SECONDARY ANALYSIS OF THE TIMSS DATA

Secondary Analysis of the TIMSS Data

Edited by

David F. Robitaille
University of British Columbia,
Vancouver, BC, Canada

and

Albert E. Beaton
Boston College,
Chestnut Hill, MA, U.S.A

KLUWER ACADEMIC PUBLISHERS
DORDRECHT / BOSTON / LONDON

A C.I.P. Catalogue record for this book is available from the Library of Congress.

ISBN 1-4020-0859-7 (HB)
ISBN 1-4020-0860-0 (PB)

Published by Kluwer Academic Publishers,
P.O. Box 17, 3300 AA Dordrecht, The Netherlands.

Sold and distributed in North, Central and South America
by Kluwer Academic Publishers,
101 Philip Drive, Norwell, MA 02061, U.S.A.

In all other countries, sold and distributed
by Kluwer Academic Publishers,
P.O. Box 322, 3300 AH Dordrecht, The Netherlands.

Printed on acid-free paper

TABLE OF CONTENTS

v

PART 6: CONCLUSION

PREFACE

Researchers who participate in IEA studies have a unique opportunity to work collaboratively with their counterparts from many different countries and disciplinary backgrounds over a period of several years on questions of shared academic interest. Once the data for a given study have been collected and the first round of international reports published, however, opportunities for that kind of collaboration tend to be much less frequent.

A major strength of IEA studies compared to other large-scale, international studies is that they are classroom based, thereby making it possible for researchers and policy makers to investigate linkages between students' achievement and a wide range of variables. Those variables could be related to instructional practices, to students' and teachers' background and attitudes, to school organizational patterns, or to opportunity to learn, to name a few. The research questions that TIMSS was designed to address make it clear that these kinds of relational, multi-variate analyses were among the major goals of the project.

The international reports of the TIMSS–95 results that were published by the International Study Center at Boston College between 1996 and 1999 were intended to provide comprehensive coverage of the basic findings of the study. They were not intended to provide in-depth analyses of research and policy issues; instead, their main purpose was to make the basic findings of the study widely available in a timely manner. This they certainly did.

The goal of the present volume is to make available the findings from a number of secondary analyses that researchers in many of the TIMSS countries have carried out since the data were collected in 1995. Thanks to the financial support provided by the U. S. National Science Foundation under Grant #REC-9815180, it has been possible to carry out some secondary analyses, and the results of those analyses are the focus of this volume. The grant made it possible to bring together 37 scholars from 10 countries for two meetings to discuss the structure of the volume and to provide feedback to them regarding their planned analyses. The grant also provided funds to provide technical support for authors in carrying out their analyses and for editing the papers they produced. Any opinions, findings, conclusions, or recommendations expressed in this book are those of the authors and do not necessarily reflect the views of the National Science Foundation.

The topics covered in this set of papers are almost as varied as the researchers who wrote them, and they illustrate the range of investigations that this kind of data makes possible. For the sake of convenience, the papers have been partitioned into several sections on the assumption that some readers would be more interested in some topics than in others. The first, or introductory section of the book includes 2 chapters and is designed to provide a brief introduction to TIMSS as a whole as well as to this volume. The second section (Chapters 3 to 8) focuses on papers related to mathematics; the third section (Chapters 9 to 12), on science; and the fourth (Chapter 13 to 19), on topics that are more cross-curricular in nature. The fifth section (Chapters 20 to 24) contains a set of papers related to measurement and methodological topics. The sixth and last section consists of closing comments from the editors regarding a number of lessons learned from TIMSS and some suggestions for further research.

The two papers in Part I provide an introduction to the volume. In Chapter 1, Hans Wagemaker, the executive director of IEA, highlights the importance of international comparisons in education and the role of IEA studies in that effort over the past 40 years. Chapter 2, written by David Robitaille from the University of British Columbia and Al Beaton of Boston College, both of whom were heavily involved in all phases of TIMSS, is a brief introduction to the study for readers who are not familiar with its scope and extent.

Part 2 consists of six chapters focusing on aspects of the mathematics component of TIMSS. John Dossey (Illinois State University), Chancey Jones (Educational Testing Service), and Tami Martin (Illinois State University) present an analysis of students' responses to constructed-response items, using the two-digit scoring codes developed for use in the study. The next paper, from David and Alan Taylor of the University of British Columbia, summarizes changes in students' achievement results over a period of about 20 years between SIMS, the second mathematics study, and TIMSS. John Dossey (Illinois State University) and Mary Lindquist (Columbus State University) discuss the influence of TIMSS on the development and dissemination of the curriculum and evaluation standards developed by the National Council of Teachers of Mathematics. Eizo Nagasaki and Hanako Senuma from the National Institute for Educational Research in Japan present an analysis of the TIMSS mathematics results from their perspective in Japan. In the next two papers, Geoffrey Howson of the University of Southampton, shares his insights about the curricular and instructional implications of the Population 2 mathematics results in Chapter 7 and of the Population 3 results in Chapter 8.

Part 3 consists of 5 chapters related to the TIMSS science results. In Chapter 9, Svein Lie and his colleagues from the University of Oslo explore students' understanding of a number of fundamental concepts in science. Chapters 10 through 13 provide reflections on the science achievement results from a range of international perspectives. These include the Czech Republic Jana Paleckova and Jana Strakova), Hong Kong (Nancy Law), Russia (Galena Kovalyova), and Scandinavia (Marit Kjærnsli and Svein Lie). In each case, the authors identify and discuss the implications of the science achievement results for informing the debate about how to improve the teaching and learning of science.

The seven chapters, Chapters 14 through 20, included in Part 4 discuss a range of issues that relate to teaching and learning, but not necessarily to mathematics or science specifically. For lack of a better term, the section is described as focusing on cross-curricular issues.

In Chapter 14, Al Beaton and Laura O'Dwyer of Boston College address separating school and classroom variance using the TIMSS data. Four scholars from UCLA discuss the correlation between students' achievement in mathematics and in science in Chapter 15. Their analysis focuses on results from the United States only. Skip Kifer from the University of Kentucky provides an analysis of the student attitude data in Chapter 16. In Chapter 17 Ina Mullis and Steve Stemler from the International Study Center at Boston College focus on an analysis of gender differences in achievement in TIMSS. Chapter 18, written by Jay Wilkins, Michalinos Zembylas, and Ken Travers, provides insight into the design of and into senior secondary students' performance on the TIMSS mathematics and science literacy study. In Chapter 19, Hans Pelgrum and Tjeerd Plomp from the University of Twente summarize finding from TIMSS having to do with the impact of technology on the teaching and learning of mathematics and science. Chapter 20 was written by Dick Wolf of Teachers' College. His paper focuses on the importance of out-of-school tutoring or coaching in various countries. Chapter 21, by Tom Kellaghan of the Educational Research Centre in Dublin and George Madaus of Boston College, use data from the TIMSS teacher questionnaires to examine the sources of teachers' information about issues related to assessment and evaluation.

Part 5 of the volume focuses on issues related research methodology. In Chapter 22, Laura O'Dwyer from Boston College discusses a new technique based for estimating between-classroom variance using what she describes as a "pseudo-classroom" approach. In Chapter 23 Dana Kelly from the American Institutes for Research describes her work on the development of international benchmarks of student achievement through scale anchoring analysis. In Chapter 24, Kadriye Ercikan and Tanya McCreith from the University of British Columbia use differential-item-functioning technology to explore the impact of translation effects on item difficulty in selected countries.

Part 6 consists of a brief concluding chapter by the editors. The goal of this chapter is not to serve as a summary of what has gone before, but rather to provide an opportunity for the editors to reflect on some of the lessons learned from TIMSS, to speculate about the kinds of research that remain to be done, and to put forward a few suggestions for the consideration of researchers who will be doing these kinds of studies in the future.

As editors of this volume, we are grateful to many individuals who helped us bring this task to a successful conclusion. We are deeply indebted to Larry Suter and the U. S. National Science Foundation for the moral and financial support provided to the authors and us throughout the process of bringing this book to publication. We are, of course, very grateful to our many authors for their patience in dealing with editorial demands and for their prompt responses to our many questions and editorial suggestions. We also grateful for the support extended to us by IEA, and particularly by its Executive Director, Hans Wagemaker, throughout the process.

On the east coast of the United States, at Boston College, the project had the support of the Center for the Study of Testing, Evaluation, and Educational Policy. Their staff members were of great assistance to us in a number of ways. We also had great technical support from Stacy Raczek, a doctoral student at Boston College, whose mastery of the software was phenomenal.

On the west coast, at the University of British Columbia, Katy Ellsworth and Bonnie Davidson took on major responsibility for getting the manuscript ready for submission to the publisher. Katy's involvement tapered off somewhat toward the end of the project as she focused on the forthcoming birth of her second child. Bonnie filled the gap admirably, and we are grateful to both of them for their contributions.

David Robitaille
Al Beaton

1 INTRODUCTION

Chapter 1

TIMSS IN CONTEXT: ASESSMENT, MONITORING, AND MOVING TARGETS

Hans Wagemaker

Writing in the early 1970s one of the architects of the early IEA studies outlined the dilemma facing practitioners and policymakers in education alike.

> At all levels in an educational system, from the teacher in the classroom, through the administrator to the policymaker, decisions have continually to be made most of the time on the basis of very little factual information. (Postlethwaite, 1974).

Educational policy is formulated and implemented at all levels of the education system even where system-level constraints such as a centralized curriculum restrict what schools and teachers might do. Discretion at the school and classroom level always remains. How and on what basis policymakers, administrators, and teachers make decisions in the educational arena is at the heart of international comparative studies of education like TIMSS. In order to more fully understand the significance of studies like TIMSS it is worth considering the way in which interest in and the impact of studies like TIMSS have evolved.

Over the last 15 years most of the developed countries of the world have initiated or experienced significant reforms in education and the wider public sector (The World Bank, 1999). Similarly, in many low- to middle-income countries, educational reform as a means of enhancing social and economic well-being has received increasing amounts of attention. This is in part attributable to the almost universal recognition that the performance of a country's educational system is a key element in establishing a nation's competitive advantage in an increasingly global economy. Education is conceived of as being implicated in a country's economic, social, and personal development and is considered one of the key means whereby inequities, social and economic, can be reduced. Perhaps the most dramatic expression of this sentiment is contained in the report from the United States, *A Nation at Risk*, in which the authors point to the threat of economic decline as supplanting the past threat of aggressor nations (United States National Commission

3

D.F. Robitaille and A.E. Beaton (eds.), Secondary Analysis of the TIMSS Data, 3–10.
© 2002 *Kluwer Academic Publishers. Printed in the Netherlands.*

on Excellence in Education, 1983). Education and the decline in educational standards were cited as the cause of economic decline in the face of intensified global competition. The authors write:

> If an unfriendly foreign power had attempted to impose on America the mediocre educational performance that exists today, we might well have viewed it as an act of war. As it stands, we have allowed this to happen ourselves ... we have, in effect, been committing an act of unthinking, unilateral educational disarmament. (United States National Commission on Excellence in Education, 1983, p.5)

Although a model which ascribes economic decline in a simplistic way to a decline in educational standards is likely to be of limited value in addressing or understanding either educational or economic policy concerns, this debate served to draw attention to real concerns about educational performance, not only in the U. S. but also in may OECD countries. It is the concern for excellence (together with concerns of equity and efficiency) that has given rise to greater intensity of focus on education and educational policy development.

While education has been receiving an increased priority in the public policy arena in many countries, it has also been facing the reality that, like many other areas of public spending, there are real limits to the amount of funding that is available for educational development. What funding is available is accompanied by increasing demands for accountability and a better understanding of the relationship between educational expenditure and educational outcomes. The fullest and perhaps most extreme expression of these concerns is reflected in publications like *Reinventing Government* (Osborne & Gaebler, 1993) in which the authors argue for an educational marketplace that should be shaped by the twin imperatives of efficiency and effectiveness. The implicit argument is that increased provision and improved instructional quality are likely to produce greater numbers of better-prepared students, which in turn will result in a more internationally competitive and better-prepared workforce. The role that TIMSS might play in such an argument is to place the focus more narrowly on the assessment of quality in mathematics and science, and presumably, therefore, on the production of more productive and high-quality scientists, mathematicians, and engineers.

In general, however, what is reflected in the kinds of concerns expressed above is a shift in focus from managing issues related to the expansion of educational systems in terms of student numbers, to one of managing issues of quality and excellence. In the case of those countries in what might be described as a less advanced stage of educational development, this has meant not surrendering to the imperatives of educational expansion at the expense of considerations of quality. The change in emphasis is noted by Tuijnman and Postlethwaite (1994) who argue that, while the history of large scale assessment dates back to the early 1960s, there was a significant development toward a more systematic focus on national monitoring with the release of reports such as *A Nation at Risk,* the release of the results of IEA's Second International Science Study, and later, again in the United States, the report from the conference of the governors of the 50 states in Charlottesville, Virginia, which sought to frame national goals for education with a strong emphasis on quality. In short, investment in education and the related policy development, it was argued, could no longer be carried out as an act of faith.

As interest in global competitiveness and local accountability has increased, so too has interest in international comparisons of educational performance. What then is the significance of, and what are the benefits of participating in international comparisons of educational achievement, and how does the Third International Mathematics and Science Study meet these expectations?

TIMSS is intended to monitor the success of mathematics and science instruction and the context in which it occurs in each of the countries that participated in the project. Three major conceptual elements drive the TIMSS design. These elements include the intended curriculum (the curriculum as described at the policy level), the implemented curriculum (the curriculum as students experience it at the school and classroom level), and the attained curriculum (the curriculum as represented by student outcomes). Through the mechanisms of a curriculum analysis, a video study, achievement tests, and background questionnaires that gathered information from schools, teachers, and students, the conceptual design was realized, providing a unique opportunity to observe and examine how mathematics and science instruction is provided in some 40 countries. What is significant is that the TIMSS design provided for an examination of those policy variables related to schooling, curriculum, and instruction that are affected by policy intervention. Furthermore, it established international benchmarks for achievement and key policy variables that allow countries to monitor their performance in an increasingly global community.

While the emphasis in comparative studies of educational achievement is often seemingly focused primarily on the achievement data, the interpretation of such system level data is not straightforward. The significance of the extensive data collected by the multiple strategies employed by TIMSS lies in the fact that countries that do not take into account the differences in the respective education systems when introducing policy reform based on comparative data risk not only disappointment, but also the possibility of developing polices that are potentially counter-productive in addressing perceived educational needs.

Moreover, the data collected through the background questionnaires allows policymakers to address particular policy needs and concerns related not only to the quantity, quality, and content of mathematics and science instruction but also to identifying factors that may be linked to achievement or to sub-populations of national importance (such as gender and ethnicity). While it is not always possible in the international context to collect data on, for example sub-groups of interest that are internationally comparable (e.g., ethnicity), the TIMSS design permits the collection of these variables as international options. For example, the TIMSS reports (Beaton et al., 1996) included information not only on such things as the characteristics of the students' home environment (e.g., books in the home), the characteristics of instructional practices (e.g., classroom organization) but also on some of the affective characteristics of the student populations (e.g., student attitudes to mathematics and science) and their relationship to achievement.

While much of what studies like TIMSS do is to describe "what is" in terms of how education is practiced in a country (the within-country perspective), the power of such studies is most fully realized when the international context they provide is considered (the between-country perspective). Given the differences in the ways in which education is organized and practiced across cultures and societies, a comparative perspective such as that provided by TIMSS not only enables an understanding of its many forms, but also serves to expand a nation's horizon as to what might be possible. As Foshay et al. (1962) noted:

> If custom and law define what is educationally allowable within a nation, the educational systems beyond one's national boundaries suggest what is educationally possible.

Identifying models or practices of education from countries around the world as a means of reflecting on one's own practice and experience is, arguably, a key function of international comparative studies like TIMSS.

TACTICAL VERSUS STRATEGIC APPROACH

While the evidence that there is an increasing interest in and awareness of the need to invest in international comparative studies of educational achievement has been presented, there is a more recent demand for the creation of international benchmarks against which a country's performance may be measured. This is associated with a growing awareness of the need to move from the process of tactical decision making and ad hoc participation in studies like TIMSS to a more strategic investment that recognizes the dynamic nature of change in the educational environment. As noted in *Education Sector Strategy* (The World Bank, 1999), market economies and a constantly changing educational, political, and economic environment prevail in countries accounting for over 80 percent of the world's population. Education is deemed vital, with those who can compete with the best having an enormous advantage in a faster paced world economy over those who are less well prepared. Globalization of markets and factors such as knowledge exacerbate these impacts. Given this dynamic it is not surprising that a demand has emerged for the regular monitoring of educational quality, particularly for those economies engaged in strategic educational reform.

Repeated monitoring over time, provided that the tests in different years can be linked, can provide evidence not only of changes in levels of performance but also of shifts in standards. Arguably, the magnitude of a country's investment in education, usually measured in billions of dollars, demands regular monitoring. What distinguishes TIMSS from many other comparative studies of educational achievement is not only the large number of countries that participated but, more importantly, that it was conceived of and designed as a trend study. That is, fundamental to the design of the assessment is the notion of regular, repeated assessments with instruments that can be linked over time.

Griffith (1989), citing the summary agreement issued by the President of the United States and the 50 governors who attended the education summit convened at the University of Virginia, noted the following extract from the report that captures the sense of this new imperative.

> Education has always been important, but never this important because the stakes have changed. Our competitors for opportunity are also working to educate their people. As they continue to improve, they make the future a moving target.

Moreover, the imperative of increasing globalization has brought with it a concern for monitoring excellence in a much broader context. The notion of performance benchmarks, borrowed from industry and introduced into the educational arena, also brought into sharp focus questions related to standards and educational performance at both the local and international levels.

IMPACT ON TEACHING CURRICULUM AND RESEARCH

While much of the discussion above has focused on placing TIMSS in the wider context of educational policy development and research in a more general sense, its impact on teaching, curriculum, and research in participating countries was and is clearly evident. Although judging the impact of any piece of research is not always a simple task, some assessment can be made as to the importance of a study like TIMSS. The available evidence to date suggests that the impact of TIMSS was far reaching. Robitaille et al. (2000) in *The Impact of TIMSS on the Teaching and Learning of Mathematics and Science* reported impact assessments from 29 countries that participated in the study. Evidence is presented which identifies the impact of the study in three key areas: curriculum, teaching, and research.

It is evident from the essays provided by the contributors that the impact on curricular reform, either by way of direct impact on materials or a more general impact in terms of additional contextual information for policy reform, was widespread. In countries as different as Iceland, Kuwait, New Zealand, Norway, Romania, and South Africa, TIMSS served as a catalyst for curricular review and change. In Iceland the information collected during the TIMSS study resulted in a recommendation for increased teaching hours for mathematics and science instruction at the elementary level. In New Zealand the TIMSS results precipitated the establishment of a task force with a focus on science and mathematics education. This task force was charged with addressing several issues, including the low expectations of success for New Zealand students held by many teachers and parents, underachievement among Maori and Pacific Island students, and teachers' lack of confidence in their ability to teach some aspects of the mathematics and science curricula.

Further analysis of TIMSS data, which was collected in a way that would allow for the identification of common misconceptions about mathematics and science, is also likely to lead to instructional and curricular refinements in many countries.

While the immediate impact on teaching was less strongly articulated, in several countries, notably Australia, Canada, Spain, Japan and in particular the Philippines, the TIMSS results were used to bring the teaching of mathematics and science into sharp focus. In the Philippines, for example, government resolve was reflected in a program to train 100,000 mathematics and science teachers over a period of five years.

Research, particularly research related to educational policy and mathematics and science learning were also identified as areas in which TIMSS had an impact. In the case of some countries, for example Macedonia, TIMSS was the first assessment of student achievement undertaken at a national level. Furthermore, all countries were able to benefit in some way from the training in large-scale student assessment that was associated with participation in the study. Some direct outcomes of the TIMSS training activities included capacity building within many countries, some of which established dedicated research institutes which used the skills and knowledge developed through the TIMSS experience to enhance their national capacity for assessment and policy development.

Finally, as evidenced by the chapters that follow, the data provided by TIMSS has spawned a large number of investigations, both national and international, directed at key policy issues as well as at developing our understanding of mathematics and science learning in a more fundamental way.

What is evident is that the release of the TIMSS results had a considerable impact in most countries, particular with respect to bringing into public focus and debate issues related to the teaching and learning of mathematics and science.

SUMMARY

Since 1959 the International Association for the Evaluation of Educational Achievement has conducted more than 16 international comparative studies intended to provide policymakers, educators, researcher, and practitioners at all levels with information about educational achievement in a range of subjects, and the contexts in which learning takes place around the world. Changing expectations over the last few decades have seen studies like TIMSS take on a new significance. Change has also meant changes in strategy in terms of investment in this type of study but also in terms of design. Regular assessments that allow for the measurement of change are now possible and are demanded. TIMSS is the most ambitious undertaking of its type to date. As Beaton et al. (1996) noted:

> Forty-five countries collected data in more than 30 different languages. Five grade levels were tested in the two subject areas, totaling more than half a million students tested around the world.

The statistics speak for themselves but the true significance of TIMSS lies in what TIMSS delivered in terms of information for policy and the platform that has been established for the assessment of trends, not only in terms of educational outcomes but also in terms of those factors that are involved in explaining educational achievement.

International studies such as TIMSS allow for comparisons among educational jurisdictions at the local, state, regional, or national level in some relative sense. They permit the examination of key policy variables relating to curriculum and instruction that are implicated in improving educational practice. They also permit the monitoring of standards in the more absolute sense, particularly where benchmarks are developed to describe desirable levels of performance. Repeated assessments based on a trend design permit the monitoring of change over time, not only in terms of between-country analyses but also in terms of within-country examinations of change related to key policy variables.

Regular monitoring of national performance in education is desirable to provide information for policy development, to identify the performance of population sub-groups to ensure that concerns for equity are not neglected, to foster public accountability, and to allow trends in performance to be identified and monitored. It is clear that concerns over monitoring achievement and altering standards cannot be addressed by a policy of ad hoc investment in comparative studies but rather, requires a design that is linked to past assessment, that captures the present situation, and that allows for future developments. By addressing these issues through its design and execution, TIMSS makes a significant contribution to our understanding of the teaching and learning of mathematics and science around the world.

REFERENCES

Beaton, A. E., Mullis, I. V. S., Martin, M. O., Gonzales, E. J., Kelly, D. L., & Smith, T. A. (1996). *Mathematics achievement in the middle school years.* Chestnut Hill, MA: TIMSS International Study Center, Boston College.

Foshay, A. W., Thorndike, R. L., Hoytat, F., Pidgeon, D. A., & Walker, D. A. (1962). *Educational achievement of thirteen-year-olds.* Hamburg: UNESCO Institute for Education.

Griffith, J. (1989). *Education indicators for quality, accountability and better practice. Indicators and Quality in Education: Second National Conference.* Surfers Paradise, Australia: Australian Conference of Directors General.

Osborne, D. & Gaebler, T. (1993). *Reinventing government: How the entrepreneurial spirit is transforming the public sector.* New York: Plume.

Postlethwaite, T. N. (1974). Introduction. *Comparative Education Review. 18*(2), 157–163.

Robitaille, D. F., Beaton, A. E., & Plomp, T. (Eds.). (2000). *The impact of TIMSS on the teaching and learning of Mathematics and Science.* Vancouver, British Columbia: Pacific Educational Press.

The World Bank. (1999). *Education sector strategy.* Washington, D.C.: The International Bank for Reconstruction and Redevelopment, World Bank.

Tuijnman, A. C. and Postlethwaite, N. T. (1994). *Monitoring the standards of education.* Oxford: Pergamon.

United States National Commission on Excellence in Education. (1983). *A nation at risk: The imperative for educational reform: A report to the nation and the Secretary of Education, United States Department of Education.* Washington, D. C.: United States Department of Education.

Chapter 2

TIMSS: A BRIEF OVERVIEW OF THE STUDY

David F. Robitaille and Albert E. Beaton

The Third International Mathematics and Science Study has been the most ambitious and arguably the most successful project ever undertaken under the auspices of IEA. In the first round of TIMSS testing in 1995, data was collected from over 500,000 students of mathematics and science at five age/grade levels, as well as from their teachers and principals in more than 40 countries around the world. A second round of TIMSS data collection involving 38 countries was carried out in 1999. That study focused on the teaching and learning of mathematics and science at the Grade 8 level, or its equivalent internationally.

A third round of TIMSS—the acronym now denotes the Trends in Mathematics and Science Study—is planned for 2003, and the preparatory work for that study is currently under way. More than 50 countries have indicated an interest in participating in that study.

A great deal has been written about TIMSS, and the list includes articles in scholarly journals, monographs, book chapters, and books. The International Study Center (ISC) for TIMSS at Boston College has published reports of the international results from both rounds of testing, as well as several technical reports. All of these reports as well as the student-level data and a partial bibliography of TIMSS-related publications are available from the ISC's website, located at www.timss.org.

Readers who wish to obtain detailed information about any aspect of TIMSS should consult the kinds of resources referred to in the previous paragraph. Our very limited goal in this chapter is to provide those readers who may be unfamiliar with the scope and extent of TIMSS with enough information about the study to make it easier for them to understand various references to the design of TIMSS that are included in virtually every chapter in this volume.

THE CONCEPTUAL FRAMEWORK FOR TIMSS

The precursors of TIMSS were four studies conducted by IEA over a period of 30 years between 1960 and 1990. Those studies were:

D.F. Robitaille and A.E. Beaton (eds.), Secondary Analysis of the TIMSS Data, 11–18.

- First International Mathematics Study, FIMS   1959-1967
- First International Science Study, FISS   1966-1973
- Second International Mathematics Study, SIMS 1976-1987
- Second International Science Study, SISS   1980-1989

The experience gained from these studies influenced the design of TIMSS in a number of important ways. In particular, the earlier IEA studies recognized the centrality of the notion of curriculum in any examination of the teaching and learning of subject matter in schools. This centrality was underscored in the conceptual framework undergirding SIMS and, later, TIMSS. Curriculum was to be examined from three viewpoints or perspectives: the curriculum as mandated at the system level (i.e. the intended curriculum), the curriculum as taught by teachers in classrooms (the implemented curriculum), and the curriculum as learned by students (the attained curriculum). That conceptual framework is summarized in Figure 1 below.

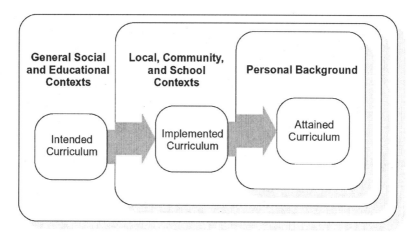

Figure 1. Conceptual framework for TIMSS.

The overall aim of TIMSS was to contribute to the improvement of the teaching and learning of mathematics and science in K–12 educational systems around the world. The intention was that policymakers, researchers, curriculum developers, and educators at all levels could use TIMSS data and findings to learn about the kinds of curriculum and instructional practices that were associated with the highest levels of achievement. In other words, educators from different national and cultural backgrounds could use the results of the study as a kind of mirror in which to study themselves, but not in isolation. Instead, TIMSS provided a unique opportunity for them to view themselves in the rich context provided by the participation of many other countries.

Four research questions guided the development and execution of the study, and these questions have been discussed in some detail in other publications (see, for example, Robitaille & Garden, 1996). It will suffice here to simply list the four questions for reference.

- How do countries vary in the intended learning goals for mathematics and science; and what characteristics of educational systems, goals, and students influence the development of those goals?

- What opportunities are provided for students to learn mathematics and science; how do instructional practices in mathematics and science vary among nations; and what factors influence these variations?

- What mathematics and science concepts, processes, and attitudes have students learned; and, what factors are linked to students' opportunity to learn?

- How are the intended curriculum, the implemented curriculum, and the attained curriculum related with respect to the contexts of education, the arrangements for teaching and learning, and the outcomes of the educational process?

DESIGN OF THE STUDY

IEA studies have traditionally employed a research design based on the use of intact classes of students. This means that the studies focus on classrooms at a particular grade level, as opposed to focusing on students of a particular age. A good case can be made for either alternative, but the prevailing opinion in IEA circles is that grade-based studies are the preferred choice for studies investigating the linkages among the intended, implemented, and attained curricula. Such studies are more likely to contribute in a significant way to our understanding of what kinds of curricula and instructional practices are associated with the highest levels of student attainment. Age-based comparisons are, in some sense, easier for the research consumer to understand, because age, unlike the grade structure of schools in different countries, is calculated in the same way everywhere. However, since students of a particular age, say 13 years, may be spread across as many as four or even five grade levels within a given system, age-based studies make the investigation of relationships between teaching practices and students' achievement virtually impossible.

TIMSS has been described, quite justifiably, as the most ambitious project ever undertaken by IEA. What is perhaps less well known is that, in its original conception, the plan for the study was even more ambitious. The first discussions about the study took place in the late 1980s and involved a number of researchers, almost all of whom had had significant involvement in the longitudinal component of SIMS. Those discussions led to the establishment of the Study of Mathematics and Science Opportunity (SMSO) which was intended to investigate teaching practices in a small number of countries, to break new ground in the investigation of opportunity to learn, to carry out the curriculum analysis component of TIMSS, and to begin the process of developing questionnaires to be administered to principals, teachers, and students participating in what became TIMSS (Schmidt et al., 1996).

The final design for TIMSS grew out of a series of discussions involving the researchers, representatives from the participating countries, and personnel from the agencies that provided the bulk of the funding required for international coordination of the project: mainly the U. S. National Science Foundation and the U. S. National Center for Education Statistics. The process of arriving at the final design of the study involved a number of compromises that affected many aspects of the study, but preserved the overall integrity of the original goals for the study.

The design of TIMSS was intended to facilitate the investigation of mathematics and science curricula (intended, implemented, and attained) from a variety of perspectives. Three populations of students—Populations 1, 2, and 3—were to be studied, and these corresponded roughly to the end of primary education, the end of compulsory and non-departmentalized education in mathematics or science in many countries, and the end of secondary education. The definitions of those populations are as follows:

•Population 1: all students enrolled in the two adjacent grades that contain the largest proportion of students of age 9 at the time of testing.

•Population 2: all students enrolled in the two adjacent grades that contain the largest proportion of students of age 13 at the time of testing.

•Population 3: all students in their final year of secondary education, including students in vocational education programs.

•Population 3 also included two sub-populations:
- Students taking advanced courses in mathematics;
- Students taking advanced courses in physics.

All countries were required to participate in the Population 2 study; participation at the other levels was optional. National samples for Populations 1 and 2 consisted of intact classes of students, and countries were encouraged to select two classes of students per grade level to facilitate later analyses. In the event, only a few countries were able to do this. Samples for Population 3 were samples of individual students, not of intact classrooms.

The Population 3 study presented a number of problems, some of them seemingly intractable. However, the degree of interest in studying student attainment at this level was so high internationally that an enormous effort was devoted to finding ways to address those problems. That effort, we believe the record shows, was largely successful.

For Population 3, there were, in fact, three studies; and countries chose to participate in as many of them as they wished. First, there was a study of mathematical and scientific literacy designed to address the question, "What mathematics and science concepts do students who are in the last year of secondary education know, and what are they able to do with that knowledge?" Students were selected for participation in the literacy component regardless of the kind of program they were enrolled in and whether or not they were studying any mathematics or science at the time the test was administered. Thus, each national sample was to include the entire range of students' backgrounds in mathematics and science: from those who were no longer studying any mathematics or science to those who were taking as many advanced science and mathematics courses as they could. Second, there was a study of "mathematics specialists," students who were taking the kinds of courses required to pursue studies in science or engineering at the post-secondary level. Third, there was a corresponding study of "physics specialists." Each student selected as part of the Population 3 sample in a given country was asked to participate in only one of the three Population 3 studies.

Data collection for all three TIMSS populations in the northern hemisphere took place in the first half of 1995. In the southern hemisphere, data collection for Populations 1 and 2 was carried out in the latter part of 1994. For Population 3 in the southern hemisphere, the data collection took place in August 1995.

THE TIMSS ACHIEVEMENT TESTS

One decision that was reached early in the development of the study was that the item pool needed to include items presented in various formats, and not be limited simply to multiple choice items. The period during which the test development occurred was one in which there was widespread support for more "realistic" forms of assessment. A great deal of effort went into the development of what was called "authentic testing" or "performance testing," and the design of the TIMSS tests reflect the spirit of that period and that discussion. The amount of testing time allocated to multiple choice, constructed response, and performance items was approximately equal.

The original plan for the study was to include four main types of items: multiple choice, short answer, extended response, and performance items. In the end, the performance-test component of the study was made optional so that countries were not required to participate in that aspect of the study. The other three types of items were used at all population levels.

For Populations 1 and 2, the mathematics and science items were distributed across eight test booklets, with some items occurring in more than one booklet. For Population 3, there were nine booklets, with some booklets designed for completion by any student, regardless of background (i.e., the literacy booklets), and others designed for students participating in the advanced mathematics or physics components. Testing time for students in Populations 1 and 2 was about an hour, and 90 minutes for Population 3.

QUALITY ASSURANCE

Any scientific study worthy of the name includes measures to ensure that the findings of the study will not be compromised or called into doubt because of a failure to implement high standards of quality control to every aspect of the study, from instrument design, to sampling, to data collection, to analysis, and finally to reporting. One of the major strengths of TIMSS is the extent to which high standards of quality assurance were proposed and implemented. The following paragraphs contain brief descriptions of the major components of that effort. More detail is available in the technical reports published by the International Study Center at Boston College (see www.timss.org).

Sampling

Not surprisingly, given the complex nature of the study and its ambitious goals, the sampling procedures for TIMSS were also rather complicated, and they held participating countries to a very high standard of compliance. Failure to meet one or more of those conditions was dealt with in several ways, depending on the degree of severity of the problem encountered in a given country. In the least serious cases, the country was footnoted in every table, with the footnote containing a brief description of how that country's sample deviated from the norm. When a more serious departure from the guidelines was found, that country was placed "below the line," in the bottom portion of each table. In the most serious cases, the country's results were either not included in the international reports at all, or were summarized briefly in an appendix to those reports. The vast majority of participating countries either met all of the sampling guidelines or were identified in footnotes as described above.

Translation

The TIMSS instruments had to be translated into more than 30 languages for use in the study. National centers were encouraged to have at least two independent translations made in each case, and to put in place a process for reconciling differences between or among the translations. They were also provided with a manual detailing procedures designed to minimize the introduction of any systematic bias to the items in the translation process. All of the translations were checked by experts centrally, and a number of statistical checks were undertaken to identify problematic translations.

Scoring of Constructed Response Items

Both the mathematics and the science item pools included large numbers of constructed response items. Some required the students to provide an answer only (short answer items) while others required students to show all of their work (extended response items). It would not have been feasible to score all of these items in one central location for all countries, so a decentralized process was developed and implemented.

Central to this scheme was the development of scoring rubrics for each of the constructed response items and the implementation of a series of training seminars for those responsible for coordinating the scoring of those items in their own country. In all, nine such seminars were held at various locations around the world. Most of the sessions were three to four days long.

Attention was also paid to the question of inter-rater reliability, both within and between countries. National centers were expected to carry out a study of inter-rater reliability on the coding of the constructed response items in their own country. The International Study Center conducted an examination of inter-rater reliability on an international level.

Quality Control Monitors

Each country was asked to identify a person to serve as a quality control monitor for their national study. The monitors attended an international training session during which they were briefed on the kinds of activities they were expected to perform. These included instructions for visiting their national center, interviewing their National Research Coordinator, visiting a number of schools participating in the study, and observing the data collection process in a number of classes. Each national Quality Control Monitor provided an independent report to the International Study Center.

TIMSS–R: A PARTIAL REPLICATION OF TIMSS

From the very beginning, TIMSS has been seen as a key component in IEA's "cycle of studies," under the terms of which the organization carries out studies in certain areas of the K–12 curriculum on a fixed schedule. The bulk of the TIMSS data was collected in 1995, and a partial replication of the study was undertaken in 1999. That study, TIMSS–R (see, for example, Martin et al., 2000), focused on the upper of the pair of grades defined as TIMSS Population 2. This was Grade 8 in almost all countries. Of the 38 countries that participated in TIMSS–R, 26 had also participated in the 1995 study.

A major goal of the new TIMSS (now denoting "Trends in Mathematics and Science Study") was to track changes in patterns of students' achievement in mathematics and science over time. A large portion of the items used in 1995 had not been released; they were kept confidential. The TIMSS–R test booklets consisted of the non-released items from the 1995 study supplemented by new items developed and field tested to parallel those that had been released.

REFERENCES

Martin, M. O., Mullis, I. V. S., Gonzalez, E. J., Gregory, K. D., Smith, T. A., Chrostowski, E. J., Garden, R. A., & O'Connor, K. M. (2000). *TIMSS 1999: International science report*. Chestnut Hill, MA: International Study Center, Lynch School of Education, Boston College.

Robitaille, D. F., & Garden, R. A. (Eds.). (1996). *TIMSS Monograph No. 2: Research questions and study design*. Vancouver: Pacific Educational Press.

Schmidt, W. H., Jorde, D., Cogan, L. S., Barrier, E., Gonzalo, I., Moser, U., Shimizu, K., Sawada, T., Valverde, G. A., McKnight, C., Prawat, R. S., Wiley, D. E., Raizen, S. A., Britton, E. D., & Wolfe, R. G. (1996). *Characterizing pedagogical flow: An investigation of mathematics and science teaching in six countries*. Dordrecht, The Netherlands: Kluwer Academic Publishers.

2 FOCUS ON MATHEMATICS

Chapter 3

ANALYZING STUDENT RESPONSES IN MATHEMATICS USING TWO-DIGIT RUBRICS

John A. Dossey, Chancey O. Jones, and Tami S. Martin

The Third International Mathematics and Science Study (TIMSS) was the first international assessment of student achievement in mathematics to make use of a large number of constructed response items in measuring and reporting student performance. The use of such items provides a great deal of data to teachers, curriculum developers, and researchers. Seeing student work allows individuals in each of these groups to examine more directly, rather than surmise about, evidence of students' misconceptions, error patterns, and problem-solving strategies. In addition to using a large number of constructed response items, TIMSS also forged new ground in employing two-digit rubrics for scoring student responses. Such scoring rubrics provide information on the correctness and methods students use to produce their work and on misconceptions noted in that work. In this chapter, we examine the patterns and findings associated with the use of such two-digit rubrics to score student responses in TIMSS.

TWO-DIGIT RUBRICS

To maximize the amount of information that might be obtained from the use of constructed response items in the TIMSS mathematics assessment, two-digit rubrics were developed to score student responses. They were devised to retain as much information as possible about student work in the codes themselves. In particular, these rubrics were developed to permit scoring student responses for correctness and capture the analytical information embedded in student responses and to:

- Be clear, distinct, and readily interpretable, and based on empirical data… so as to account for the most common correct responses, typical errors, and misconceptions.

- Be capable of encoding the adequacy of an explanation, justification, or strategy as well as the frequency with which it is used.

D.F. Robitaille and A.E. Beaton (eds.), Secondary Analysis of the TIMSS Data, 21–45.

- Be simple, in order to get high reliability and not to impose unreasonable time or resource burden.

- As far as possible, allow for nuances of language and idiosyncratic features of various countries, but avoid being so complex that coders are overwhelmed and tend to limit themselves to a few stereotypic codes.

- Have a number of codes that is not excessive, but sufficient to reduce coding ambiguity to a minimum (Lie, Taylor, & Harmon, 1996).

To achieve these goals, TIMSS used the same general two-digit rubric structure across all parts of the study. The ten's digit of the two-digit code represents the number of score points assigned to the response, based on its level of correctness. The unit's digit, when combined with the ten's digit, provides a code number to identify specific approaches or strategies by credit level, or, in the case of student difficulty with the item, common errors or misconceptions.

Figure 1 displays item V2 from the Population 2 assessment. This item, with the same rubric also was used as part of the Population 3 Mathematical Literacy Assessment as item A12. The rubric was developed through analyzing student field trial responses and in the actual administration of the TIMSS assessment. Student performance on this item for students in Population 2 and the Population 3 Literacy study will be reviewed later in the chapter.

In general, the structure of the two-digit rubric might be conceptualized as shown in Table 1. Variations of this general scheme were developed for problems worth 3, 2, and 1 points (Lie, Taylor, & Harmon, 1996). The overall aim was to develop a means of giving students credit for partially correct solution; and, at the same time, develop documentation of the kinds of approaches used by students from different countries. The latter kind of information could be invaluable to curriculum developers and teachers wishing to develop instructional materials that take into account the kinds of errors made by students on such items.

Because of the immense scope of the coding effort required to score the constructed response items across the countries participating in TIMSS, training sessions were designed and delivered at an international level. A series of 4-day training sessions were developed to train representatives of national centers who, in turn, had the responsibility for training personnel in their respective countries to reliably apply the two-digit codes. A total of 10 of these training sessions were held around the globe to prepare country-level teams to train their national coders (Mullis, Jones, & Garden, 1996). While there were concerns about whether this process would work, the final reliability results concerning the coding indicated that it did. An analysis of random samples of scored work indicated that the actual achieved average reliability coefficients ranging from 0.95 to 0.97 for the first-digit correctness codes and from 0.87 to 0.93 for the second-digit diagnostic codes (Mullis et al., 1997; Beaton et al., 1996; Mullis et al., 1998).

V2. The following two advertisements appeared in a newspaper in a country where the units of currency are *zeds*.

BUILDING A	BUILDING B
Office space available	Office space available
85 - 95 square meters	35 - 260 square meters
475 *zeds* per month	90 *zeds* per square meter
	per year
100 - 120 square meters	
800 *zeds* per month	

If a company is interested in renting an office of 110 square meters in that country for a year, at which office building, A or B, should they rent the office in order to get the lower price? Show your work.

Note: There is no distinction made between responses with and without units.

Code	Response
Correct Response	
30	Building A. Correct calculation of rents for both buildings. 9600 yearly/800 monthly and 9900 yearly/825 monthly, OR 825 to compare with 800 given.
39	Other correct.
Partial Response	
20	Building A. Correct calculation of rent for Building A OR B but not both.
21	Building B OR building is not named. Correct calculation of rents for both buildings.
Minimal Response	
10	Building A. Calculations or explanation are incorrect or inadequate.
11	Building A. No work shown.
12	Building B, OR building is not named. Correct calculation of rent for Building A OR B but not both.
16	Building A. Explanation is given only in the form of extracts from the advertisements.
19	Other minimal.
Incorrect Response	
70	Building B. Incorrect or inadequate calculations.
71	Building B. No work shown.
79	Other incorrect.
Nonresponse	
90	Crossed out/erased, illegible, or impossible to interpret.
99	BLANK

Figure 1. Population 2 mathematics item V2 and rubric.

Table 1. TIMSS two-digit coding scheme for a two-point item

Code	Meaning
20	Correct response, answer category or method #1
21	Correct response, answer category or method #2
22	Correct response, answer category or method #3
29	Correct response, other answer category used
10	Partially correct response, answer category or method #1
11	Partially correct response, answer category or method #2
12	Partially correct response, answer category or method #3
19	Partially correct response, other answer category used
70	Incorrect response, common misconception, error #1
71	Incorrect response, common misconception, error #2
72	Incorrect response, common misconception, error #3
76	Incorrect response, information in stem repeated
79	Incorrect response, other misconception/error noted
90	Work crossed out, erased, illegible, or impossible to interpret
99	Blank student paper

Two-Digit Coding as an Investigative Tool in International Assessments

In order to examine the use of two-digit codes in scoring items in an international context, a selection of countries and items was made. To do this, the national level student response data on constructed response items were subjected to cluster analyses by population. The data were clustered using the variables of the percent of fully or partially correct responses and the percent of items omitted by students in each country. Five clusters of similarly performing countries were formed for each population using Ward's algorithm for cluster formation (JMP Statistical Software, 2000; Hair, Anderson, Tatham, & Black, 1998). Then, within each population and cluster, representative countries were selected for further analysis. The countries selected, shown in bold in Table 2, were chosen on the basis of their appearance across populations and, to some degree, in an effort to provide some variation in curricular types and structures (Robitaille, 1997). No countries were selected for further study from Cluster 5 with the exception of the United States in the Population 3 advanced mathematics portion of the study.

The five clusters of countries identified in Table 2 are listed, within each TIMSS population, in order of the overall average percent correct or partially correct for the cluster on the constructed response items for their respective populations. Table 3 provides the cluster performance averages for the percent of constructed response items scored as being correct or partially correct, incorrect, or omitted by population level. It is interesting to note that even though the clusters within populations were ordered by the percent of items scored as correct or partially correct, the percents of items scored as incorrect or omitted responses are not, in general, ordered in the

Table 2. Clusters by TIMSS population with comparison counties highlighted

Population	Cluster 1	Cluster 2	Cluster 3	Cluster 4	Cluster 5
1	Korea **Japan** **Singapore**	Australia Austria Czech Rep. Hong Kong **Hungary** Ireland **Netherlands** Slovenia **United States**	**Canada** Cyprus Great Britain Iceland Israel Latvia (LSS)* New Zealand **Norway** Thailand Scotland	**Greece** Iran Portugal	Kuwait
2	Hong Kong, **Japan,** Korea, **Singapore**	Australia Austria Belgium (Fl)* **Canada** Czech Rep. **France** Great Britain Iceland Ireland **Hungary** **Netherlands** New Zealand **Norway** Scotland Slovak Rep. Spain Sweden Switzerland Slovenia Thailand **United States**	Belgium (Fr)* Bulgaria Denmark **Germany** Israel Lithuania Romania Russian Fed.	Cyprus **Greece** Iran Latvia (LSS)*, Portugal	Columbia Kuwait South Africa
3-Literacy	Australia Austria **Canada** **Netherlands** New Zealand **Sweden**	Denmark **France** **Norway** Slovenia Switzerland	Cyprus **Czech Rep.** **Germany** **Hungary** Iceland Italy Lithuania Russian Fed.	**United States**	South Africa
3-Advanced	**France** **Greece** Russian Fed. Switzerland	Cyprus **Czech Rep.** Denmark Lithuania Slovenia **Sweden**	Australia **Canada**	Austria **Germany** Italy	**United States**

* The Flemish and French educational systems in Belgium participated separately in TIMSS. Latvia is annotated LSS for Latvian Speaking Schools only.

same way. This indicates that there is something to look for in the student response patterns that are identified by two-digit codes.

Table 3. Percent of constructed response items coded correct or partially correct, incorrect, and omitted by cluster within Populations 1, 2, 3 Literacy, and 3 Advanced

Population	Correct/Partially	Incorrect	Omit
Population 1			
Cluster 1	73.9	24.2	1.9
Cluster 2	60.3	31.2	8.5
Cluster 3	49.9	37.1	13.1
Cluster 4	34.0	38.5	27.5
Cluster 5	15.3	42.3	42.3
Population 2			
Cluster 1	75.1	23.7	1.2
Cluster 2	59.9	36.6	3.5
Cluster 3	57.0	34.0	8.9
Cluster 4	47.9	45.6	6.5
Cluster 5	29.5	60.0	10.5
Population 3 Literacy			
Cluster 1	65.5	28.3	6.3
Cluster 2	64.0	25.0	11.0
Cluster 3	54.7	29.4	15.9
Cluster 4	53.7	39.8	6.5
Cluster 5	29.9	55.3	14.8
Population. 3 Advanced Math			
Cluster 1	48.3	24.9	26.8
Cluster 2	39.5	28.9	31.6
Cluster 3	33.6	38.3	28.1
Cluster 4	29.3	34.5	36.2
Cluster 5	27.8	54.8	17.5

INTERPRETING STUDENT PERFORMANCE USING THE RUBRICS

The following sections contain selected analyses of students' work on constructed response items using data resulting from the use of the two-digit coding. For each of the four TIMSS populations, the percentages of students in the selected comparison countries, as well as the international percentages, are given for each of the rubric scoring codes. In the tables containing this data, the comparison countries are ordered by performance within their clusters from 1 to n.

Population 1 Examples: A Look at Number and Number Sense

To grasp the nature of student performance on constructed response items in Population 1, data on the performance of students from Canada, Greece, Hungary,

Japan, Netherlands, Norway, Singapore, and the United States were selected for analysis. The analyses of students' responses to selected constructed response items across these eight countries revealed several differences. Foremost among these was the level of response or non-response to the items themselves. The amount of non-response can be further subdivided into the percentage of students who made some form of uninterpretable response (Code 90) and the percentage of students who left their papers blank (Code 99). In no instance did over five percent of the students in the Population 1 comparison countries receive a Code 90. However, there were substantial percentages of students from Greece, Hungary, and Norway who left their papers blank (Code 99). Examine these percentages as you view the response rates provided with the following sample items.

The two Population 1 items selected for in-depth analysis deal with a number sentence item and a number sense situation. These items give a picture of emerging student understanding of whole number relationships and place value.

Item S02: Complete Number Sentence. Item S02 is a short-answer item that measures students' competencies in performing routine procedures in the content area of fractions and number sense. This item, shown in Figure 2, requires students to find the number that satisfies a given equation. Table 4 shows the varied student performances in the Population 1 comparison countries as well as the mean international performance percentages for all countries in Population 1. Although 62.6 percent of students internationally responded correctly to the item, performances in the comparison countries ranged from 45.2 (Norway) to 92 percent (Singapore) correct.

Table 4. Student performance percents for item S02

	Cluster	Full Credit	Incorrect	Non-response
Singapore	1	92.0	7.5	0.5
Japan	1	85.9	11.7	2.4
Netherlands	2	83.3	14.9	1.8
Hungary	2	76.0	11.9	12.1
United States	2	57.7	38.6	3.7
Canada	3	65.4	28.1	6.5
Norway	3	45.2	44.4	10.4
Greece	4	51.7	26.5	21.8
International		62.6	26.7	10.7

Analysis of the codes for incorrect and omitted responses, shown in Table 5, shows that the most frequent error made by students in most countries was answering 7 instead of 700 (Code 70). This perhaps reflects an inability of students in several countries to recognize the difference between the place value of a number and its face value. Note, however, the very low percentage of this occurrence among students in Singapore and Japan. Several students in Greece (4.2 percent) confused the place value of 700 with that of 70 (Code 72). The larger percentages in the column for Code 73 represent the use of other numbers that can be found embedded in the number 2739, such as 739.

S2.   Here is a number sentence.

$$2000 + \square + 30 + 9 = 2739$$

What number goes where the $\square$ is to make this sentence true?

Answer: _____

Code	Response
Correct Response	
10	700 or written out as "seven hundred."
Incorrect Response	
70	7
71	43
72	70
73	Gives other numbers made by digits in 2739 such as 73, 30, 9, 39, 739, 2739,...
79	Other incorrect.
Nonresponse	
90	Crossed out/erased, illegible or impossible to interpret.
99	BLANK

Figure 2. Population 1 mathematics item S02 and rubric.

The examination of the last two columns in Table 5, those for Codes 90 (non-interpretable response) and 99 (blank) provide other information. While the number of non-interpretable responses was low, there were large differences among the comparison countries on the percentages of papers left blank. Note also that the three countries with over 10 percent of students' papers blank represent different performance clusters. Thus, the non-response data reflects something about how these countries' students approach responding to constructed response items. Note that there were also differences in the overall correct performance of students from these three countries.

Table 5. Percents of correct, incorrect, and non-response for item S02

	Cluster	Scoring Code							
		10	70	71	72	73	79	90	99
Singapore	1	92.0	0.9	0.0	0.1	2.1	4.2	0.2	0.5
Japan	1	85.9	3.0	0.0	0.6	1.9	5.4	0.9	2.4
Netherlands	2	83.3	8.4	0.0	0.5	3.1	2.5	0.4	1.8
Hungary	2	76.0	6.7	0.0	0.6	0.9	3.4	0.3	12.1
United States	2	57.7	15.7	0.2	1.8	5.6	14.8	0.6	3.7
Canada	3	65.4	13.9	0.1	2.7	1.8	8.6	1.1	6.5
Norway	3	45.2	32.7	0.0	2.7	2.1	4.1	2.8	10.4
Greece	4	51.7	11.8	0.0	4.2	4.2	5.2	1.0	21.8
International		62.6	13.1	0.1	1.9	2.9	7.7	1.0	10.7

Item T02: Make Smallest Whole Number. Item T02 assessed students'
competencies in solving problems in the content area involving number sense. The
item called on students to employ their knowledge of place value in whole numbers
to find the smallest whole number that could be written with the four digits 4, 3, 9,
and 1. This item was a significantly more difficult item for Population 1 students
than the previous item, as the international mean percentage correct was only 43.4.

T2. What is the smallest whole number that you can make using the digits
4, 3, 9 and 1? Use each digit only once.

Answer: _____

Code	Response
Correct Response	
10	1349
Incorrect Response	
70	1,3,4,9
71	1
72	4
73	17
74	Any four-digit number with digits 4,3,9 and 1, other than 1349
75	13 OR "1 and 3" OR "3 and 1"
79	Other incorrect.
Nonresponse	
90	Crossed out/erased, illegible or impossible to interpret.
99	BLANK

Figure 3. Population 1 mathematics item T02 and scoring rubric.

Again, Table 6 shows that there were considerable differences in the performances of students from the comparison countries. The percent correct ranged from 28.6 (Norway) to 72.4 (Singapore).

Table 6. Student performance percents for item T02

	Cluster	Full Credit	Incorrect	Non-response
Singapore	1	72.4	26.5	1.3
Japan	1	52.9	41.3	5.8
Hungary	2	57.9	29.9	12.2
United States	2	49.7	46.2	4.2
Netherlands	2	43.1	54.4	2.6
Canada	3	48.5	44.3	7.4
Norway	3	28.6	50.1	21.3
Greece	4	31.5	35.9	32.6
International		43.4	43.7	12.9

The analysis of errors, summarized along with correct responses in Table 7, shows that the most common error reflected a possible misunderstanding of the item (Code 71). These students selected the smallest of the four individual whole numbers given and simply responded with the number 1. Sizeable numbers of students from the Netherlands and Norway responded incorrectly with either 4 (Code 72) or 17 (Code 73). At the same time, some students in Canada, Greece, Hungary, Norway, and the United States responded with incorrect four-digit numbers (Code 74). This error indicates potential misunderstandings involving place value and whole number comparisons based on place value.

Table 7. Patterns of correct, incorrect, and non-response percents for item T02

		Scoring Codes								
COUNTRY	Cluster	10	70	71	72	73	74	75	79	90	99
Singapore	1	72.4	3.7	7.5	1.3	0.4	3.7	0.1	9.8	0.4	0.9
Japan	1	52.9	3.7	19.8	2.4	0.2	2.0	1.9	11.3	1.2	4.6
Hungary	2	57.9	0.3	4.0	1.7	0.8	9.0	1.5	12.6	0.0	12.2
United States	2	49.7	2.6	8.4	2.7	5.2	7.3	1.3	18.7	0.3	3.9
Netherlands	2	43.1	1.7	16.1	5.3	7.2	2.7	4.1	17.3	0.1	2.5
Canada	3	48.5	1.8	9.7	2.8	3.4	7.3	1.6	17.7	0.7	6.7
Norway	3	28.6	1.7	13.3	4.7	10.3	4.9	2.5	12.7	1.9	19.4
Greece	4	31.5	3.4	9.6	1.1	0.7	4.9	0.4	15.8	4.4	28.2
International		43.4	2.3	11.7	2.5	3.7	5.7	1.3	16.7	0.9	12.0

The same patterns existed in the countries having more than 10 percent of their students with blank responses, but this time the percents were substantially higher for Greece and Norway.

Population 2 Examples: Equations and Consumer Sense

The 10 comparison countries selected from Population 2 were Canada, France, Greece, Germany, Hungary, Japan, Netherlands, Norway, Singapore, and the United States. An analysis of Population 2 student performance on constructed response items indicated two items that had a large amount of variability in response patterns. These problems related to applications of the students' knowledge of algebra and knowledge of number sense in a consumer data interpretation setting. Neither of these items required students to do more than perform routine procedures.

Item L16: Solve for x. Item L16 required Population 2 students to solve a linear equation with variables appearing on both sides of the equals sign. Table 8 shows the wide variety of performance levels on this item. Students in Canada (26.9) and the Netherlands (21.0) exhibited some of the weakest performance whereas students in Japan (85.7) and Singapore (78.8) performed the best. The Netherlands (35.0) and Germany (31.4) had high student non-response rates, while Singapore (4.4) and Japan (3.0) had low student non-response rates.

L16. Find x if $10x - 15 = 5x + 20$

Answer: _____

Code	Response
Correct Response	
10	7
Incorrect Response	
70	1 OR 2.33.. OR 3
71	Other incorrect numeric answers.
72	Any expression or equation containing x.
79	Other incorrect
Nonresponse	
90	Crossed out/erased, illegible, or impossible to interpret.
99	BLANK

Figure 4. Population 2 mathematics item L16 and scoring rubric.

Table 8. Student performance percents for item L16

	Cluster	Full Credit	Incorrect	Non-response
Japan	1	85.7	11.2	3.0
Singapore	1	78.8	16.8	4.4
Hungary	2	72.0	19.1	8.8
France	2	50.8	32.3	16.9
Norway	2	38.6	41.9	19.5
United States	2	31.4	54.4	14.2
Canada	2	26.9	51.6	21.6
Netherlands	2	21.0	44.0	35.0
Germany	3	42.5	26.2	31.4
Greece	4	42.6	32.7	24.7
International		45.3	34.5	20.2

The percent of students reporting the correct solution, 7 (Code 10), is shown in Table 9. The percent of students coded as having one of the various types of incorrect responses or non-responses is also shown in Table 9. Code 70 corresponded to incorrect solutions of 1 or 2.333... or 3. A student might have obtained $x = 1$ by adding and subtracting incorrectly on each side of the equation to obtain $25x = 25$. A similar error may have resulted in an intermediate equation of $5x = 15$, leading to the solution $x = 3$. If the student had incorrectly added $5x$ to the left side and subtracted it from the right side, and then performed other algebraic steps correctly, he or she would have obtained $x = 2.33...$(Code 70). Code 71 represents other incorrect numerical answers. Code 72 represents solutions or equations involving x. The error that may have occurred here was students leaving the variables on separate sides of the equation. As a result, the solution would be x equal to some expression involving x along with other coefficients or constants. Code 79 represents other incorrect solutions.

Table 9. Credit patterns, incorrect responses, and non-response percents for item L16

	Cluster	Code 10	Code 70	Code 71	Code 72	Code 79	Code 90	Code 99
Japan	1	85.7	2.3	6.4	0.3	2.2	0.7	2.3
Singapore	1	78.8	6.0	6.7	1.4	2.7	0.2	4.2
Hungary	2	72.0	5.9	11.1	1.3	0.8	0.2	8.6
France	2	50.8	3.0	18.5	8.8	2.0	0.6	16.3
Norway	2	38.6	6.2	21.1	12.5	2.1	5.9	13.6
United States	2	31.4	3.4	32.6	8.0	10.4	4.0	10.2
Canada	2	26.9	3.4	35.2	8.7	4.3	5.9	15.7
Netherlands	2	21.0	3.2	16.6	12.7	11.5	20.6	14.4
Germany	3	42.5	4.1	11.4	6.3	4.4	2.6	28.8
Greece	4	42.6	5.3	21.4	4.8	1.2	2.7	22.0
International		45.3	3.9	15.6	9.3	5.7	3.2	17.0

For this exercise, the most common error across the ten comparison countries, was other incorrect numeric answers (Code 71). Students in Canada and the United States provided the most incorrect answers of this type. Students in the Netherlands and Norway provided the greatest number of incorrect answers with terms involving x (Code 72). An analysis of the illegible or noninterpretable answers (Code 90) shows a high value of 20.6 percent of the responses in the Netherlands. Germany and Greece had over 20 percent of their students leaving pages blank.

Item V2: Price of Renting Office Space. Item V2, which was shown earlier in Figure 1 along with its rubric, required students to consider representations in the context of data representation, analysis, and probability. Specifically, students had to compare prices (with a currency unit of *zeds*) for renting a specified office size, given the rental rates in two buildings. For this problem, full credit was awarded three points, partial response was awarded two points, and minimal response was awarded one point.

The analysis of the percentage of students earning full, partial, minimal, incorrect, and non-response scores on Item V2 at the Grade 8 level showed that 45.8 percent earned at least one point, 25.1 percent earned at least two points, and 19.1 percent for students earned three points. As this item was also given as part of the Population 3 literacy assessment, further analysis of student performance at the Population 2 level is included there in contrast to the performance noted at that level.

Population 3 Mathematics Literacy: Consumer Mathematics and Data Sense

The literacy items given to the Population 3 students were similar in content and difficulty to many of the items given to the Population 2 students. In this section, we report data for a group of nine focus countries: Canada, the Czech Republic, France, Germany, Hungary, the Netherlands, Norway, Sweden, and the United States. The two items selected for analysis for the Population 3 Literacy portion are drawn from consumer literacy (the building rent problem that was examined in Population 2) and from statistical literacy.

Item A12: Price of Renting Office Space. Item A12, is identical to the item shown in Figure 1 that was given to Population 2 students (V2). It requires students to compute the prices (in zeds) of office space in two buildings and to determine the lower price. For this problem, full credit was awarded three points, partial response was awarded two points, and minimal response was awarded one point. The international average of students earning full credit was 49.1 percent. This compares to an international average of 19.1 percent for Population 2, but one has to be careful in making such comparisons, as the countries participating in TIMSS at the two levels of schooling differ in their membership. Restricting the data for full credit at the Population 2 level data to exactly the same countries participating in the Mathematical Literacy Study for Population 3, we find that the restricted Population 2 average performance would have been 25 percent. In either case, students had made significant progress, but were still performing below a level that indicates strong consumer skills.

Table 10 shows summary data for full credit, partial credit, minimal credit, incorrect, and non-response data for the eight focus countries for Population 3 Literacy. Correct (3 point) responses included both the correct price computations and the correct comparison. The partial and minimal (2 points and 1 point, respectively) responses had either the comparison or the price computation correct or partially correct.

Table 10. Student performance percents for item A12

	Cluster	Full Credit	Partial Credit	Minimal Credit	Incorrect	Non-response
Netherlands	1	70.3	6.5	12.5	6.6	4.1
Sweden	1	69.6	5.5	10.4	5.5	9.1
Canada	1	57.3	7.2	11.0	9.4	15.0
France	2	66.0	6.5	7.9	3.7	15.9
Norway	2	59.6	7.0	11.7	7.2	14.4
Germany	3	45.4	8.4	11.4	10.2	2.8
Hungary	3	39.4	5.8	13.8	12.5	28.4
Czech Republic	3	38.4	6.8	14.2	10.2	30.3
United States	4	39.1	8.7	16.8	17.1	18.5
International		49.1	6.4	12.0	10.0	22.6

For Population 3, the best performance on item A12 in the nine-country sample was by students in the Netherlands. The Netherlands also had the lowest rate of non-response, which was generally consistent with what was reported for other items in other TIMSS populations. The Czech Republic, Hungary, and the United States all had similarly poor performance on this item. The Czech Republic and Hungary were the two countries with the highest rates of non-response.

Seven countries were included in both the Population 2 and the Population 3 focus groups. These were Canada, France, Germany, Hungary, the Netherlands, Norway, and the United States. For all of these countries, the Population 3 sample did considerably better than the Population 2 sample. The biggest difference in performance between the two populations was in the Netherlands where the percentage of students earning full credit was 45.9 percentage points higher for the Population 3 sample than for the Population 2 sample. In contrast, the least difference in performance was in Hungary, where the average percent of students earning full credit in Population 3 was only 19.1 percentage points higher than the percent of students earning full credit from the Population 2 sample. Table 11 shows the distribution of more specific categories of correct, partial, minimal, incorrect, and non-responses to item A12. The codes for correct, partial, minimal, and non-response are as given in Figure 1.

Table 11. Credit pattern percents for student work on item A12

	Cluster	Scoring Code								
		30	39	20	21	10	11	12	16	19
Canada	1	56.6	0.7	5.9	1.3	3.8	2.6	4.4	0.0	0.2
Czech Rep.	3	36.1	2.3	5.4	1.4	3.2	7.2	3.6	0.1	0.1
France	2	65.6	0.4	5.0	1.5	3.3	1.3	3.2	0.0	0.1
Hungary	3	39.1	0.3	3.4	2.4	2.1	2.7	8.5	0.3	0.2
Netherlands	1	69.9	0.4	4.3	2.2	3.0	3.4	5.4	0.4	0.3
Germany	3	43.9	1.5	5.6	2.8	2.3	5	3.9	0.1	0.1
Norway	2	58.4	1.2	4.1	2.9	3.4	3.0	5.1	0.2	0.0
Sweden	1	67.9	1.7	3.6	1.9	3.5	2.2	4.1	0.5	0.1
United States	4	38.9	0.2	8.0	0.7	3.8	5.9	6.2	0.9	0.0
International		48.1	1.0	4.8	1.6	3.3	3.6	4.5	0.4	0.2

Table 12. Patterns of incorrect response and non-response percents for item A12

	Cluster	Scoring Codes				
		70	71	79	90	99
Canada	1	4.1	3.8	1.5	1.1	13.9
Czech Rep.	3	3.1	5.8	1.3	0.9	29.4
France	2	1.5	1.5	0.7	0.6	15.3
Hungary	3	5.3	4.7	2.5	1.3	27.1
Germany	3	13.1	9.9	0.4	1.6	4.1
Netherlands	1	2.4	2.9	1.3	1.0	3.1
Norway	2	3.6	2.9	0.7	1.4	13.0
Sweden	1	2.5	2.4	0.6	0.4	8.7
United States	4	4.3	9.5	3.3	1.5	17.0
International		4.0	4.3	1.7	0.9	21.7

For all countries in the focus group, within the classification of partial credit, more students correctly said that Building A had the lower price, but reported the correct rent calculation for only one of the buildings (Code 20) than calculated both rents correctly but said that Building B was the lower rental (Code 21). In the classification of minimal response, the specific errors that included calculation errors or no calculations were more prevalent than the error of extracting components of advertisement for a justification. The specific incorrect responses did not seem to follow a clear pattern for all countries nor for any specific countries, although the percentages of incorrect responses were lower in Population 3 than the percentages for Population 2.

Item D17: Graph with Robberies per Year. Item D17 shows an interrupted bar graph with a statement claiming a huge difference in the values represented by the two bars. Students were asked whether the statement was a reasonable interpretation of the graph and to explain their response. For this problem, a correct response was awarded two points, a partial response was awarded one point, and incorrect responses were not awarded any points. The international average of students responding correctly was 19 percent. Figure 5 shows item D17 and its rubric.

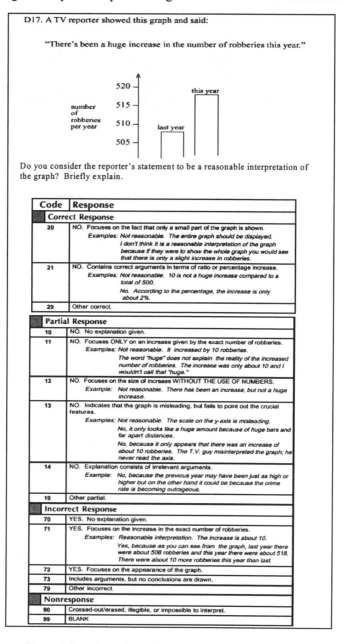

Figure 5. Population 3 mathematics item D17 and scoring rubric.

The full credit responses to this item include a claim that the interpretation of the graph is not reasonable and appeal to the misleading nature of the interrupted scale or to the specific values. Partial credit responses claim that the statement is not a reasonable interpretation of the graph, but the explanation is incomplete or irrelevant. The incorrect responses claim that the statement is a reasonable interpretation of the graph, with various types of explanations. Table 13 shows the summary distribution of the eight-country focus group within the categories of full credit, partial credit, incorrect, and non-response.

Sweden was the country with the best performance on this item, with 37.4 percent of students being awarded full credit. Czech Republic and Hungary both had very poor performance with only 5.8 percent and 4.3 percent of students being awarded full credit, respectively. The Netherlands had, by far, the lowest rate of non-response. The Czech Republic, France, and Hungary all had very high rates of non-response.

Table 13. Student performance percents for item D17

	Cluster	Full Credit	Partial Credit	Incorrect	Non-response
Canada	1	22.6	34.6	15.4	27.6
Czech Rep.	3	5.8	26.0	31.1	37.2
France	2	21.6	24.7	9.7	44.0
Hungary	3	4.3	25.5	17.2	53.1
Germany	3	20.5	26.0	12.7	41.0
Netherlands	1	29.8	26.7	34.4	9.3
Norway	2	34.3	24.5	12.5	28.9
Sweden	1	37.4	28.7	13.3	20.8
United States	4	14.4	40.6	29.3	15.5
International		18.8	25.1	20.5	35.8

Tables 14 and 15 provide more detailed information about the distribution of more specific categories of full, partial, incorrect, and non-responses to Item D17. Of the specific types of full-credit responses, there was a consistent pattern across all countries. Students provided an explanation about why the statement was not reasonable based on the specific numbers that could be read from the graph much more frequently than they appealed to the general problem of having an interrupted vertical axis giving a misleading visual representation of the relative heights of the bars.

Likewise, there was a consistent pattern of preferred partial credit responses for all eight countries in the focus group. The most frequently observed partial credit response was assigned Code 12. This explanation claimed that the increase was not as large as the statement made it seem, but did not appeal to specific numbers from the graph. Codes 10 or 11 were the least frequently occurring partial credit responses for all countries except Hungary. Code 10 corresponds to a response that omits an explanation. Code 10 was the least frequently used code for all countries except Canada. Code 11 corresponds to an explanation that includes the difference in number of robberies, not the proportional difference, which would be required to justify a claim of huge difference. The strategies associated with the incorrect responses did not fall into the same type of consistent international pattern that the full and partial credit responses did. For most countries these strategies were somewhat uniformly distributed. However, this was not the case for the Netherlands and the United States. In these countries, there were greater numbers of students that were misled by the relative heights of the bars shown in the diagram. These students claimed that the statement was reasonable and their explanations were based on the appearance of the graph.

Table 14. Credit pattern percents for student work on item D17

		Scoring Code							
	Cluster	20	21	29	10	11	12	13	14	19
Canada	1	4.9	11.5	4.4	1.8	0.8	22.5	4.2	4.4	2.7
Czech Rep.	3	1.2	2.8	1.2	0.6	2.6	13.4	5.1	2.5	2.4
France	2	1.7	15.2	3.9	0.8	2.0	12.4	2.3	6.6	1.4
Hungary	3	1.2	2.8	0.2	0.1	1.5	18.7	4.4	0.5	0.4
Germany	3	2.7	15.2	1.1	1.5	3.5	10.6	5.3	4.8	1.8
Netherlands	1	5.3	18.7	4.8	1.0	1.2	8.2	7.1	7.5	2.7
Norway	2	7.9	16.3	10.0	0.1	1.6	15.1	3.4	4.2	0.2
Sweden	1	8.0	17.8	10.7	0.9	2.1	16.4	3.8	4.2	2.2
United States	4	1.7	8.4	4.3	0.0	1.6	26.3	6.1	3.6	3.0
International		3.9	10.8	3.2	0.8	1.7	14.0	4.1	3.3	2.0

Table 15. Patterns of incorrect responses and non-response percents for item D17

		Scoring Code						
	Cluster	70	71	72	73	79	90	99
Canada	1	1.4	3.3	2.7	0.3	7.7	0.1	27.5
Czech Rep.	3	7.9	6.1	6.2	7.4	3.5	0.4	36.8
France	2	2.2	3.0	1.8	0.8	1.9	2.9	41.1
Germany	3	3.8	1.6	2.8	1.3	3.2	1.4	39.6
Hungary	3	4.4	6.8	3.0	0.7	2.3	0.2	52.9
Netherlands	1	4.2	6.4	18.6	1.0	4.2	1.3	8.0
Norway	2	3.9	1.4	4.1	1.4	1.7	0.6	28.3
Sweden	1	2.5	1.9	2.8	2.5	3.6	0.1	20.7
United States	4	2.4	5.5	11.6	1.2	8.6	0.8	14.7
International		4.1	5.1	5.0	2.2	4.0	0.7	35.1

Population 3 Advanced Mathematics Problem Solving and Proving

The differing set of countries participating in the Population 3 advanced mathematics portion of TIMSS necessitated choosing another group of comparison countries. As at the other levels of the assessment, a cluster analysis was performed on the response patterns of students to multiple choice, short answer, and extended response items. After examining the various clusters formed, seven countries were selected to both represent the groups as well as to maintain some of the same countries contained within the comparison groups for other portions of this chapter. This resulted in the choices of Canada, the Czech Republic, France, Germany, Greece, Sweden, and the United States.

Item K14: String on Cylinder. Item K14, a constructed response item measuring students' ability to determine the length of a string wound around a cylinder, is perhaps one of the most frequently exhibited items from the advanced mathematics item set. Student performance on the item indicates that Item K14 was the most difficult item on the advanced mathematics examination. It was developed with the intention of measuring students' problem solving competencies involving spatial and geometric reasoning. As Table 16 shows, there was considerable variability among the focus countries on the item with slightly less than 10 percent of students internationally receiving full credit for the item. The average high non-response rate of 42 percent may indicate that the item was considered by students to be unique and residing outside the curriculum they had experienced.

Table 16. Student performance percents for item K14

	Cluster	*Full Credit*	*Partial Credit*	*Incorrect*	*Non-response*
Greece	1	5.0	0.9	22.8	71.5
France	1	4.1	1.8	27.7	66.4
Sweden	2	23.5	1.0	52.1	23.3
Czech Rep.	2	8.3	4.4	40.4	46.7
Canada	3	12.2	0.7	72.6	14.4
Germany	4	7.9	0.8	45.4	46.0
United States	5	3.8	0.5	80.8	14.9
International		9.7	1.8	46.2	42.4

The performances of students receiving full credit for the item in the comparison countries ranged from a high of nearly 24 percent in Sweden to a low of less than 4 percent in the United States. The performance of the Swedish students was an outlier in the distribution of performances. Only in the Czech Republic did a sizeable portion of other students receive partial credit for their work on the item. Students receiving full or partial credit for the item most often approached the item through an application of the Pythagorean theorem. The differences in these solution methods are the ways in which students visualize the unwrapping of the string from the cylinder to form rectangles.

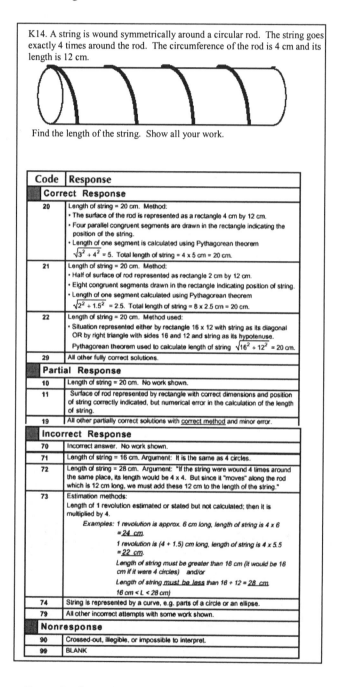

K14. A string is wound symmetrically around a circular rod. The string goes exactly 4 times around the rod. The circumference of the rod is 4 cm and its length is 12 cm.

Find the length of the string. Show all your work.

Code	Response
Correct Response	
20	Length of string = 20 cm. Method: • The surface of the rod is represented as a rectangle 4 cm by 12 cm. • Four parallel congruent segments are drawn in the rectangle indicating the position of the string. • Length of one segment is calculated using Pythagorean theorem $\sqrt{3^2 + 4^2} = 5$. Total length of string = 4 x 5 cm = 20 cm.
21	Length of string = 20 cm. Method: • Half of surface of rod represented as rectangle 2 cm by 12 cm. • Eight congruent segments drawn in the rectangle indicating position of string. • Length of one segment calculated using Pythagorean theorem $\sqrt{2^2 + 1.5^2} = 2.5$. Total length of string = 8 x 2.5 cm = 20 cm.
22	Length of string = 20 cm. Method used: • Situation represented either by rectangle 16 x 12 with string as its diagonal OR by right triangle with sides 16 and 12 and string as its hypotenuse. Pythagorean theorem used to calculate length of string $\sqrt{16^2 + 12^2} = 20$ cm.
29	All other fully correct solutions.
Partial Response	
10	Length of string = 20 cm. No work shown.
11	Surface of rod represented by rectangle with correct dimensions and position of string correctly indicated, but numerical error in the calculation of the length of string.
19	All other partially correct solutions with <u>correct method</u> and minor error.
Incorrect Response	
70	Incorrect answer. No work shown.
71	Length of string = 16 cm. Argument: It is the same as 4 circles.
72	Length of string = 28 cm. Argument: "If the string were wound 4 times around the same place, its length would be 4 x 4. But since it "moves" along the rod which is 12 cm long, we must add these 12 cm to the length of the string."

73	Estimation methods:
	 Length of 1 revolution estimated or stated but not calculated; then it is multiplied by 4.
	    Examples: 1 revolution is approx. 6 cm long, length of string is 4 x 6
	       = <u>24 cm</u>.
	      1 revolution is (4 + 1.5) cm long, length of string is 4 x 5.5
	       = <u>22 cm</u>.
	      Length of string must be greater than 16 cm (it would be 16 cm if it were 4 circles) and/or
	      Length of string <u>must be less</u> than 16 + 12 = <u>28 cm</u>.
	      16 cm < L < 28 cm)

74	String is represented by a curve, e.g. parts of a circle or an ellipse.
79	All other incorrect attempts with some work shown.
Nonresponse	
90	Crossed-out, illegible, or impossible to interpret.
99	BLANK

Figure 6. Advanced Mathematics item K14 and rubric.

An analysis of the ways in which students approached the item, represented in Table 17, shows that most of the Swedish students who earned full credit relied on two distinct strategies (Codes 20 and 22) for obtaining their solution. Czech students tended to rely more consistently on one strategy (Code 20) as did the students in the other comparison countries.

Table 17. Credit pattern percents for student work on item K14

| | Cluster | Scoring Code | | | | | | |
		20	21	22	29	10	11	19
Greece	1	0.6	0.9	0.8	2.7	0.9	0.0	0.0
France	1	2.7	1.4	0.0	0.0	0.0	1.3	0.5
Sweden	2	14.5	1.5	7.5	0.0	0.2	0.6	0.2
Czech Republic	2	6.3	1.2	0.8	0.0	1.5	0.0	2.9
Canada	3	9.3	0.4	1.1	1.4	0.4	0.2	0.1
Germany	4	4.2	0.3	3.0	0.4	0.8	0.0	0.0
United States	5	3.2	0.0	0.6	0.0	0.2	0.2	0.1
International		6.5	1.0	1.5	0.7	0.7	0.2	0.9

Table 18. Patterns of incorrect responses and non-response percents for item K14

| | Cluster | Scoring Code | | | | | | | |
		70	71	72	73	74	79	90	99
Greece	1	8.8	0.6	0.0	12.5	0.0	0.9	6.7	64.8
France	1	1.5	8.9	2.5	2.4	0.0	12.4	2.7	63.7
Sweden	2	1.3	6.8	21.0	2.0	0.8	20.2	2.4	20.9
Czech Republic	2	5.7	8.7	6.3	4.0	1.3	14.4	9.6	37.1
Canada	3	0.6	17.7	11.5	0.6	0.4	41.8	4.1	10.3
Germany	4	1.5	10.7	7.9	2.3	0.4	22.6	17.5	28.5
United States	5	2.3	12.2	9.7	0.3	0.1	56.2	3.0	11.9
International		3.4	8.4	7.6	2.8	1.4	22.5	6.3	36.1

Students who worked on the item, but who did not receive any credit for their work varied a great deal in their approaches. Across the various codes for incorrect work, each one indicates that students in one or more of the countries tried to solve the problem using a method that would fall under at least one of the patterns in the rubric. The code of 71 that reflected that the students felt that the length would be the same as four circle circumferences, ignoring the movement of the string down the cylinder, was a favorite in many of the countries. An adjustment to the argument to include the 12 cm length of the rod also was attempted by many students as shown by Code 72. Greece was unusual in that these two error patterns were less likely to be encountered than some of the others included in the list of incorrect approaches observed.

Item K18: Proving ΔABC is Isosceles. An interesting constructed response item was K18, a geometry item measuring the ability of students to justify and prove. K18, shown in Figure 7 along with its rubric, asked students to construct a proof that triangle ABC is isosceles based on the angle information given. Considerable differences existed in the performances of the students in the seven countries observed (See Table 19). In the top performing countries, France and Greece, approximately 65 percent of the students received full or partial credit on the item. In comparison, 55 percent of Canadian and 50 percent of Swedish students earned full or partial credit, whereas only 19 percent of U. S. students earned credit at either level. Internationally, 48.4 percent of students earned full or partial credit.

Table 19. Student performance percents for item K18

	Cluster	Full Credit	Partial Credit	Incorrect	Non-response
Greece	1	64.9	0.6	3.8	30.7
France	1	53.4	11.1	17.7	17.9
Sweden	2	40.8	9.2	24.1	26.0
Czech Republic	2	33.8	10.3	27.9	28.0
Canada	3	34.4	21.1	38.4	6.2
Germany	4	21.1	10.0	19.2	49.7
United States	5	9.7	9.3	63.3	17.8
International		34.0	14.4	24.8	26.8

Examining the data for the awarding of full and partial credit gave rise to some interesting patterns. The Greek students tended to get full credit as compared to the combinations of full and partial credit patterns noted in other countries. By taking a deeper look into the distribution of full credit (two points) strategies, shown in Table 20, we see that among the students receiving full credit for the item, almost all students in each of the seven countries gave a fairly traditional formal proof based on angle sums and the isosceles triangle theorem (Code 20).

Table 20. Credit pattern percents for student work on item K18

		Scoring Code					
	Cluster	20	21	29	10	11	19
Greece	1	64.9	0.0	0.0	0.6	0.0	0.0
France	1	51.1	2.0	0.3	6.7	0.9	3.5
Sweden	2	37.8	2.2	0.8	6.8	1.6	0.8
Czech Republic	2	31.4	1.1	1.3	6.9	2.1	1.3
Canada	3	33.1	0.6	0.7	8.2	2.0	10.9
Germany	4	18.2	1.4	1.5	7.2	0.4	2.4
United States	5	9.1	0.6	0.0	7.4	0.5	1.4
International		31.4	2.0	1.5	10.7	1.3	3.2

K18. In the ΔABC the altitudes BN and CM intersect at point S. The measure of ∠MSB is 40° and the measure of ∠SBC is 20°. Write a PROOF of the following statement:

"ΔABC is isosceles."

Give geometric reasons for statements in your proof.

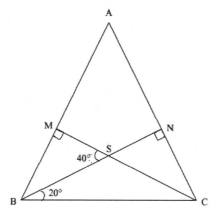

Note: To be considered correct, all responses must include mention of all geometric facts used, all calculations made, and a conclusion

Code	Response
Correct Response	
20	Correct proof. Proves that ∠B = ∠C using the following facts: • the sum of angles in any triangle is 180°. • if two angles of a triangle are equal, the triangle is isosceles. and possibly also uses: • vertically opposite angles are equal. • supplementary angles add to 180°. The concept of congruence is not used.
21	As code 20 but somewhere in the proof uses the fact that some triangles: e.g. triangles BCM and CBN, OR triangles BMS and CNS, are congruent.
29	All other fully correct and complete proofs.
Partial Response	
10	As in codes 20-21 shows ∠B and ∠C are equal giving steps in logical order, but omits one step or one reason or gives one incorrect reason.
11	As in codes 20-21 shows ∠B and ∠C are equal, states correct geometric facts but not in a logically correct order.
19	Other responses with minor errors.
Incorrect Response	
70	Shows measures of angles correctly on figure but no geometric facts mentioned or argumentation given.
71	Incorrect argumentation and/or includes more than one incorrect geometric fact, step, or reason.
72	"Proof" is circular; makes use of statements which are equivalent to what is to be proven.
79	Other incorrect responses.
Nonresponse	
90	Crossed-out, illegible, or impossible to interpret.
99	BLANK

Figure 7. Advanced Mathematics item K18 and scoring rubric.

A look at the codes for partial credit (1 point) shows that the patterns were somewhat similar in all countries with the exception of Canada, where a frequently used partial justification focused on something other than angles B and C and their measures.

An examination of the incorrect and non-response data for item K18, summarized in Table 21, shows that the United States had the highest total with 81 percent of its students in the two categories. These data indicate the vast differences in national predispositions to respond. Greek students, for example, tended not to respond unless they knew something about the item. Students from the United States and Canada tended to give some form of incorrect response rather than leave an item with a non-response. Students in the other three countries tended to have about as many students giving an incorrect response as giving a non-response. The most common form of incorrect work was noted by Code 71. It indicated a pattern of incorrect arguments including incorrect geometric facts, steps, or reasons.

Table 21. Patterns of incorrect responses and non-response percents for item K18

		Scoring Code					
	Cluster	*70*	*71*	*72*	*79*	*90*	*99*
Greece	1	2.9	0.0	0.0	0.9	3.8	30.7
France	1	2.8	5.1	0.5	9.3	17.7	17.8
Sweden	2	9.6	3.5	0.0	11.0	24.1	25.9
Czech Republic	2	6.5	8.6	4.3	8.5	27.9	28.0
Canada	3	4.5	12.6	1.1	20.2	38.4	6.1
Germany	4	6.8	1.8	0.6	10.0	16.9	32.8
United States	5	3.8	33.6	3.7	22.2	63.3	17.7
International		8.0	6.1	1.6	8.9	2.9	22.6

CONCLUSION

The examination of the data related to student performance at the item level as done in this chapter provides mathematics educators and policy researchers with a variety of findings. The first is the varied response rates of students across countries. Here the data show that students in some countries may have a propensity to respond to items whether they know the answers or not. In other countries, students may tend to respond only if they feel confident about the correct answer.

A second finding is the illumination of students' thinking provided by the two-digit rubrics in scoring items and in examining the results of students' work. Although the two-digit coding does not provide detailed information about students' thinking on every problem, it does provide an extra lens through which we may view differences in student performance across countries. In particular, by coding several different error patterns and coding several different solution strategies, we can link patterns of misconceptions and predominant solution methods to various groups of students. These patterns, then, provide some insight into the strengths and weaknesses of the groups, which, in this case, were the school programs of different countries. For example, on the string on a cylinder item, Item K14 in Advanced Mathematics for Population 3, we see that, in contrast to many other countries, students in Sweden were able to engage in the three-dimensional visualization required to solve the problem. This may indicate that geometric visualization is a strong component of Sweden's geometry curriculum. Similarly, two-digit coding enhanced the information obtained from Items S02 and T02 for Population 1. The coding scheme enabled us to identify which countries had students who struggled with place value, as well as identify the frequency with which students made the various types of errors.

These findings, and others, highlight the valuable information that may be obtained using a two-digit scoring rubric in large scale assessments. This type of scoring method can also add greatly to national and regional assessments of student performance in mathematics. If students are to learn mathematics from a conceptual and constructive standpoint, it is important that the scope and nature of that understanding is probed in a fashion that illuminates both the depth and breadth of students' understanding. The two-digit rubric scoring approach provides one such method for making these observations.

REFERENCES

Beaton, A. E., Mullis, I. V. S., Martin, M. O., Gonzalez, E. J., Kelly, D. L., & Smith, T. A. (1996). *Mathematics achievement in the middle school years.* Boston, MA: TIMSS International Study Center.

Hair, J. F., Jr., Anderson, R. E., Tatham, R. L., & Black, W. C. (1998). *Multivariate data analysis* (5th Edition). Upper Saddle River, NJ: Prentice Hall.

JMP Statistical Software. (2000). *JMP® Statistics and graphics guide.* Cary, NC: SAS Institute, Inc.

Lie, S., Taylor, A., & Harmon, M. (1996). Scoring techniques and criteria. In M. O. Martin & D. L. Kelly (Eds.), *Third International Mathematics and Science Study technical report: Volume 1: Design and development.* Chestnut Hill, MA: Center for the Study of Testing, Evaluation, and Educational Policy, Boston College.

Mullis, I. V. S., Jones, C. O., & Garden, R. A. (1996). Training sessions for free-response scoring and administration of performance assessment. In M. O. Martin & D. L. Kelly (Eds.), *Third International Mathematics and Science Study technical report: Volume 1: Design and development* (pp. 10-1 to 10-11). Chestnut Hill, MA: Center for the Study of Testing, Evaluation, and Educational Policy, Boston College.

Mullis, I. V. S., Martin, M. O., Beaton, A. E., Gonzalez, E. J., Kelly, D. L., & Smith, T. A. (1997). *Mathematics achievement in the primary school years.* Boston, MA: TIMSS International Study Center.

Mullis, I. V. S., Martin, M. O., Beaton, A. E., Gonzalez, E. J., Kelly, D. L., & Smith, T. A. (1998). *Mathematics and science achievement in the final year of secondary school.* Boston, MA: TIMSS International Study Center.

Robitaille, D. F. (Ed.). (1997). *National contexts for mathematics and science education.* Vancouver, B.C.: Pacific Educational Press.

Chapter 4

FROM SIMS TO TIMSS: TRENDS IN STUDENTS' ACHIEVEMENT IN MATHEMATICS

David F. Robitaille and Alan R. Taylor

Over the past four decades, IEA has carried out four investigations of students' achievement in mathematics. The first survey, known as FIMS or the First International Mathematics Study, was conducted in 1964 with 12 countries participating. The second, SIMS or the Second International Mathematics Study, was carried out in 1980–82 with 20 participating countries or educational systems. In the 1990's two rounds of TIMSS were carried out—TIMSS-95 and TIMSS-99— with 41 countries participating at the Grade 8 level in the former, and 38 in the latter.

An earlier volume included a paper by the authors (Robitaille & Taylor, 1989) describing changes in patterns of students' achievement in mathematics between FIMS and SIMS. The goal of the present paper is to extend that analysis by examining changes between SIMS and the two rounds of TIMSS.

Each of IEA's four mathematics studies included a population of 13-year-old students who were enrolled in what, in almost all of the participating countries, was their eighth year of formal schooling. The operational definitions for that population of students in the four studies are listed below for comparison purposes. In the remainder of this paper, we will refer to this as the Grade 8 population.

In other words, all four studies included a Grade 8 population, and that population was defined in almost exactly the same way in each of the studies. Part of the reason for the similarity is that this age or grade level was considered in all four studies to be the highest at which one could be reasonably sure that most of the children included in the age cohort were still in school in most countries and were still studying mathematics. It is also the case that in many countries all students at this level take the same mathematics course. That is to say, in most countries tracking or streaming of students into different mathematics courses or different sections of the same course on the basis of some measure of aptitude or ability has yet to begin in earnest.

D.F. Robitaille and A.E. Beaton (eds.), Secondary Analysis of the TIMSS Data, 47–62.

Table 1. IEA mathematics studies: Populations of 13-year old students

IEA Study	Population Description
FIMS Population A	All students who were 13 years old at the time of testing, or who were in the grade where the majority of 13-year-old students were to be found. (Husén, 1967)
SIMS Population A	All students in the grade (year level) where the majority have attained the age of 13.00 to 13.11 years by the middle of the school year. (Travers, Garden, & Rosier,1989)
TIMSS–95 Population 2 (upper grade)	All students in the [upper of the] pair of adjacent grades that contained the most students who were 13 years old at the time of testing. (Robitaille & Garden, 1996)
TIMSS–99 Population 2 (upper grade)	All students in the [upper of the] pair of adjacent grades that contained the most students who were 13 years old at the time of testing. (Mullis et al., 2000)

The focus of this chapter is on changes in students' mathematics achievement between SIMS and the two rounds of TIMSS. One factor influencing this choice was our desire to maximize the numbers of countries included in the comparisons. Since few countries have participated in all four studies to date, we decided not to include the FIMS data in this analysis. A second factor was the fact that there were comparatively few achievement items used in FIMS, where all students responded to the same test booklet. On the other hand, the item pools for each of the later studies contained more than 150 mathematics items: 157 in SIMS, 151 in TIMSS–95, and 162 in TIMSS–99. The greater the number of items, especially when those items are spread across a number of content areas or strands, the more confidence one may have in making generalizations about students' achievement in mathematics in general.

Unfortunately, there are no items that appeared in exactly the same form in all four studies. Taken pairwise, the studies did share significant numbers of items. There were 35 items common to FIMS and SIMS, 23 for SIMS and TIMSS–95, and 48 for TIMSS-95 and TIMSS-99. However, in the process of developing the item pool for each study, most of the items that might have been repeated underwent some degree of change. The TIMSS policy of holding about half of the items secure and not releasing them also limited the numbers of common items available for publication.

Our comparisons of students' performance across the several studies, therefore, are based not on comparisons of performance on the same items, but on sets of items similarly labelled. Thus we compare, among others, changes in students' performance in algebra. The items making up the algebra subtest differed in each study, but each of those sets of items was deemed to be a valid measure of achievement in algebra. Moreover, our analysis focuses on the relative performance of students and not on the raw scores obtained. This helps control for variation in the difficulty of the items used in the different studies.

Over the almost four decades that have elapsed since the FIMS data were collected, several influential waves of educational reform focusing on the teaching and learning of mathematics have occurred. These have included "New Math," "Back to Basics," student-centered approaches, and most recently standards-based approaches. Each such reform wave has presumably left its mark on the systems involved, but no attempt has been made in the IEA studies or in the analyses undertaken in the preparation of this paper to account for such impact.

In this chapter we focus on comparisons of students' achievement in mathematics in 12 educational systems from 11 countries at three points in time: SIMS, TIMSS–95, and TIMSS–99. The jurisdictions are listed below, with asterisks indicating those that also participated in FIMS.

Belgium (Flemish)*	Israel*
Canada (BC)	Japan*
Canada (Ontario)	Netherlands*
England*	New Zealand
Hong Kong	Thailand
Hungary	United States*

Of the 12, nine represent entire countries. However, in Belgium and Canada only parts of each country participated in all three studies. In Belgium, the Flemish school system participated in all three studies; in Canada, the provinces of British Columbia and Ontario did so. In this paper, each of the 12 jurisdictions is referred to as a "country" in order to simplify the text. The authors are aware that the term "country" is used somewhat inaccurately in this context.

A final caveat concerns the quality of the national samples selected to participate in each study. The standards for sampling in IEA studies are very high and the quality of the samples selected is also very high, especially in the more recent studies. However, there is some variation in sample quality between countries in the same study, and within a given country between studies. Detailed information on sampling quality is available in the international reports published for each study.

STUDENT DEMOGRAPHICS

As was noted above, all three studies focused on Grade 8 students. Tables 2 and 3 present a summary of some basic demographic information about the students who participated in the three studies for comparison purposes.

As expected, the mean age of Grade 8 students participating in the three studies from a given country remained fairly constant across studies and across countries. Three exceptions occurred in the cases of Japan, Hong Kong, and Ontario where the mean age of their students in SIMS was about a year less than the mean age elsewhere. In Hong Kong and Japan, a deliberate decision had been made to use Grade 7 students rather than Grade 8 students, and that accounts for the discrepancy in their cases. In Ontario, the students were in Grade 8, but the figure reported may have been the ages of the students at the beginning of the year rather than at the time of testing, which was some nine months later.

Table 2. Mean age of students participating in the IEA mathematics studies (in years)

	SIMS	TIMSS-95	TIMSS-99
Belgium (Flemish)	14.2	14.1	14.1
Canada (BC)	14.0	13.9	13.9
Canada (Ontario)	13.4	14.0	13.9
England	14.1	14.0	14.2
Hong Kong	13.2	14.2	14.2
Hungary	14.2	14.3	14.4
Israel	14.0	14.1	14.1
Japan	13.5	14.4	14.4
Netherlands	14.4	14.3	14.2
New Zealand	14.0	14.0	14.0
Thailand	14.3	14.6	14.5
United States	14.1	14.2	14.2

Table 3 summarizes the gender composition of the national samples of students by reporting what percent of each country's sample was girls. There are a few anomalous results which may be related to sampling issues—for example, the large proportion of girls in the Thai sample for TIMSS-95—but the overall pattern is one of nearly equal number of boys and girls from each country in each study.

Table 3. Gender of students participating in the IEA mathematics studies (percent girls)

	SIMS	TIMSS-95	TIMSS-99
Belgium (Flemish)	52	50	50
Canada (BC)	50	49	51
Canada (Ontario)	50	50	50
England	54	48	49
Hong Kong	49	46	49
Hungary	52	51	50
Israel	49	51	51
Japan	49	48	50
Netherlands	49	50	52
New Zealand	50	48	49
Thailand	48	61	54
United States	52	49	50

TRENDS IN STUDENTS' ACHIEVEMENT IN MATHEMATICS

In each of the three studies, a relatively large number of test items was used to assess the level of students' performance in mathematics. Also, in each case, the items were divided into a number of discrete subtests covering major curricular strands such as algebra, geometry, and measurement. About half the items used in TIMSS–95 and TIMSS–99 were identical, but only a small proportion of the SIMS items were used in either of the later studies. That is to say, the tests were not identical but they were comparable in the sense that they were both designed to assess students' performance in mathematics in general as well as in a number of well-defined curricular strands.

The scores from the three studies are not directly comparable. SIMS used percent correct scores; TIMSS–95 reported Rasch scores; and TIMSS–99, scaled scores using a three-parameter IRT model. Because of the incomparability of these scores in absolute terms, we decided to compare students' performance across the three studies in relative terms using z-scores, with a mean of 0 and a standard deviation of 1. These z-scores were based on the international average and standard deviations of results from among the countries being compared for each study. Comparisons were made using total test scores as well as five subtest scores: algebra, geometry, number and number sense, measurement, and data analysis.

Total Test Comparisons

The graphs in Figure 1 summarize changes in patterns of students' achievement over the three studies. The three scores for each country have been connected in order to make it easier for the reader to track changes country by country. Across the 12 jurisdictions, the z-scores for SIMS range from a low of –1.5 for Thailand to a high of 2.1 in Japan. In the case of TIMSS–95, the spread is from –1.2 in the United States to 1.9 in Japan; and for TIMSS–99, from –1.5 in Thailand to 1.6 in Hong Kong.

The graph shows some consistent patterns across the three studies. Japanese students did extremely well in all three studies, with each of their total-test z-scores being in the neighborhood of 2.0. Their performance on SIMS is all the more remarkable when one considers that the Japanese students were a year younger than those from almost all the other countries. The Hong Kong students in SIMS were also a year younger, and their SIMS result was considerably weaker than those achieved by their Grade 8 counterparts in the two rounds of TIMSS.

The performance of students in Belgium (Flemish) and those in Hong Kong is particularly noteworthy. These are the only two jurisdictions whose scores, relative to those for students in other countries, improved consistently from study to study.

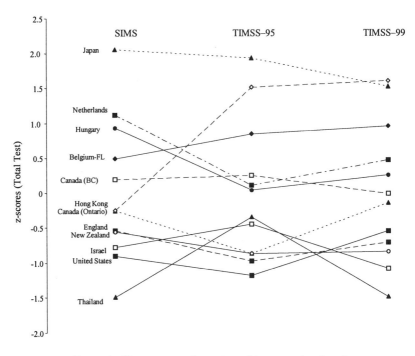

Figure 1. Changes in mathematics achievement (total test).

Table 4 summarizes the total-test scores in a somewhat different way. In that table, the countries are grouped within each study into three groups. A country was classified as high-achieving if its total score was more than half a standard deviation above the mean, and it was classified as low-achieving if its total score was more than half a standard deviation below the mean. Otherwise, the country was classified as mid-range. In the table, the countries are listed alphabetically within each cell.

Table 4. Relative achievement levels for total test score

	SIMS	*TIMSS–95*	*TIMSS–99*
High	Hungary	Belgium (Flemish)	Belgium (Flemish)
	Japan	Hong Kong	Hong Kong
	Netherlands	Japan	Japan
Mid-range	Belgium (Flemish)	Canada (BC)	Canada (BC)
	Canada (BC)	Hungary	Canada (Ontario)
	Canada (Ontario)	Israel	Hungary
	England	Netherlands	Netherlands
	Hong Kong	Thailand	United States
Low	Israel	Canada (Ontario)	England
	New Zealand	England	Israel
	Thailand	New Zealand	New Zealand
	United States	United States	Thailand

Viewed in this way, the three best-performing countries within this group of 12 were Belgium (Flemish), Hong Kong, and Japan. The total-test scores of students from those countries were in the high category in at least two of the three rounds of testing. Overall scores for England, Israel, New Zealand, Thailand, and the United States were in the low category in at least two of the three rounds.

Subtest Comparisons

For each of the three studies, the item pool was partitioned into a number of subtests. The names of these subtests varied somewhat across the studies, but for the purpose of this analysis we have labelled them as follows: algebra, geometry, number and number sense, measurement, and data analysis. Table 5 presents a summary of students' achievement on these subtests across studies. As in the previous table, the scores have been categorized as high (denoted by ⇑), low (denoted by ⇓), or mid-range (denoted by ⇔).

Table 5. Comparative achievement levels by subtest [a]

	Algebra			Geometry			Number & Number Sense			Measurement			Data Analysis		
	SIMS	TIMSS–95	TIMSS–99	SIMS	TIMSS–95	TIMSS–99	SIMS	TIMSS–95	TIMSS–99	SIMS	TIMSS–95	TIMSS–99	SIMS	TIMSS–95	TIMSS–99
Belgium (Fl)	⇑	⇑	⇑	⇔	⇔	⇑	⇑	⇑	⇑	⇑	⇑	⇑	⇔	⇑	⇑
Canada (BC)	⇔	⇔	⇔	⇔	⇔	⇔	⇑	⇑	⇔	⇔	⇔	⇔	⇔	⇑	⇔
Canada (Ontario)	⇓	⇓	⇔	⇔	⇓	⇔	⇔	⇓	⇔	⇔	⇓	⇔	⇔	⇓	⇔
England	⇓	⇓	⇓	⇔	⇓	⇓	⇓	⇓	⇓	⇔	⇔	⇔	⇔	⇔	⇔
Hong Kong	⇔	⇑	⇑	⇔	⇑	⇑	⇔	⇑	⇑	⇔	⇑	⇑	⇓	⇑	⇑
Hungary	⇑	⇑	⇑	⇑	⇔	⇔	⇑	⇔	⇔	⇑	⇔	⇑	⇔	⇔	⇔
Israel	⇔	⇔	⇓	⇓	⇔	⇓	⇔	⇓	⇓	⇓	⇓	⇓	⇔	⇓	⇓
Japan	⇑	⇑	⇑	⇑	⇑	⇑	⇑	⇑	⇑	⇑	⇑	⇑	⇑	⇑	⇑
Netherlands	⇑	⇔	⇔	⇑	⇔	⇔	⇑	⇔	⇑	⇑	⇔	⇑	⇑	⇑	⇑
New Zealand	⇓	⇓	⇓	⇑	⇓	⇓	⇓	⇓	⇓	⇓	⇓	⇔	⇔	⇔	⇓
Thailand	⇓	⇓	⇓	⇓	⇓	⇔	⇔	⇓	⇓	⇓	⇓	⇓	⇔	⇓	⇓
United States	⇓	⇔	⇔	⇓	⇔	⇓	⇓	⇓	⇔	⇓	⇔	⇓	⇓	⇓	⇔

[a] ⇑ indicates high achieving; ⇓, low achieving; and ⇔, mid range.

The table illustrates that countries whose scores placed them in the high-performing group on the basis of their total test scores tended to do well on all or almost all of the subtests. Japan was in the high-scoring group for each of the 5 subtests in all three rounds of testing, for a perfect result of 15 high achieving scores and 0 low achieving scores. The result for Belgium (Flemish) was 12 high scores and 0 low scores; for Hong Kong, it was 10 and 1, the latter being a low achieving score in data analysis on SIMS.

At the other end of the distribution, Thailand had 12 low scores and no high scores. The result for New Zealand was 11 low scores and 1 high score. For Israel it was 10 low scores and 0 high ones; for the United States, 9 and 0; for England, 8 and 0.

Achievement in Algebra

The algebra subtest covered similar topics in all three studies. The SIMS algebra subtest included items dealing with integers, rational numbers, formulas, expressions, exponents, relations and functions, linear equations and inequalities, and sets. Topics in the algebra subtest for TIMSS included linear equations, expressions, and patterns and relationships. There were 30 items on the SIMS algebra subtest; 27 on the TIMSS–95 subtest; and 35 on the TIMSS–99 subtest.

Students from three of the participating countries did consistently well on the algebra subtest. These were Belgium (Fl), Hungary, and Japan. On the other hand, students from England, New Zealand, and Thailand had consistently lower scores on this subtest than those from the other countries included in these comparisons.

$$\frac{x}{2} < 7 \text{ is equivalent to}$$

A. $x < \dfrac{7}{2}$ D. $x > 5$

B. $x < 5$ E. $x > 14$

C.* $x < 14$

* indicates the correct response.

Figure 2. SIMS Item 118, TIMSS-95 Item K4: Algebra.

Table 6. Percent Correct on SIMS Item 118, TIMSS–95 Item K4: Algebra

	SIMS	TIMSS–95
Belgium (Fl)	43	57
Canada (BC)	38	46
Canada (Ontario)	38	40
England	33	33
Hong Kong	41	61
Hungary	36	53
Israel	30	52
Japan	45	69
Netherlands	40	49
New Zealand	34	35
Thailand	43	–[a]
United States	45	52

[a]Data not available.

A sample algebra item that was included in the algebra subtest for both SIMS and TIMSS–95 is shown above. This item, along with all of the other sample items included in this paper, was among the items released for public use after TIMSS–95. None of the released items was used again in TIMSS–99; they were replaced by items that were parallel in form and content.

Performance on this item improved by five percentage points or more between SIMS and TIMSS-95 in 8 of the 11 countries for which complete data were available, and no country showed a decrease. Particularly large increases in performance were registered by students in Hong Kong, Israel, and Japan. Option A was the most popular distractor, chosen by between 25 and 35 percent of students in most countries in both studies.

Less consistent results were found on an item involving substitution into an equation. Students were asked to solve for a variable in a formula by substituting values for the other variables. The item is shown below.

$P = LW$. If $P = 12$ and $L = 3$, then W is equal to

A. $\dfrac{3}{4}$ D. 12

B. 3   E. 36

C.* 4

Figure 3. SIMS Item 17, TIMSS–95 Item Q7: Algebra.

Table 7. Percent Correct on SIMS Item 17, TIMSS–95 Item Q7: Algebra

	SIMS	*TIMSS–95*
Belgium (Fl)	61	77
Canada (BC)	73	77
Canada (Ontario)	73	65
England	56	53
Hong Kong	70	91
Hungary	68	66
Israel	56	75
Japan	81	88
Netherlands	71	75
New Zealand	60	59
Thailand	57	73
United States	68	71

Results in five countries increased by five or more percentage points on this item, and only one showed a decrease of more than five points. In the other countries, little change in performance was noted. The most popular incorrect answer choice was A.

Achievement in Geometry

The geometry subtest in SIMS included items dealing with properties of plane figures, congruence and similarity, coordinates, transformations and Pythagorean triangles. In TIMSS, topics included transformations, two-dimensional geometry, and three dimensional shapes and surfaces. Many of the subtopics included within each of these categories were the same in both studies. The geometry subtest consisted of 39 items in SIMS, 23 in TIMSS–95, and 21 in TIMSS–99.

Students from Japan did well on the geometry subtest in all three studies, while those from Hong Kong showed strong performance in geometry in two of the three studies. Students from England, New Zealand, Thailand, and the United States had the weakest overall results in geometry, with z-scores that were at least half a standard deviation below the average on two of the three occasions.

Students found the following geometry item rather difficult. It involved the concept of congruence and required students to identify corresponding angles and sides in a pair of congruent triangles.

These triangles are congruent. The measures of some of the sides and angles of the triangle are shown.

What is the value of x?

A. 52
B.* 55
C. 65
D. 73
E. 75

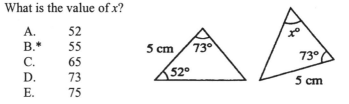

Figure 4. SIMS Item 25, TIMSS–95 Item K8: Geometry.

Table 8. Percent Correct on SIMS Item 25, TIMSS–95 Item K8: Geometry

	SIMS	TIMSS–95
Belgium (Fl)	23	43
Canada (BC)	30	20
Canada (Ontario)	30	21
England	37	31
Hong Kong	34	61
Hungary	48	39
Israel	31	43
Japan	49	69
Netherlands	30	21
New Zealand	25	26
Thailand	23	33
United States	20	17

Less than 50 percent of students answered this item correctly in either study. In the case of SIMS, Japanese students recorded the highest percent correct on this item with a score of 49. They also had the highest score among this group of countries on this item in TIMSS–95, but that score was also relatively low at 69. In quite a few cases, *p*-values for this item were close to or below chance level.

Not surprisingly, the most frequently chosen incorrect response was "A." About half the students selected that response. Those students may have been misled by the diagram since the lengths of the sides are rather similar in both triangles. In any case, they failed to identify the appropriate corresponding side in the triangle on the right.

The geometry item that showed the greatest improvement in students' performance involved the concept of supplementary angles. That item is shown below.

In this figure, AB is a straight line.

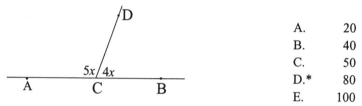

A.	20
B.	40
C.	50
D.*	80
E.	100

What is the measure, in degrees, of angle BCD?

Figure 5. SIMS Item 155, TIMSS–95 Item M7: Geometry.

Table 9. Percent Correct on SIMS Item 155, TIMSS–95 Item M7: Geometry

	SIMS	TIMSS–95
Belgium (Fl)	53	85
Canada (BC)	63	85
Canada (Ontario)	69	76
England	52	67
Hong Kong	66	83
Hungary	74	86
Israel	62	67
Japan	80	87
Netherlands	68	74
New Zealand	65	73
Thailand	63	79
United States	51	61

Performance on this item increased in every one of the 12 jurisdictions by at least five percentage points. In several cases, performance increased by more than 20 points to *p*-values between 85 and 90 percent.

Achievement in Numbers and Number Sense

This subtest was called "Arithmetic" in SIMS and "Fractions and Number Sense" in the two rounds of TIMSS. In the case of SIMS, this subtest covered topics such as whole numbers, fractions and decimals, ratio, proportion, percent, and estimation. In TIMSS the material on this subtest included common and decimal fractions, estimation, and number sense. The SIMS arithmetic subtest consisted of 46 items. The TIMSS–95 subtest in this area included 51 items; and the TIMSS–99 subtest, 61.

Scores for Belgium (Flemish) and Japan were among the highest on this subtest in all three studies. Canada (BC), Hong Kong, and the Netherlands were among the highest in two of the three. Scores for England and New Zealand were among the lowest on all three occasions, while those for Israel, Thailand, and the United States were among the lowest on two out of three.

Only two items common to SIMS and TIMSS–95 were among the TIMSS items that were released for publication. The first involved division with decimals, and the second relationships between ordered pairs. These are discussed below.

$$\text{Divide:} \quad 0.004\overline{)24.56}$$

$$
\begin{array}{ll}
\text{A.} & 0.614 \\
\text{B.} & 6.14 \\
\text{C.} & 61.4 \\
\text{D.} & 614 \\
\text{E.*} & 6140
\end{array}
$$

Figure 6. SIMS Item 108, TIMSS–95 Item J14: Number Sense.

Table 10. Percent Correct on SIMS Item 108, TIMSS–95 Item J14: Number Sense

	SIMS	TIMSS–95
Belgium (Flemish)	41	56
Canada (BC)	53	59
Canada (Ontario)	61	47
England	29	19
Hong Kong	37	70
Hungary	65	68
Israel	34	55
Japan	59	71
Netherlands	50	34
New Zealand	24	23
Thailand	23	34
United States	60	46

None of the scores from either study were very high on this item, and this is rather surprising given the fact that the item is solely about where to place the decimal point in the quotient. Six countries showed some improvement, while performance in four others declined. Scores in England and New Zealand were considerably lower than those in other countries.

The second sample item from the numbers subtest involved ordered pairs. This item produced particularly low scores, which for most countries were close to or below chance. Students found this item difficult. In no country did more than half the students obtain the correct answer in either study, and in most cases results were below the chance level. Between the studies, five countries improved their results by 5 or more percentage points, with the greatest improvements occurring in British Columbia and Belgium (Flemish).

The table shows the values of x and y,
where x is proportional to y.

x	y
3	7
6	Q
P	35

A. P = 14 and Q = 31
B. P = 10 and Q = 14
C. P = 10 and Q = 31
D. P = 14 and Q = 15
E.* P = 15 and Q = 14

What are the values of P and Q?

Figure 7. SIMS Item 142, TIMSS–95 Item L14: Number Sense.

Table 11. Percent Correct on SIMS Item 142, TIMSS–95 Item L14: Number Sense

	SIMS	TIMSS–95
Belgium (Fl)	18	33
Canada (BC)	18	38
Canada (Ontario)	18	22
England	22	18
Hong Kong	24	38
Hungary	25	24
Israel	26	17
Japan	50	48
Netherlands	22	29
New Zealand	17	19
Thailand	26	39
United States	22	20

Performance in Measurement

The measurement subtest for SIMS consisted of 24 items covering topics such as units of measurement, approximation, area and perimeter, and volume. The TIMSS–95 test included 18 measurement items covering similar topics; and the TIMSS–99 test, 24 items. There were no measurement items that appeared in exactly the same form in SIMS and TIMSS-95, so no sample items are discussed here. Students from Belgium (Fl), Hong Kong, Hungary, Japan, and the Netherlands did comparatively better on this subtest than students from other countries. Those from Israel, New Zealand, Thailand, and the United States had poorer results.

Performance in Data Analysis

The SIMS test included a subtest with the title, "Statistics." It covered topics such as data collection, organization of data, data representation, and data interpretation, as well as specific topics such as mean, median, and mode. In the case of TIMSS the subtest was called "Data Representation, Analysis, and Probability." It covered interpretation and analysis of graphs, graphs of functions and relations, probability, and sampling. For this report, these subtests have been labelled data analysis. There were 18 items on the SIMS subtest in this area, and 21 for both TIMSS–95 and TIMSS–99. There were three items on this subtest that were common to SIMS and TIMSS-95, but none were released for publication.

Countries with the best performance in data analysis across the three rounds of testing were Belgium (Fl), Hong Kong, Japan, and the Netherlands. Israel, Thailand, and the United States had the lowest scores among these countries.

CONCLUSION

However one looks at this trend data, it is clear that the mathematics achievement of Grade 8 students in some countries was consistently strong in all three studies when compared to the achievement of students from other countries in the comparison group. In particular, the data show that students from the Flemish-speaking sector of Belgium, from Hong Kong, and Japan were almost always among the highest-scoring countries in SIMS, TIMSS-95, and TIMSS-99 whether one looks at the overall score or at subtest scores.

There is, quite naturally, a great deal of interest internationally in trying to ascertain what can be learned about "what works" in successful schools and school systems, so that successful practices might be emulated in other environments. One thing that is clear from the IEA studies is that no single variable (class size, teacher qualifications, hours of instruction, …) and no simple combination of such variables differentiates among systems in a consistent way. Each such variable may well have a role in explaining differences across systems; but so do factors such as the socio-cultural milieu of the schools, societal and parental support for and involvement in schools, and a host of others that are largely beyond the immediate influence of the schools.

Some relevant information is available from the "TIMSS Encyclopedia" (Robitaille, 1997). For that volume, authors from each of the countries participating in TIMSS were asked to write about mathematics and science education in their countries. In particular, they were asked to comment on recent changes in areas such as curriculum, examinations, and the use of technology. This information provides some contextual background that can be used to interpret the achievement results.

The chapter about Belgium in the TIMSS Encyclopedia (Monseur & Brusselmans-Dehairs, 1997) deals with the country as a whole. The authors report that major changes in the mathematics curriculum have been introduced, and that "a purely modern approach has been replaced by a compromise between modern and traditional mathematics, with more stress on algorithms and geometry and less attention to set theory, groups, and vector spaces." All students follow the same course of study for mathematics until the end of Grade 8. The authors also point out that the kinds of comparisons that are drawn between countries on the basis of studies such as these are taken quite seriously by policy makers in Belgium.

The paper about Hong Kong (Leung & Law, 1997) reports that few changes had been made to the mathematics curriculum there, and that "the impact of modern technology on the curriculum has been minimal." All students follow the same course of study for mathematics through Grade 9, although some remedial programs are offered.

The authors of the chapter related to Japan (Miyake & Nagasaki, 1997) make several points about the mathematics education of Japanese students. They point out that there is no streaming or tracking of students for mathematics at this level; all Grade 8 students follow the same, compulsory mathematics course. A revised curriculum for mathematics was introduced during the 1990s, and the goals of that revision were very similar to those in other countries around the world. They also say that, although mathematics teachers are encouraged to use calculators and computers in their teaching, very few do so.

REFERENCES

Husén. T. (Ed.). (1967). *International study of achievement in mathematics, Vols. 1 and 2.* New York: John Wiley.

Leung, F. K. S., & Law, N. W. Y. (1997). Hong Kong. In David F. Robitaille (Ed.), *National contexts for mathematics and science education: An encyclopedia of the education systems participating in TIMSS.* Vancouver, BC: Pacific Educational Press.

Miyake, M., & Nagasaki, E. (1997). Japan. In David F. Robitaille (Ed.), *National contexts for mathematics and science education: An encyclopedia of the education systems participating in TIMSS.* Vancouver, BC: Pacific Educational Press.

Monseur, C., & Brusselmans-Dehairs, C. (1997). Belgium. In David F. Robitaille (Ed.), *National contexts for mathematics and science education: An encyclopedia of the education systems participating in TIMSS.* Vancouver, BC: Pacific Educational Press.

Mullis, I. V. S., Martin, M. O., Gonzalez, E. J., Gregory, K. D., Garden, R. A., O'Connor, K. M., Chrostowski, S. J., & Smith, T. A. (2000). *TIMSS 1999 international mathematics report.* Chestnut Hill, MA: International Study Center, Lynch School of Education, Boston College.

Robitaille, D. F., & Garden, R. A. (Eds.). (1989). *The IEA study of mathematics II: Contexts and outcomes of school mathematics.* Oxford: Pergamon Press.

Robitaille, D. F., & Taylor, A. R. (1989). *Changes in patterns of achievement between the first and second mathematics studies.* In David F. Robitaille & Robert A. Garden (Eds.), *The IEA study of mathematics II: Contexts and outcomes of school mathematics,* Oxford: Pergamon Press.

Robitaille, D. F., & Garden, R. A. (1996). *Design of the study*. In D. F. Robitaille & R. A. Garden (Eds.), *TIMSS monograph No. 2: Research questions and study design*. Vancouver, BC: Pacific Educational Press.

Robitaille D. F. (Ed.). (1997). *National contexts for mathematics and science education: An encyclopedia of the education systems participating in TIMSS*. Vancouver, BC: Pacific Educational Press.

Travers, K. J., Garden, R. A. & Rosier, M. (1989). *Introduction to the study*. In D. F. Robitaille & R. A. Garden (Eds.), *The IEA study of mathematics II: Contexts and outcomes of school mathematics*. Oxford: Pergamon Press.

Chapter 5

THE IMPACT OF TIMSS ON THE MATHEMATICS STANDARDS MOVEMENT IN THE UNITED STATES

John A. Dossey and Mary M. Lindquist

Discussions of the findings of large-scale international assessments and of the health of mathematics education in the United States have been integrally intertwined since the release of the results of the Second International Mathematics Study (SIMS) in the late 1980s. These results, along with those from the National Assessment of Educational Progress (NAEP) in mathematics, an assessment carried out by the U. S. Department of Education, have been the markers against which the mathematical progress of U. S. students have been judged. During the last fifteen years of the 20th century there was a virtual explosion of new assessments, as almost all of the 50 states instituted their own assessment programs (American Federation of Teachers, 1998).

It is in this context that efforts in the United States to reform the teaching and learning of school mathematics have been taking place. Spurred by the release of the National Council of Teachers of Mathematics' *Curriculum and Evaluation Standards for School Mathematics* (NCTM, 1989) and its companion pieces for instruction and assessment (NCTM, 1991, 1995), almost all states either instituted or revised their curricular goals for school mathematics (Blank et al., 1997). This process continues, with the focus in most states on aligning state and local curricular documents to bring them into line with the recently released NCTM *Principles and Standards for School Mathematics* (NCTM, 2000). There is no doubt that both the Standards movement and the growth of large-scale assessments have had a great influence on mathematics education in the United States. Present reforms and governmental legislation show the ongoing belief that politicians and the public have in the link between standards in general and the results of assessments of student achievement. This belief in testing and in high standards for school programs seems to be even more tightly held when the assessment results come from international studies. However, the interaction of the two, especially where international studies are involved, is an understudied topic. Both authors have been deeply involved with the Standards as well as with the international studies and NAEP. Thus, this chapter is partly their reflection on this interaction.

D.F. Robitaille and A.E. Beaton (eds.), Secondary Analysis of the TIMSS Data, 63–79.
© 2002 *Kluwer Academic Publishers. Printed in the Netherlands.*

THE EVOLUTION OF TIMSS IN RELATION TO MATHEMATICS REFORM IN THE U. S.

Any analysis of the impact of TIMSS findings on mathematics education in the United States must begin with the fit of the TIMSS framework to extant curriculum documents or recommendations for change in school mathematics. Work on the TIMSS framework began shortly after the release of the 1989 version of the *Standards* and work on the *Principles* began shortly after the collection of the TIMSS student achievement data in 1995-96.

The TIMSS framework evolved from a number of activities. The initial work on the framework was carried out by the Study of Mathematics and Science Opportunities, which was funded by the National Science Foundation. This study, chronicled in *Characterizing Pedagogical Flow* (Schmidt et al., 1996), was a small-scale precursor to TIMSS structured to help characterize the major variables involved in describing and studying mathematics and science curricula in an international context. In the end, the curricular framework for TIMSS grew from that used in previous IEA studies, most prominently that of SIMS (Travers & Westbury, 1989). The initial drafts of the curricular framework were vetted at conferences attended by international groups of mathematics educators, one held in the spring of 1991 in Washington, D.C. and the second, sponsored by the International Commission on Mathematical Instruction, in Calonge, Spain (Robitaille et al., 1993).

The resulting framework included three aspects of mathematics curricula: content, performance expectations, and perspectives. The content areas included in the mathematics framework were numbers; measurement; geometry: position, visualization, and shape; geometry: symmetry, congruence, and similarity; proportionality; functions, relations, and equations; data, probability, and statistics; elementary analysis; validation and structure; and an omnibus "other" content area consisting basically of topics dealing with computer science and logic. This framework provided enough latitude to measure patterns of content coverage from the primary grades through the end of secondary school with sufficient depth to discern similarities and differences in national curricula. The content portion of the framework also was neutral enough to capture the content irrespective of whether the curriculum was constructed about a spiraled or totally integrated model.

In addition, the framework allowed nations' textual materials, videos, and other curricular artifacts to be coded with respect to the intentions with which they were developed. These codings were called performance expectations and they described what the student was supposed to do or to gain from the experiences provided by the materials. In mathematics, these student performance expectations were five in number: knowing, using routine procedures, investigating and problem solving, mathematical reasoning, and communicating. The final dimension of the framework consisted of the perspectives: ancillary topics that may receive special emphasis in one country or another. These perspectives were: attitudes towards mathematics, careers in mathematics, participation in mathematics by underrepresented groups, mathematics topics to increase interest, and mathematical habits of mind (Robitaille et al., 1993).

The National Council of Teachers of Mathematics published the *Curriculum and Evaluation Standards in School Mathematics* in 1989, about the time that the work on the TIMSS framework was beginning. The standards address what students should know and be able to do in K-12 mathematics. The 1989 NCTM Standards grew out of a number of factors, but most central was the realization that U. S. mathematics education in the middle grades was still rooted in the study of arithmetic and repetition of skills with whole numbers. An examination of U. S. curricula as a result of SIMS had shown that the mathematics curricula of the majority of other countries were already looking forward to algebra at this level. In addition, the results of SIMS indicated that the U. S. school mathematics curriculum contained too many topics at any given grade level. It needed greater focus (McKnight et al., 1987). Given that the locus of responsibility for the setting of educational standards rests at state and local levels in the United States, the original NCTM Standards took great care not to appear as a move toward a national curriculum. Rather it was an attempt to stimulate national discussion about what students should be expected to know and be able to do in three grade ranges: K-4 (ages 5-9); 5-8 (ages 10-13), and 9-12 (ages 14-17).

The 13 or 14 content recommendations made by the 1989 Standards at each grade-level interval were organized and named differently. The content standards at any grade level could easily be grouped into number, algebra, geometry, measurement, and probability and data analysis. For example, at grades K-4 (ages 5-9) there are standards on estimation, number sense and numeration, concepts of whole number operations, and whole number operations that are included in the single content area of number in TIMSS. There were also four process-oriented standards that were named the same across the grade-bands. These were mathematics as problem solving, mathematics as communication, mathematics as reasoning, and mathematical connections. These process standards are quite similar in nature to the performance expectations of TIMSS.

During the 1990s, mathematics educators in the United States were actively involved in reform efforts. Curriculum materials at all levels were being developed to meet the intent of the *Curriculum and Evaluation Standards in School Mathematics*. Professional development efforts were launched around the Standards, including more emphasis on technology and teaching for meaning. Assessment programs were aligning their tests with the proposed changes in curriculum. Surveys of states' curricula documents show that most states changed their mathematics curricula in the period from 1989 to 1995 to conform to the NCTM's recommendations (Blank et al., 1997). However, even with a great amount of activity, the amount of real change in mathematics classrooms was minimal. Ten years of effort produced a level of awareness, but there was still a need for another look at the Standards. The National Council of Teachers of Mathematics undertook the effort to reexamine and update the Standards shortly after the administration of TIMSS.

The document resulting from this effort, *Principles and Standards for School Mathematics* (NCTM, 2000), narrowed the content focus to the five content areas of number and operations, algebra, geometry, measurement, and data analysis and probability. In addition, the document delineated more explicit performance expectations for students in the following grade/age intervals: Pre-K–2 (ages 4-7), 3–5 (ages 8-10), 6–8 (ages 11-13), and 9–12 (ages 14-18). These Standards added representation to the previous four process standards: problem solving, reasoning and proof, communication, connections. In addition, the 2000 NCTM document discussed global principles that must be present in support and delivery structures to assure that the students can achieve the content and process-related goals listed.

THE FIT OF THE TIMSS FRAMEWORK TO U. S. CURRICULA

The question of the degree of fit between the TIMSS curricular framework and the U. S. curriculum is difficult to answer. Given that there are over 15,000 school districts in the United States and that each of these local entities has some ability to create its own mathematics curriculum for the students it serves, any attempt to describe a fit is problematic. This great diversity is one of the major problems in shaping and moving reform in school mathematics in the United States. Perhaps the easiest way of exploring the nature of a fit is a comparison of the TIMSS framework with that of the mathematics framework used for the 1996 NAEP (National Assessment Governing Board, 1994). NAEP is often referred to as "The Nation's Report Card," in that it is the only regularly scheduled national examination of student performance in mathematics carried out on a regular basis. In addition to collecting student achievement data, it also collects data on students, teachers, schools, and instruction in mathematics classrooms (National Center for Educational Statistics, 1999). As such, it is perhaps the best set of data at a national level about the current status of mathematics learning and teaching at a national level.

Population 1 Comparisons

Table 1 shows the comparison of the related Population 1 content topics areas as a percent of each of the assessments (National Assessment Governing Board, 1994; Mullis et al., 1997; Mitchell et al., 1999).

Although the data in the table present some notion of the match between the TIMSS and U. S. programs, the comparison is not straightforward. The percentages of items in each category are about the same, but the nature of an assessment rests in the items and the finer descriptions of the content. However, both the TIMSS and NAEP frameworks lack this level of specificity with regard to the populations and grade-levels of students, respectively. So, it is really the items on the student achievement portions of the assessments that actually define the alignment of the programs.

Table 1. Comparison of the TIMSS test and the 1996 NAEP Grade 4 test

Content Area	Number of Items							
	Multiple Choice		Short Answer		Extended Response		Total	
	TIMSS	NAEP	TIMSS	NAEP	TIMSS	NAEP	TIMSS	NAEP
Number & Operation	34	35	7	22	5	2	46	59
Measurement	16	19	3	7	1	0	20	26
Geometry	12	10	2	13	0	2	14	25
Data & Chance	8	9	2	5	2	2	12	16
Algebra & Functions	9	8	1	9	0	1	10	18
Total	79	81	15	46	8	7	102	144

The majority of items in the TIMSS Population 1 pool were reflective of the concepts and skills expected in the U. S. curricular recommendations for Population 1 students. However, there are several notable exceptions in the TIMSS item pool.

TIMSS included very few estimation items. Estimation skills are difficult to test in a large-scale assessment format, but the items included were based only on rounding to find an estimation of a computation. There was also no assessment of students' ability to produce reasonable answers to word problems.

The test for TIMSS Population 1 placed more emphasis on proportional reasoning items than what was called for in the NAEP framework. This was partly due to the decision that proportional reasoning was to be a highlighted strand for Population 2 in TIMSS. About ten, or one-tenth, of the items for Population 1 involved some notion of proportional reasoning or depended upon students seeing comparisons as multiplicative, rather than additive, in nature. United States students had not had much classroom experience by age 9 in solving such problems, and data from the NAEP assessment show that few students were successful when asked to respond to such tasks.

The measurement items in TIMSS used non-standard units such as squares, paper clips, or triangles (5 items), metric units (12 items), or non-metric units (3 items) such as hours or degrees. Although United States students are expected to know metric units, they also have the added burden of learning the traditional system of measurement (feet, cups, and pounds). Of course, none of these customary units were used in the international assessment items.

The format of most items of the TIMSS Population 1 was multiple-choice, a format which provided little opportunity for obtaining evidence of students' understanding or ability to communicate their thinking and reasoning processes. However, a real effort was made to include among the distractors for such items responses that would provide some degree of insight into the thinking processes employed by the students.

Population 2 Comparisons

The TIMSS tests for Population 2 were similar in structure to those proposed for the U. S. Grade 8 students by NAEP. Table 2 shows the comparison of the related content topics for this group of students (National Assessment Governing Board, 1994; Beaton et al., 1996; Mitchell et al., 1999).

Table 2. Comparison of the TIMSS test and the 1996 NAEP Grade 8 test

Content Area	Multiple Choice		Short Answer		Extended Response		Total	
	TIMSS	NAEP	TIMSS	NAEP	TIMSS	NAEP	TIMSS	NAEP
Number & Operation	49	31	11	13	2	2	62	46
Measurement	13	15	3	10	2	2	18	27
Geometry	22	15	1	15	0	1	23	31
Data & Chance	19	13	1	9	1	3	21	25
Algebra & Functions	22	21	3	12	2	0	27	33
Total	125	95	19	59	7	8	151	162

As with the Grade 4 comparisons of TIMSS and NAEP, the percentage of items in each content area is quite similar for both assessments. One area with some difference is that of number. This area was emphasized more for Population 2 students in TIMSS than it was for Grade 8 students in NAEP. There is, however, a substantial difference in the type of item format used. The TIMSS test contained more multiple choice items and fewer short answer items. Many of the short answer items in NAEP required a brief student-constructed explanation of their work in contrast to the TIMSS assessment that required only answers for such items.

A study sponsored by the National Center for Educational Statistics (NCES) gives more insight into the differences between TIMSS and NAEP (Nohara & Goldstein, 2001). This study compared the item pools for TIMSS-R in 1999 and NAEP as well as the Organization for Economic Cooperation and Development's (OECD) Program of International Student Assessment (PISA).

This comparison of items from the three assessments is germane, as the TIMSS-R assessment is basically a clone of the TIMSS test given in 1995. The results of study indicate that computation items comprised 34 percent of the TIMSS-R assessment compared to 27 percent on the NAEP test. Closer examination of these items showed that the computation required in TIMSS-R items focused on operations with fractions and integers. Very few of these items were to be found in NAEP 1996. The NCES report also found that 41 percent of NAEP items required multi-step reasoning in contrast to 31 percent of the TIMSS-R items. Thus, in many ways, the TIMSS-R test was less demanding cognitively than the NAEP test, yet often involved more advanced content relative to the typical U. S. curriculum at Grade 8.

Population 3 Literacy Comparisons

While the overall curriculum framework for TIMSS represented broad coverage of the potential mathematics that could be studied, the mathematics content selected for inclusion in the actual TIMSS assessments varied somewhat from one population level to another. The selection and inclusion of items was a function of the initial item pools developed and the performance of items in the field tests. As a result, the items that comprised the mathematical literacy assessment given to students enrolled in the final year of secondary school focused on items from number sense (fractions, percents, and proportions), elementary algebra, measurement, geometry, and some data representation and analysis. As such, the item coverage fell short of the content mastery expectations that most states currently hold for their students in algebra and geometry. In 1999, a survey of state officials overseeing mathematics education (Blank et al., 1999) reflected that, in over 80 percent of the states, over 50 percent of students were completing at least one year of geometry and two years of algebra study at the high school level.

Table 3. Comparison of the TIMSS literacy test and the 1996 NAEP Grade 12 test

Content Area	Number of Items							
	Multiple Choice		Short Answer		Extended Response		Total	
	TIMSS	NAEP	TIMSS	NAEP	TIMSS	NAEP	TIMSS	NAEP
Number & Operation	11	26	2	12	0	1	13	39
Measurement	11	12	1	11	0	0	12	23
Geometry	3	12	1	14	0	1	4	27
Data & Chance	4	14	1	18	2	3	7	35
Algebra & Functions	5	29	3	10	0	3	8	42
Total	34	93	8	65	2	8	44	166

The items on the TIMSS mathematical literacy assessment were, content wise, well within the competencies that American students should have achieved as a result of their study of mathematics at the middle and secondary school levels. However, in many cases, it is possible that their mathematics classes focused on a more formal study of the content, one less suited to performance on the more literacy-oriented level of items found in the TIMSS mathematical literacy assessment.

Population 3 Advanced Mathematics Comparisons

The content of the Population 3 assessment of advanced students exceeded what many of the United States' students had opportunities to learn. In order to get a sufficient number of students to meet the sample size requirements for participation in the advanced mathematics study, the United States had to include students who had not studied calculus. At the time of the assessment, only 14 to 15 percent of

U. S. students were taking an advanced mathematics class intended for students in their final year of secondary school. Slightly less than one half of these students had exposure, at least in any significant way, to the concepts of calculus. Thus, the U. S. sample for the advanced mathematics assessment represented about 14 percent of the age cohort still in school in the final year. Of these, only about one-half (7 percent of the age cohort) had any significant exposure to many of the topics tested (Takahira, Gonzales, Frase, & Salganik, 1998). A 2000 study replicating TIMSS with students in Advanced Placement Calculus classes found that their performance on the advanced TIMSS test in mathematics was significantly higher than that of any country, other than France, participating in the advanced Population 3 study (Gonzalez, O'Connor, & Miles, 2001). Thus, when advanced mathematics students in the United States have had an opportunity to learn the calculus, their performance is quite acceptable. This finding parallels others that suggest that the U. S. mathematics curriculum lags at least one year behind that observed in other countries from the middle grades forward.

While there was no direct comparison with NAEP for the TIMSS advanced mathematics assessment, an examination of the TIMSS content showed that the items were divided into content areas as follows: numbers and equations (26 percent), geometry (35 percent), probability and statistics (11 percent), validation and structure (5 percent), and calculus (23 percent). An examination of the items shows that the focus of the content in the test on calculus went beyond the content seen by approximately 50 percent of the students in the United States sample. This was one factor in the low achievement level of U. S. students on the assessment. However, in examining student performance data, it appears that other factors were at work as well. Besides topics from calculus, the other areas assessed were geometry; numbers, equations, and functions; and probability and statistics. The performance of United States' students was low in these areas as well, even though students had opportunities to study content from these areas as part of their secondary school mathematics curricula.

REFLECTIONS ON THE PERFORMANCE OF U. S. STUDENTS

The analyses by Dossey, Jones, and Martin elsewhere in this volume (Chapter 3) indicate that all three groups of United States students involved in TIMSS could be described as having response patterns that reflected moderate levels of correct responses, relatively high levels of incorrect responses, and low levels of omitted items. In short, United States' students are willing to try to answer items, whether or not they know the answers. This in itself may be symptomatic of the cluttered curricula they have experienced. Such exposure to many topics, but with few studied in any depth, may have left them with a wide exposure to concepts, but with an inability to decisively perform in almost any area of the mathematics curriculum.

This pattern is also seen when one compares the performance of United States' students (upper grades at Populations 1 and 2) with their peers in other countries that participated in the Population 1, Population 2, and Population 3 literacy assessments of TIMSS.

Table 4. Performance of countries participating in Population 1, 2, and 3-Literacy TIMSS

	Population 1	*Population 2*	*Population 3 (Literacy)*
Above the Mean (more than 1 s.d. above mean)	Netherlands Austria Hungary Australia United States	Netherlands Austria Hungary Australia Canada	Netherlands New Zealand Hungary
At the Mean (within ± 1 s.d. of mean)	Canada	New Zealand United States	Canada Australia Austria
Below the Mean (more than 1 s.d. below mean)	New Zealand		United States

A review of the TIMSS achievement tests for Populations 1 and 2 and expectations set forth for Grades 3-5 and Grades 6-8, "concluded that, for the most part, *Principles and Standards* is more rigorous than TIMSS in terms of what is outlined for mathematics teaching" (Kelly et al., 2000). They also attribute the discrepancy between expectations and performance to the lack of successful implementation of what is articulated in the Standards in U. S. classrooms.

In examining the various items on the Population 3 assessments, some items stand out in what they tell us about United States students' performances. In the mathematical literacy assessment, they performed better than average on Item D-16a (see Figure 1) which asked students to estimate (to the nearest minute) the total time required to record five songs, each of which was measured in minutes and seconds.

Teresa wants to record 5 songs on tape. The length of time each song plays for is shown in the table.

Song	Length of Time
1	2 minutes 41 seconds
2	3 minutes 10 seconds
3	2 minutes 51 seconds
4	3 minutes
5	3 minutes 32 seconds

Estimate to the nearest minute the total time taken for all five songs to play and explain how this estimate was made.

Estimate: _____

Figure 1. Population 3 Mathematical Literacy Item D16-a.

Full credit was given for estimates of either 15 or 16 minutes. While the international average correct for the item was 39 percent, United States students' average was 41.4 percent correct.

An analysis of student work showed that about 64 percent of the United States' students receiving full credit gave a response of 15 minutes. The remaining students getting full credit answered 16 minutes. Only 13 percent of U. S. students did not attempt to answer the question. Internationally, 25 percent of students failed to attempt the item. An examination of the data shows that 11.5 percent of U. S. students found the correct numerical answer, 15 minutes and 14 seconds, but failed to round to correctly answer the item as directed, compared to 9.5 percent of the international sample.

U. S. students' responses to the second part of the item asking for an explanation of how they reached their answer showed a variety of explanations. Most prominent of among these were the following: added 3 five times (5 percent), rounded first then added (1.7 percent), responded correctly about rounding with no written calculation work (11.3 percent), added correctly to the total time then rounded (7.2 percent), or used another approach (2.1 percent). The answers to both parts of this item show that U. S. students know how to perform estimates in a variety of ways reflecting that the emphasis placed on estimation and approximation has had some effect on their overall performance. Their overall performance on this literacy level item was slightly above the international level for providing a correct answer to the first portion and slightly below the international average in explaining the processes used.

An item that typifies U. S. performance on the advanced mathematics assessment for Population 3 was Item K18, shown in Figure 2 on the next page. This item asked students to prove that a given triangle was isosceles. This item, measuring the content area of geometry and the expectation of justifying and proving, had varied patterns of performance by students from the countries participating in this section of the assessment. Internationally, 34 percent of students received full credit for this item with another 14.4 percent of the students getting some form of partial credit. Only 9.7 percent of U. S. students earned full credit, while another 9.3 percent received some partial credit for the item. The data show that while 63.3 percent of U. S. students were marked as having incorrect responses, only 17.8 percent of them failed to attempt the item. The international percentages for these categories were 24.8 percent incorrect and 26.8 percent nonresponding.

The multiple choice item L7 shown on Page 74 points out a lack of a good conceptual grasp of the meaning of the definite integral in terms of the area under a curve among U. S. advanced students. Only 27 percent of American students selected C as the correct answer. Thirty-five percent of international students responded correctly. The fact that only slightly better than a chance level proportion of students could correctly respond to the item, while 50 percent of the students have had adequate exposure to such an interpretation of the integral is disappointing.

K18. In the ΔABC the altitudes BN and CM intersect at point S. The measure of ∠MSB is 40° and the measure of ∠SBC is 20°. Write a PROOF of the following statement:

"ΔABC is isosceles."

Give geometric reasons for statements in your proof.

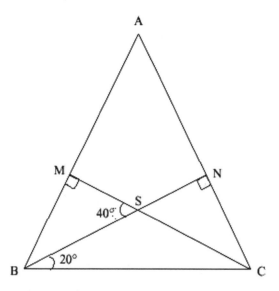

Figure 2. Population 3 Advanced Mathematics Item K18.

The performance of U. S. students at the Population 3 level has been a point of almost crisis-like concern for many in the United States. Few, including mathematics educators, remember that this level of performance is quite similar to that observed for advanced students in SIMS in the early 1980s (McKnight et al., 1987). Here in TIMSS, as then in SIMS, the entire group of U. S. students in advanced studies had lower achievement performances in areas dealing with analysis and geometry than their Population 3 peers in other participating countries.

Unfortunately, the results do not reflect the achievement levels of the group of students who have completed the College Board's Advanced Placement Calculus test that is equivalent to a full year of university calculus. The performance of these students would be quite high, but they only reflect a small portion of those students included in the Advanced Mathematics sample for the U. S. (Gonzalez, O'Connor, & Miles, 2001).

The lower level of performance for the advanced mathematics group as a whole is due to less opportunity to learn and, perhaps, to less desire to do well on the TIMSS assessments. The first factor, reduced opportunity to learn results from a cluttered curriculum in which too many topics are covered annually to the exclusion of any real mastery of a smaller number of more core, and mathematically important, topics. The second issue, one hard to quantify but one easily observed in many U. S. schools, is the fact that the TIMSS test is a low-stakes assessment. Researchers come, the test is given, but there is no individual feedback or outcome for which the students are personally held accountable. TIMSS, NAEP, and many state assessments all share in this problem. By the final year of secondary school, many students do not care about completing another assessment and, thus, the resulting observed levels of performance might be a partial reflection of this attitude.

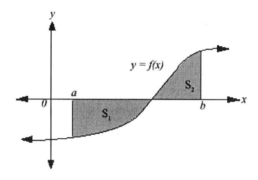

This figure shows the graph of $y = f(x)$.
S_1 is the area enclosed by the x- axis, $x = a$ and $y = f(x)$;
S_2 is the area enclosed by the x- axis, $x = b$ and $y = f(x)$;
where a < b and $0 < S_2 < S_1$.

The value of $\int_a^b f(x)dx$ is

A. $S_1 + S_2$

B. $S_1 - S_2$

C. $S_2 - S_1$

D. $|S_1 - S_2|$

E. $\frac{1}{2}(S_1 + S_2)$

Figure 3. Population 3 Advanced Mathematics Item L7.

OTHER ASPECTS OF THE TIMSS COLLECTION OF STUDIES

Along with the achievement and background questionnaire studies, the United States participated, along with Germany and Japan, in the contextual background studies and a video study of classroom teaching at the Population 2 level. While little national attention has been given to the results of the contextual studies (Stevenson & Nerison-Low, undated), the video studies (Stigler, Gonzales, Kawanaka, Knoll, & Serano, 1999) have received a great deal of attention in the United States. While the teaching observed in the tapes reflects random samples of U. S. and German classrooms, Japanese school officials, in many cases, selected the Japanese teachers. This practice may have led to a selection process that resulted in something different than a random sampling process would have given (Stigler et al., 1999).

In any case, the analyses of the videotapes have pointed to the lack of lesson focus in U. S. Population 2 classrooms. Other analyses point to the lack of justification and reasoning in these lessons as well (Manaster, 1998). Other analyses point to the lack of content in the problems considered in U. S. classrooms, compared to the problems considered in lessons from Germany and Japan (Neubrand & Neubrand, 2001). Continued analysis of these video tapes, along with data from the broader set of classrooms and countries being video-taped as part of TIMSS-R, will perhaps open new vistas of understanding for U. S. mathematics educators.

IMPACT OF INTERNATIONAL STUDIES ON U. S. MATHEMATICS EDUCATION

The negative slope of U. S. performance over populations against the TIMSS achievement scale has caught the attention of U. S. educators and policy makers involved in the design, delivery, assessment, and modification of the mathematics curriculum in U. S. schools. Educational policy makers across the United States have moved to calling for requiring more algebra of all students, both at an earlier level and across more years of the curriculum. Early study of algebra is seen as one of the most obvious differences between U. S. student opportunities in mathematics and those given to students in other countries. Data resulting from the examination of those students in Population 2 that were completing a year of algebra by the end of Grade 8 showed that these students were performing, on average, at a level equivalent to students in many of the top performing countries (Schmidt et al., 1999).

Part of the discussion of the TIMSS results has focused on the use of technology in U. S. mathematics classrooms. While surface-level analyses of the data do not immediately point to calculator usage as a causal factor, secondary analyses are needed to examine the degree to which having learned content with the aid of a calculator, but being tested without it, may have hampered students accustomed to making heavy use of calculators in problem solving and computationally complex situations. While they may have the procedural skills required to do the work, the fact that they do not use them on a regular basis may have added to the difficulty of the test for them.

Although TIMSS is credited with "having a major impact on the way the U. S. thinks about mathematics and science education" (Welch, 2000, p. 161), there is little systematic reporting of such influence. However, if one examines recent frameworks and standards documents, there is evidence of such influence. For example, it is clear from conversations with the writers of the *Principles and Standards for School Mathematics* that findings from TIMSS were influential in forming their recommendations. Likewise, background research papers developed for the Standards often cited the results of TIMSS (Kilpatrick, Martin, & Shifter, in progress).

Indeed, there are several references to TIMSS in the Standards. As part of the general argument that there was a need to continue to improve the mathematics education of U. S. students (NCTM, 2000), the document notes the performance of our students on TIMSS. The Curriculum Principle of the Standards calls for a coherent curriculum from Pre-kindergarten through Grade 12, in line with the findings of the TIMSS studies of curricula and the results of the implemented curriculum survey. Not only did the Standards call for focus on coherence across the curriculum, it references the video study of TIMSS in making the case for internal coherence within a lesson.

The Standards are concerned with the support that mathematics teachers receive in working to make the changes necessary to provide our students with coherent and focused programs of study in mathematics. The case for professional development was informed and supported by the experiences of teachers in some of the high-performing TIMSS countries. In particular, the findings that teachers in those countries are provided time and are expected to work with colleagues to develop lessons was central to the recommendations concerning professional development (Ma, 1999; Lewis, 1995).

TIMSS was an important reference in the overall and content recommendations of the *Standards*. At each level of schooling, major ideas were identified to help counter the interpretation of TIMSS results that indicated a lack of focus in the U. S. curriculum. For example, at Grades 3-5, the focus is on multiplicative reasoning, equivalence, and computational fluency with whole numbers. Since geometry and algebra are often the focus in Grades 6 to 8 in other countries, these content areas were given more prominence in the new recommendations. At each of the grade-level bands, the expectation of what students should know when entering these grades is summarized. The discussion then moves forward to new topics that should be emphasized in that grade band. This is an attempt to move away from the present United States practice of repeating topics each year as was evidenced by the TIMSS curriculum analyses. Although many of the texts used in this analysis were developed before the original *Curriculum and Evaluation Standards* (NCTM 1989) which also called for more focus and coherence, the TIMSS curriculum results reinforced the notion that focus and repetitiveness needed to be addressed in the current version of the Standards.

Overall, TIMSS added little new information for those who had been long involved in international comparisons of student achievement in mathematics. The findings about lack of focus, the over-abundance of topics, and the shortcomings of American students' performances in geometry and measurement had all been noted in SIMS. However, TIMSS provided solid data support for up these points. While they had been noted in earlier studies, the databases for those studies were insufficient for providing solid support for the generalizations formed. Besides doing this, TIMSS provided those interested in curricular comparisons with a broader understanding of the relative roles of policy, curriculum, instruction, social context, and their relationship to the observed student achievement in mathematics across a broad spectrum of schooling. The inclusion of Population 1 students and the two groups of students in Population 3 gave perhaps the most comprehensive views to date of mathematics education internationally. These views have fueled discussion, debate, and decisions. But, most importantly, they have provided mathematics educators from the United States with important information and points of support as they work to improve the learning and teaching of mathematics for students in school mathematics.

REFERENCES

American Federation of Teachers. (1998). *Making standards matter: 1998*. Washington, DC: Author.

Beaton, A. E., Mullis, I. V. S., Martin, M. O., Gonzalez, E., Kelly, D. L., & Smith, T. A. (1996). *Mathematics achievement in the middle school years: IEA's Third International Mathematics and Science Study*. Chestnut Hill, MA: TIMSS International Study Center, Boston College.

Blank, R. K., & Langesen, D. (1999). *State indicators of science and mathematics education:1999*. Washington, DC: Council of Chief State School Officers.

Blank, R. K., Langesen, D., Bush, M., Sardina, S., Pechman, E., & Goldstein, D. (1997). *Mathematics and science content standards and curriculum frameworks*. Washington, DC: Council of Chief State School Officers.

Gonzalez, E. J., O'Connor, K. M., & Miles, J. A. (2001). *How well do Advanced Placement students perform on the TIMSS advanced mathematics and physics tests?* Chestnut Hill, MA: TIMSS International Study Center, Boston College.

Kelly, D. L., Mullis, I. V. S., & Martin, M. O. (2000). *Profiles of student achievement in mathematics at the TIMSS international benchmarks: U. S. performance and standards in an international context.* Chestnut Hill, MA: TIMSS International Study Center, Boston College.

Kilpatrick, J., Martin, G., & Shifter, D. (in press). *A research companion to the Principles and Standards*. Reston, VA: National Council of Teachers of Mathematics.

Lewis, C. C. (1995). *Educating hearts and minds: Reflections on Japanese preschool and elementary education*. New York, NY: Cambridge University Press.

Ma, L. (1999). *Knowing and teaching elementary mathematics: Teachers' understanding of fundamental mathematics in China and the United States*. Hillsdale, NJ: Erlbaum.

Manaster, A. B. (1998). Some characteristics of eighth-grade mathematics classes in the TIMSS Videotape study. *The American Mathematical Monthly, 105*, 793- 805.

McKnight, C.C., Crosswhite, F. J., Dossey, J. A., Kifer, E., Swafford, J. O., Travers, K. J., & Cooney, T. J. (1987). *The underachieving curriculum: Assessing U. S. school mathematics from an international perspective.* Champaign, IL: Stipes.

Mitchell, J. H., Hawkins, E. F., Jakwerth, P. M., Stancavage, F. B., & Dossey, J. A. (1999). *Student work and teacher practices in mathematics.* Washington, DC: National Center for Education Statistics.

Mullis, I. V. S., Martin, M. O., Beaton, A. E., Gonzalez, E., Kelly, D. L., & Smith, T. A. (1997). *Mathematics achievement in the primary school years: IEA's Third International Mathematics and Science Study.* Chestnut Hill, MA: TIMSS International Study Center, Boston College.

National Assessment Governing Board. (1994). *Mathematics framework for the 1996 and 2000 National Assessment of Educational Progress.* Washington, DC: U. S. Department of Education.

National Center for Education Statistics. (1999). *The NAEP guide.* Washington, DC: U. S. Department of Education.

National Council of Teachers of Mathematics. (1989). *Curriculum and evaluation standards for school mathematics.* Reston, VA: Author.

National Council of Teachers of Mathematics. (1991). *Professional teaching standards for school mathematics.* Reston, VA: Author.

National Council of Teachers of Mathematics. (1995). *Assessment standards for school mathematics. Reston, VA: Author.*

National Council of Teachers of Mathematics. (2000). *Principles and standards for school mathematics. Reston, VA: Author.*

Neubrand, J., & Neubrand, M. (2001). *The role of problems in the mathematics lessons of the TIMSS-Video-Study.* Paper presented at the meeting of the American Education Research Association, Seattle, Washington.

Nohara, D., & Goldstein, A. (2001). *A comparison of the National Assessment of Educational Progress (NAEP), the Third International Mathematics and Science Study (TIMSS), and the Programme for International Student Assessment (PISA).* (Report Number: NCES 2001-07). Washington, DC: National Center for Education Statistics.

Robitaille, D. F., Schmidt, W. H., Raizen, S., McKnight, C. C., Britton, E., & Nichol, C. (1993). *Curriculum frameworks for mathematics and science.* Vancouver, Canada: Pacific Educational Press.

Schmidt, W. H., Jorde, D., Cogan, L. S., Barrier, E., Gonzalo, I., Moser, U., Shimizu, K., Sawada, T., Valverde, G. A., McKnight, C., Prawat, R. S., Wiley, D. E., Raizen, S. A., Britton, E. D., & Wolfe, R. G. (1996). *Characterizing pedagogical flow: An investigation of mathematics and science teaching in six countries.* Dordrecht, The Netherlands: Kluwer Academic Publishers.

Schmidt, W. H., McKnight, C. C., Cogan, L. C., Jakwerth, P. M., & Houang, R. T. (1999). *Facing the consequences: Using TIMSS for a closer look at U. S. Mathematics and science education.* Dordrecht, The Netherlands: Kluwer Academic Publishers.

Stevenson, H. W., & Nerison-Low, R. (undated) *To sum it up: Case studies of education in Germany, Japan, and the United States.* Available from the national Institute on Student Achievement, Curriculum and Assessment (http://www.ed.gov/offices/OERI/SAI).

Stigler, J. W., Gonzales, P., Kawanaka, T., Knoll, S., & Serrano, A. (1999). *The TIMSS Videotape classroom study: Methods and findings from an exploratory research project on eighth-grade mathematics instruction in Germany, Japan, and the United States.* Washington, DC: National Center for Education Statistics.

Travers, K. J., & Westbury, I (Eds.). (1989). *The IEA Study of Mathematics I: Analysis of mathematics curricula.* Oxford: Pergamon Press.

Takahira, S., Gonzales, P., Frase, M., & Salganik, L. H. (1998). *Pursuing excellence: A study of U. S. twelfth-grade mathematics and science achievement in international context.* Washington, DC: National Center for Education Statistics.

Welch, C. M. (2000). United States. In D. F. Robitaille, A. E. Beaton, & T. Plomp (Eds.), *The impact of TIMSS on the teaching and learning of mathematics and science.* Vancouver, Canada: Pacific Eductional Press.

Chapter 6

TIMSS MATHEMATICS RESULTS: A JAPANESE PERSPECTIVE

Eizo Nagasaki and Hanako Senuma

Japan has been a member country of IEA since the organization was first established and has participated in all of its mathematics studies: the first in 1964 (FIMS), the second in 1981 (SIMS), the third in 1995 (TIMSS), and most recently in 1999 (TIMSS-R). In addition, Japan has participated in several IEA studies of science education and computer education. Both the results and the methodology of the IEA studies (Robitaille et al., 1993; Robitaille & Garden, 1996) have had a strong impact on Japanese education.

In this paper, the impact of the TIMSS results on the Japanese mathematics education are interpreted and discussed. Readers may find it helpful to read more on this topic in the chapter by Miyake and Nagasaki on Japan in *The Impact of TIMSS on the Teaching and Learning of Mathematics and Science* (Robitaille, Beaton, & Plomp, 2000).

LEARNING FROM THE FORMER IEA STUDIES OF MATHEMATICS

Mathematics education has a long history in Japan (JSME, 2000). In recent years, IEA's international mathematics studies have impacted on mathematics education in Japan in a number of ways. Japanese mathematics educators have paid particular attention to students' outcomes in the affective domain and to background information as well as achievement in the cognitive domain. FIMS revealed that higher-achieving Japanese students tended to think of mathematics as stable knowledge to be memorized (NIER, 1967). In response to this, national courses of study developed during the 1970s and 1980s stressed the developmental aspects of mathematics, now more commonly known as "mathematical ways of thinking." SIMS revealed that Japanese students did not like mathematics and that they thought of mathematics as being irrelevant to society. It further revealed that calculators were not used in mathematics classrooms (NIER, 1981; NIER, 1982; NIER, 1983; NIER, 1991). The national courses of study developed in the 1990s therefore

D.F. Robitaille and A.E. Beaton (eds.), Secondary Analysis of the TIMSS Data, 81–93.
© 2002 *Kluwer Academic Publishers. Printed in the Netherlands.*

stressed an appreciation of the meaning of mathematics. The TIMSS findings and their implications, discussed later in this chapter, had a similar impact in Japan.

Participation in TIMSS

Japan participated in TIMSS Populations 1 and 2 and the main survey data was collected in February 1995. After the survey, NIER published a national report (NIER, 1996) and two international reports in Japanese (NIER, 1997; NIER, 1998). The content of the two Japanese international reports was almost the same as those published by the International Study Center (Beaton et al., 1996; Mullis et al., 1997).

The TIMSS results were reported widely in the Japanese press. The emphasis in most stories was on the high achievement of Japanese students and their dislike of mathematics. The Japanese reports analyzed the intended, implemented, and attained curricula. In addition, we, as members of TIMSS team, have written in journals, participated in symposia, and given lectures on TIMSS. However, the analysis based on the three curricular levels was not reported in the general media.

The broader education community in Japan showed a great deal of interest in the results. TIMSS has been often referred to in theoretical research in education and mathematics education, as well as in professional development programs for teachers of mathematics. After TIMSS, Japan participated in TIMSS-R Population 2 in 1999. In this paper, the results of TIMSS-R (NIER, 2001) are also included.

In this paper, we interpret Japanese mathematics education in line with the three levels of curriculum: attained, intended and implemented.

JAPANESE INTERPRETATION OF THE ATTAINED CURRICULUM

TIMSS and TIMSS-R revealed many characteristics of the attained curriculum through the main surveys (Beaton et al., 1996; Mullis et al., 1997; Mullis et al., 2000). These were analyzed from the Japanese standpoint (NIER, 1997; NIER, 1998; NIER, 2001; Nagasaki, 1999a; Nagasaki, 1999b; Nagasaki & Senuma, 2000).

The Attained Curriculum in Japan

Characteristics of the attained curriculum of Japan are classified into four categories as shown in Table 1. These are the cognitive domain, the affective domain, background information, and relationships between attained curriculum and attributes.

The affective domain rather than cognitive domain has been a major focus of Japanese interpretations of TIMSS results on the attained curriculum. Japanese students' attitudes toward mathematics are not as positive as those of students in other countries. However, attitudes toward mathematics had been recognized as important issues in FIMS and SIMS, too. A priority to improve mathematics education is being given to enhancing students' attitudes toward mathematics.

Table 1. Characteristics of the Japanese attained curriculum

I. Cognitive domain

1. Mathematics achievement is higher than in other participating countries.
2. Mathematics achievement in problem solving is relatively lower than in other categories of performance expectation.
3. Mathematics achievement in extended response problem types is relatively lower than in other types.
4. Variation in mathematics achievement among lower secondary students is greater in other participating countries.
5. The proportion of students who reached the top 10 percent level in mathematics achievement is higher than in other participating countries.
6. Mathematics achievement during the past 40 years has been stable.

II. Affective domain

1. Students do not like mathematics and do not enjoy mathematics.
2. Students tend to think that mathematics is not so important in their lives.
3. Students do not have confidence in doing well in mathematics.
4. Students tend to think that having fun is more important than doing well in mathematics.
5. Students tend to think that they do not want a job that uses mathematics

---

III. Background information

1. Time for study outside school tends to be decreasing.
2. Time for TV or computer games is highest among participating countries.

IV. Relationships between attained curriculum and attributes

1. There are gender differences in mathematics achievement and in attitudes toward mathematics.
2. There is a positive correlation between mathematics achievement and liking mathematics.
3. There is a positive correlation between mathematics achievement and believing that one is doing well in mathematics.
4. There is a positive correlation between mathematics achievement and overall attitudes toward mathematics.
5. There is a positive correlation between mathematics achievement and the number of books in the home.

Cognitive Domain in the Attained Curriculum

In the cognitive domain, Japanese scores were significantly higher than the international mean. However, there is a great deal of interest in Japan in examining the results at a more detailed level. Therefore we analyzed the achievement results in the following way.

The TIMSS data were analyzed by categories of performance expectation—knowing, performing routine procedures, using complex procedures, and solving problems—for Populations 1 and 2. The reason why we analyzed the performance expectation data was that results of a national survey on mathematics conducted by the Ministry of Education (Elementary & Secondary Bureau, 1997; Elementary & Secondary Bureau, 1998) showed that students' achievement in higher-order skills—the category called "mathematical ways of thinking"—was low. "Mathematical ways of thinking" can be viewed as equivalent to "solving problems" in TIMSS performance expectation terminology. Japanese mean achievement and the international mean achievement for each category of TIMSS were calculated and analyzed by z-scores based on mean achievement for all categories. This is shown in Table 2.

Table 2. Mean achievement by performance expectation in TIMSS (z-score)

	Japanese		International	
Performance Expectation	*Grade 4*	*Grade 8*	*Grade 4*	*Grade 8*
Solving problems	-1.15	-1.62	-1.49	-1.28
Complex procedures	-0.86	1.03	-0.37	0.73
Routine procedures	1.13	0.37	1.06	-0.54
Knowing	0.81	0.39	0.74	1.24

The z-scores for Japanese students on "problem solving" in TIMSS—the category similar to "mathematical ways of thinking" of Japan—were lower than their z-scores in other categories. This TIMSS result shows the same trends as the results of the national survey. However, the results on the right side of Table 2 suggest that not only Japanese students but also students around the world registered lower mean achievement on higher-order skills. In the process of the analysis, another interesting fact was found that almost every country has the same achievement order of performance expectations, namely, from solving problems to procedures and knowing.

The TIMSS data were also analyzed by categories of item format—extended response items, short answer items, and multiple-choice items—for Populations 1 and 2. Japanese mean achievement and the international mean achievement for each category of item format were calculated and analyzed by z-scores based on a mean achievement for all categories as shown in Table 3.

Table 3. Mean achievement by item format in TIMSS (z-score)

Item Format	Japanese		International	
	Grade 4	Grade 8	Grade 4	Grade 8
Extended response items	-1.13	-1.12	-0.99	-1.05
Short answer items	0.36	0.32	-0.01	0.11
Multiple choice items	0.77	0.80	1.01	0.94

Though Japanese students get lower z-scores on extended response items than on other formats, students in other countries also get lower z-scores on extended response items.

Another issue in the cognitive domain was whether or not there had been a decline in students' mathematics achievement over the period from 1964 to 2000. Using data from the IEA studies enabled us to give a clear picture. We used anchor items to equate mathematics achievement in FIMS, SIMS, TIMSS, and TIMSS-R. The anchor items were not necessarily representative of the overall tests they were part of, so one needs to exercise caution before drawing any firm conclusions. This approach has obvious methodological weaknesses, but this data suggests that mathematics achievement was at much the same level in 1999 as it had been in 1964, and that standards have not fallen. This theme will be referred to again later.

Affective Domain and Background Information in the Attained Curriculum

Japanese students have less positive attitudes toward mathematics and the learning of mathematics than students in other countries. These trends were also evident in SIMS. In particular, Japanese students do not like mathematics and do not enjoy mathematics. They tend to think that mathematics is not so important in everyone's life and they do not have confidence in doing well in mathematics. They tend to think that having a fun is more important than doing well in mathematics.

Another interesting finding concerns changes in students' life style. Students now spend more time watching TV or playing computer games and less time studying in their homes. This change is clear from both TIMSS and TIMSS-R.

Relationships between the Attained Curriculum and Attributes

There are some attributes that contribute to mathematics achievements: e.g., attitudes toward mathematics and home condition. Liking mathematics, doing well in mathematics, and overall attitudes toward mathematics have positive correlations with mathematics achievement. For example, the more students like mathematics, the better the students achieve in mathematics.

Boys have higher achievement and are more positive toward mathematics than girls. Having books in the home correlates positively with mathematics achievement in TIMSS-R. In other words, social and economic conditions have an effect on mathematics achievement.

JAPANESE INTERPRETATION OF THE INTENDED AND IMPLEMENTED CURRICULA

TIMSS revealed many characteristics of the intended and implemented curricula through the study itself, (Beaton et al., 1996; Mullis et al., 1997; Robitaille, 1997; Schmidt et al., 1997; Mullis et al., 2000), the textbook case study (Howson, 1995) and the video study (Schumer, 1998; Schmidt et al., 1997; Stigler & Hiebert, 1997). We analyzed those reports from the Japanese standpoint (NIER, 1997; NIER, 1998; NIER, 2001; Nagasaki, 1999a; Nagasaki, 1999b; Nagasaki & Senuma, 2000).

The Intended Curriculum in Japan

In order to study the intended curriculum in Japan, we analyzed the educational system, national courses of study, and textbooks in international terms using TIMSS data as shown in Table 4.

Table 4. Characteristics of the Japanese intended curriculum

I. The educational system
1. Decision-making for the mathematics curriculum is centralized.
2. The number of students in a class is larger than in most other countries.
3. The time allotted to mathematics is less than in most other countries.

II. The national courses of study for mathematics
1. The amount of mathematics content has been decreased.
2. Performance expectations are described more generally.
3. Mathematics topics in each grade are divided into several larger content areas.
4. Mathematics topics are introduced not spirally but linearly.
5. Equations and functions are introduced earlier.

III. Mathematics textbooks used in lower secondary schools
1. There is no consideration for different achievement levels.
2. Some content areas, such as percent, do not appear in elementary school mathematics textbooks.
3. Algebra is heavily emphasized.
4. Formal proofs are emphasized.
5. Specific topics relating to calculators are not included.
6. The history of mathematics is included.
7. Topics related to students' interests, attitudes, or careers in mathematics rarely appear.

The Implemented Curriculum in Japan

In order to study the implemented curriculum in Japan, we analyzed the teachers' instructional practices in general, teachers' practices in lower secondary schools, and teachers' perceptions of mathematics in international terms by using TIMSS data as shown in Table 5. Teachers' practices in the lower secondary schools were the focus of the TIMSS video survey.

Table 5. Characteristics of the Japanese implemented curriculum

I. Teachers' practices in general
1.   Teachers almost always refer to textbooks in teaching.
2.   The whole-class teaching method is most common.
II. Teacher's practices in lower secondary school
1.   Teachers tend not to relate the lesson contents to daily life.
2.   Teachers tend not to use calculators and computers.
3.   Teachers tend to adopt problem-solving approaches.
4.   Teachers urge students to think from various points of view and to provide explanations for problems.
5.   The time allowed for student discussions is limited.
6.   More individual learning than group learning takes place in a whole-class setting.
7.   Teachers tend not to assign homework.
III. Teachers' perception
1.   Teachers think that being able to think creatively is important for finding success in mathematics.
2.   Teachers do not think that understanding how mathematics is used in the real world is important to be good at mathematics.

The main characteristic of mathematics teaching in Japan is whole-class instruction based on textbooks. Teachers also tend to adopt problem-solving approaches where students are requested to think individually before discussion. Teachers do not want to use calculators and computers for mathematics education. Lower secondary school teachers tend not to assign homework. Teachers do not think that it is important to connect mathematics to the real world. The proportion of teachers who think that remembering formulas and procedures is important to be good at mathematics has decreased from 60 to 45 percent between 1995 and 1999.

POSSIBLE EXPLANATION OF THE ATTAINED CURRICULUM IN TERMS OF THE INTENDED AND IMPLEMENTED CURRICULA

Some of the Japanese achievement results could be explained from the perspective of the intended and implemented curricula. Higher achievement in mathematics during these 40 years may be due to the fact that the curriculum framework for compulsory mathematics education has not drastically changed: namely, four operations on whole numbers, decimal fractions, and fractions have continued to be taught in elementary schools; algebraic expressions and proof in geometry are still taught in lower secondary schools. On the other hand, this framework may have had an effect on students' attitudes toward mathematics: students do not enjoy mathematics and do not feel confident about it. We need to know whether the objectives of the mathematics curricula at the various grade levels are attainable by most students. If the objectives are easily attainable, students are likely to enjoy mathematics and feel confident about it. The trade-off may be that many students will not achieve their potential. Ideally, the curriculum and assessment should provide enough challenge for students to achieve well without having to put in an unreasonably great amount of effort.

Students tend to think that mathematics is not so important in their lives. Teachers do not think applications of mathematics in the real world are important and they tend not to relate their mathematics lesson contents to daily life. Teachers' attitudes on the relationship between mathematics and the real world might have an effect on students' attitudes. This tendency of Japanese mathematics educators is characterized as "theory-oriented" as opposed to "practice-oriented" (Shimada, 1980). The fact that calculators and computers are not used in mathematics classrooms can also be explained in terms of this orientation.

The importance of doing well in school may be changing. More Japanese students reported that having time to have fun is important than reported that doing well in mathematics is important, especially in lower secondary school. It appears that students are not as focused on studying and doing well as they used to be. The fact that the amount of content in national courses of study has decreased and that topics regarding perspectives are rare in textbooks may also affect students' attitudes. Today, even if students do not study hard, they can still enter upper secondary school or university. This is due, in no small measure, to constantly decreasing numbers of students. Students appear to be losing sight of the reasons for learning.

The amount of time spent studying out of school is decreasing. We may attribute this to the fact that Japanese teachers are moving away from assigning homework, especially in lower secondary schools. Since about 50 percent of lower secondary school students attend *Juku* (Central Council for Education, 1996), teachers may hesitate to assign too much homework.

The TIMSS data presented some crucial results and their possible explanations for Japanese mathematics education. We think that these relationships should be investigated.

THE IMPACT OF TIMSS ON NATIONAL EDUCATIONAL POLICY

A number of educational improvement schemes were already underway in Japan prior to the release of TIMSS results in November 1996. The beginning of this educational reform is related to a report by the Central Council for Education published in July 1996 (Central Council, 1996). The Council discussed what a life-long learning society should be, how to cope with issues related to entrance examinations, memory-centered learning, and students' passive attitudes. They made several recommendations, including a five-day school week and a new motto for education, "Zest for Living." A new subject, "Integrated Learning," would be introduced, which would replace the prior structure of separate subject teaching. In upper secondary schools, "Fundamentals of Mathematics" would be introduced, in order to enhance student interest and to increase perceptions of the relevance of mathematics to society.

However, the TIMSS results were widely discussed in the Curriculum Council's debates on setting national curriculum standards. The Curriculum Council was established to embody the recommendations of the Central Council of Education. The indicators in the affective domain were noted, in particular that Japanese students did not like mathematics, did not think that mathematics was relevant to their lives, and did not think that they wanted a job that used mathematics. There seems to be a certain level of agreement within the education community that much of the mathematics content included in the old curriculum was too difficult or too advanced for most students (see Central Council, 1996), and that their resulting frustration led to their negative comments about the subject on the TIMSS survey.

The New National Courses of Study for Mathematics and TIMSS

The policy governing revisions to the national courses of study in mathematics published by the Curriculum Council in July 1998 amended the general objectives for the subject, altered the time allocated to the subject, and shifted content areas.

The new courses of study were published in December 1998 and March 1999. They will be implemented in elementary and lower secondary schools in 2002, and in upper secondary schools in 2003.

The new objectives seek to foster thinking abilities and the development of positive attitudes toward using mathematics. "Enjoyment" was explicitly expressed as an objective in elementary and lower secondary school levels. At these levels, the objectives are achieved through mathematical activities. At the upper secondary level, "cultivating basic creativity" was added. These additions to the general objectives were directly related to the Curriculum Council's reaction to the TIMSS results.

About 20 percent of the time allotment has been deleted from each subject in order to introduce a five-day school week and a new subject, "Integrated Learning." For example, the number of mathematics periods for Grades 1 to 6 were reduced from 175 to 150 per year, and for Grades 7 to 9, from about 140 to 105. A class period is ordinarily 45 minutes in elementary school and 50 minutes in lower secondary school. A corresponding quantity of content was deleted from the curriculum.

In spite of these changes, the essential framework of the mathematics curriculum for elementary and lower secondary schools remained the same. Arithmetic is completed in elementary school, although the difficulty level has been reduced. In lower secondary school, algebra is taught, along with formal proofs in geometry. In upper secondary school, calculus is still the summit of mathematics.

TIMSS and the New Evaluation Standards

Following the revision of the national content standards, the Curriculum Council reported on evaluation. The changes included a shift from norm-referenced to criterion-referenced evaluation. They also identified a need for continuous, nation-wide surveys of student achievement. The report stressed that the TIMSS results had had an impact on the revision of the national courses of study, and that TIMSS methodology should be used in nation-wide surveys of students' achievement.

TIMSS and National Educational Policy

The TIMSS results were also discussed at the National Commission on Educational Reform in April, 2000. Furthermore, two reports published in 2000 referred to the results, *Articulation among Elementary, Secondary and Tertiary Education* by the Central Council for Education (2000) and *Japanese Government Policies in Education, Science, Sports and Culture* by the Ministry of Education (2000). Also, TIMSS was referred to several times in the National Diet in relation to views on the decline of students' achievement.

In 1999, several mathematicians reported on university students' poor computational skills (Okabe et al., 1999). They expressed opposition to the deletion of content in the revised national courses of study. Some of them started a movement to oppose the implementation of the new national courses of study in 2000.

Interestingly, the TIMSS results were used by both sides as a basis for their positions throughout the controversy. This was because TIMSS was the most recent survey and was thought to be the most reliable educational survey in Japan. Both sides recognize that the proportion of Japanese students holding negative attitudes toward mathematics was the highest among participating countries. In addition, opponents of the new course of study focused on the fact that the time allocated to classroom mathematics in Japan was lower than the international mean. Therefore they insisted that more time should be allotted to mathematics. Advocates of the new courses of study stressed that Japanese students' mean correct response rate was the same across all the three IEA studies. Therefore they insisted that the Integrated Learning course be more encouraged in order to improve students' attitudes.

CONCLUSION

The results of TIMSS have had an undeniably strong impact on mathematics education in Japan. Specially, results in the affective domain rather than cognitive domain have been heavily emphasized. Among the members of the academic society of mathematics education, TIMSS is considered basic knowledge. In addition, the methodology of TIMSS is frequently referred to. Beyond educational researchers, many classroom teachers are also familiar with TIMSS.

Within the general society, interest in international surveys such as TIMSS is increasing. Since TIMSS results were used by both sides in the controversy over the decline in student achievement, educational researchers have expressed greater interest in international, quantitative studies. The Ministry of Education regards the TIMSS results positively and discusses them extensively in their reports.

As we mentioned earlier, TIMSS has been welcome in Japan. However, there are few in-depth reports based on TIMSS data in Japan. Empirical research in education and in-depth investigation are necessary to improve our education further.

REFERENCES

Beaton, A. E., Martin, M. O., Mullis, I. V. S., Gonzalez, E. J., Kelly, D. L., & Smith, T. A. (1996). *Mathematics achievement in the middle school years.* Chestnut Hill, MA: Center for the Study of Testing, Evaluation, and Educational Policy, Boston College.

Central Council for Education. (1996). *The first report for the 15th term.* Tokyo.

Central Council for Education. (2000). *Articulation among elementary, secondary and tertiary education.* Tokyo.

Elementary & Secondary Education Bureau, Ministry of Education, Sports and Culture. (1997). *Report of comprehensive survey on implementation of national curriculum, elementary school mathematics.* (in Japanese)

Elementary & Secondary Education Bureau, Ministry of Education, Sports and Culture.. (1998). *Report of comprehensive survey on implementation of national curriculum, lower secondary school mathematics.* (in Japanese)

Howson, G. (1995). *Mathematics textbooks: A comparative study of Grade 8 texts. TIMSS monograph. No. 3.* Vancouver, B.C.: Pacific Educational Press.

Japan Society of Mathematical Education (JSME). (2000). Mathematics education in Japan during the fifty-five years since the War: Looking towards the 21st century. *Journal of JSME, 82*(7–8).

Ministry of Education, Sports and Culture. (2000). *Japanese government policies in education, science, sports and culture.*

Miyake, M, & Nagasaki, E. (2000). Japan. In D. F. Robitaille, A. E. Beaton, & T. Plomp (Eds.), *The impact of TIMSS on the teaching & learning of mathematics & science.* Vancouver, B.C.: Pacific Educational Press.

Mullis, I. V. S., Martin, M. O., Beaton, A. E., Gonzalez, E. J., Kelly, D. L., & Smith, T. A. (1997). *Mathematics achievement in the primary school years.* Chestnut Hill, MA: Center for the Study of Testing, Evaluation, and Educational Policy, Boston College.

Mullis, I. V. S., Martin, M. O., Gonzalez, E. J. , Gregory, K. D. , Garden, R. A. , O'Connor, K. M. , Chrostowski, S. J., & Smith. T. A. (2000). *TIMSS 1999 International Mathematics Report Findings from IEA's Repeat of the Third International Mathematics and Science*

Study at the Eighth Grade . Chestnut Hill, MA: Center for the Study of Testing, Evaluation, and Educational Policy, Boston College.

Nagasaki, E. (1999a). Curriculum studies based on international comparison of mathematics Education (in Japanese). In *Toward the reform of curriculum in mathematics.* Japan Society of Mathematical Education.

Nagasaki, E. (1999b). Is Japanese students' achievement to apply mathematics really low? (in Japanese). In *Education and information.* The Ministry of Education Science, Sports, and Culture.

Nagasaki, E., & Senuma, H. (2000). Japan's mathematics achievement in IEA studies (in Japanese). *Bulletin of the National Institute for Educational Research of Japan, 129,* 43-77.

National Institute for Educational Research (NIER). (1967). *International study on mathematics education, Report by the IEA Japanese National Commission* (in Japanese). Tokyo: Author.

National Institute for Educational Research. (1981). *Student achievement in mathematics in lower secondary and upper secondary schools, interim report for the Second International Mathematics Study.* (in Japanese). Tokyo: Daichihouki Shuppansha.

National Institute for Educational Research. (1982). *Student achievement in mathematics & various conditions in lower secondary and upper secondary schools, national report for The Second International Mathematics Study.* (in Japanese). Tokyo: Daichihouki Shuppansha.

National Institute for Educational Research. (1983). *Student achievement in mathematics & teachers' instruction in lower secondary schools, national report for the Second International Mathematics Study.* (in Japanese). Tokyo: Daichihouki Shuppansha.

National Institute for Educational Research. (1991). *International comparison of mathematics education, the final report for the Second International Mathematics Study.* (in Japanese). Tokyo: Daichihouki Shuppansha.

National Institute for Educational Research. (1996). *Student achievement in mathematics & science in Japanese elementary and lower secondary schools, national interim report for the Third International Mathematics & Science Study.* (in Japanese). Tokyo: Toyokan Shuppansha.

National Institute for Educational Research. (1997). *International comparison of mathematics & science education in lower secondary schools, report for the Third International Mathematics & Science Study.* (in Japanese). Tokyo: Toyokan Shuppansha.

National Institute for Educational Research. (1998). *International comparison of mathematics & science education in elementary schools, the final report for the Third International Mathematics & Science Study.* (in Japanese). Tokyo: Toyokan Shuppansha.

National Institute for Educational Policy Research (NIER). (2001). *International comparison of mathematics & science education, the report for the Third International Mathematics & Science Study-Repeat.* (in Japanese). Tokyo: Goyousei.

Okabe, T., Tose, N., & Nishimura, K. (1999). *University students who cannot compute fractions.* (in Japanese). Tokyo: Toyokeizaishinposha.

Robitaille, D. F., Schmidt, W. H., Raizen, S., McKnight, C., Britton, E., & Nicol, C. (1993). *Curriculum frameworks for mathematics and science: TIMSS monograph. No. 1.* Vancouver, B.C.: Pacific Educational Press.

Robitaille, D. F., & Garden, R. A. (1996). *Research questions & study design: TIMSS monograph. No. 2.* Vancouver, B.C.: Pacific Educational Press.

Robitaille, D. F. (Ed.). (1997). *National contexts for mathematics and science education: An encyclopedia of the education systems participating in TIMSS.* Vancouver, B.C.: Pacific Educational Press.

Robitaille, D. F., Beaton, A. E., & Plomp, T. (Eds.). (2000). *The impact of TIMSS on the teaching & learning of mathematics & science*. Vancouver, B.C.: Pacific Educational Press.

Schumer, G. (1998). Mathematics education in Japan, the United States and Germany: Some observations on the difficulty of evaluating schools. Studien zur Wirtshafts-und Erwachsenenpadagogik aus der Humboldt-Universitat zu Berlin. Band 13.3.

Schmidt, W. H. (Ed.). (1997). *Many visions, many aims, Volume 1: A cross-national investigation of curricular intention in school mathematics*. Dordrecht: Kluwer Academic Publishers.

Shimada, S. (1980). *Calculators in Schools in Japan*. Mathematics Education Information Report. International Calculator Review.

Stigler, J. W. & Hiebert, J. (1997). Understanding and improving classroom mathematics instruction: An overview of the TIMSS video study. *Phi Delta Kappan*, 14–21.

Chapter 7

TIMSS, COMMON SENSE, AND THE CURRICULUM

Geoffrey Howson

In this paper we shall look at the data collected by TIMSS in its Population 2 study in order to investigate two particular questions. The first concerns the mathematics that students learn outside the mathematics classroom; the second, curriculum design and, in particular, the benefits of the early study of probability and of what is now sometimes referred to as "pre-algebra."

In recent years, I have become increasingly interested in what might be termed common sense mathematics, if by this we mean the appreciation of mathematics that the student has gained outside the classroom: from the media, leisure activities, or other activities within the social community. In many cases there is a need to help students structure this knowledge before we move on to extend it, for example, to develop the primary school child's understanding of fractions. In other cases, as the TIMSS data show, common sense may run counter to mathematical thought and there is a need to identify and correct misapprehensions. What guidance can the TIMSS findings provide on these aspects of teaching and learning mathematics?

An interest in curriculum construction and in comparative curriculum studies recently led to three publications (Howson, 1991, 1995, 1998[1]) which, respectively, contrasted fourteen national mathematics curricula, Grade 8 textbooks drawn from eight countries, and the primary/elementary school curricula and texts of six countries. These studies raised serious doubts in my mind about the design of, indeed, the philosophy underlying, some recently implemented (or planned) curricula, particularly as they related to the teaching of probability and algebra. Did TIMSS offer any evidence that would either reinforce or alleviate my misgivings?

THE METHODOLOGY

Simply taking averages of the results of students coming from 40 or more disparate educational systems is unlikely to produce any guidance of great value. In this paper, then, we concentrate on students in Grades 7 and 8[2] and on a subset of countries having roughly similar social and economic backgrounds, yet representative of a variety of educational traditions. Moreover any inferences are

95

D.F. Robitaille and A.E. Beaton (eds.), Secondary Analysis of the TIMSS Data, 95–111.
© 2002 *Kluwer Academic Publishers. Printed in the Netherlands.*

based upon a generous sample of items, considered individually and selected on clear criteria prior to the detailed inspection of students' responses, rather than on a handful of items chosen *a posteriori* to help support pre-determined beliefs.

The nine educational systems selected for study were Belgium (Flemish), Canada, England, France, Hungary, Japan, the Netherlands[3], Switzerland and the United States. Each country had been studied in at least one of the reports referred to above and some in all three. Items were selected based on the findings of the Test-Curriculum Matching Analysis (see Beaton et al., 1996, and below), and were those which were "on the fringe of the curricula." That is, in every case some countries claimed they were "appropriate" for their median students and others said they were not. In many instances items were thought "inappropriate" (i.e., had not been taught) at Grade 7, but "appropriate" at Grade 8. This allows the reader to compare "taught" versus "common sense" approaches to items, and gives some indication of how progress from year to year was affected by specific teaching, although in a non-longitudinal study any attempts to measure "progress" are extremely hazardous.

WHO WAS TAUGHT WHAT?

What does it mean to say that an item was appropriate, that it had been taught? Answering this question in relation to, say, Japanese students is relatively straightforward. The curriculum in Japan is determined by grade and is followed by all students who, with extremely few exceptions, are in the grade determined by their age. Schools have a choice of officially authorized textbooks and these tend to follow the same ordering of topics through the year. By contrast, in the Netherlands schooling is multilateral, students do not all share the same mathematics syllabus, and the relatively loosely defined curriculum is not laid down by grade but by a "stage" comprising several years. In England, students are expected to progress through the curriculum at different rates. At the time of TIMSS, its national curriculum was still relatively new and it was by no means easy to say what mathematics students in mid "key stage" would have studied. The United States presents an even greater problem; there is no national curriculum and the encyclopedic textbooks for each grade would appear to be designed to cater for pupils at very different levels of attainment. It is impossible to infer from reading them what, say, an average student would have been taught.

TIMSS then, in order to judge the appropriateness of the test items, had eventually to settle for the Test-Curriculum Matching Analysis. This is described in detail in Appendix B of Beaton et al. (1996) and essentially depended upon each participating country establishing a panel to decide whether or not an item was appropriate for median Grade 7 and Grade 8 students, i.e., had probably been taught to them. As will be seen from the TIMSS report and also from the data reproduced below, there is a general level of consistency of results suggesting that the exercise was taken seriously and that the panels had normally assessed matters correctly. There is, however, an obvious anomaly in the Tables of Appendix B concerning the

United States. We have noted how difficult it would be to estimate the appropriateness of an item for a "nation-wide median student." Nevertheless, simply asserting that every mathematics and science item would have been covered by the median Grade 7 student by the time the test was taken would not seem a helpful response. In the tables that follow, therefore, it would be unwise to take the reported U. S. coverage at face value. Rather, the tables can be used to cast light on what the "median" U. S. student is likely to have covered.

THE OVERALL RESULTS

When considering how students performed on a particular topic, especially, one which they had not been taught, it is important to know what overall mathematical attainments those students possessed: how well did they perform on all the items? Table 1, therefore, gives the mean scores for students from Grades 7 and 8 of the countries studied. The rankings of the various countries are given in parentheses. Also included are the average ages of students in the two grades and data concerning the estimated percentage of 13-year olds who had not reached Grade 7 (as a result of countries requiring low-attainers to repeat years) and those who had already been accelerated beyond Grade 8. It is impossible to estimate how these differences influence national means. However, if TIMSS set out to test the attainments of 13-year olds, then it is likely that the scores shown for countries, other than England and Japan, are, to differing degrees, inflated.

Table 1. TIMSS Data for Selected Countries, Population 2

	Mean Grade 7	Percent below Grade 7	Mean age Grade 7	Mean Grade 8	Mean age Grade 8	Percent above Grade 8
Belgium (Fl)*	65 (2)	5.4	13.0	66 (2)	14.1	0.2
Canada	52 (6)	8.1	13.1	59 (7)	14.1	0.6
England	47 (9)	0.6	13.1	53 (8=)	14.0	0.5
France	51 (7)	20.5	13.3	61 (5)	14.3	1.3
Hungary	54 (4)	10.5	13.4	62 (3=)	14.3	2.0
Japan	67 (1)	0.3	13.4	73 (1)	14.4	0.0
Netherlands	55 (3)	9.8	13.2	60 (6)	14.3	0.4
Switzerland	53 (5)	8.3	13.1	62 (3=)	14.2	0.2
United States	48 (8)	9.0	13.2	53 (8=)	14.2	0.2

* The Flemish (Fl) and French (Fr) educational systems in Belgium participated separately.

PROBABILITY

It is only since the 1950s that the teaching of probability has become commonplace at the secondary school level. However, in recent years there have also been attempts to introduce probability even earlier in the school system, despite the fact that mathematical probability is by no means an easy topic. Thus, for example, England included it in its national curriculum for primary schools when that was first developed in the late 1980s although a recent revision of the curriculum has seen the disappearance of quantitative probability at that level.[4] Yet other countries still seem influenced by the kind of thinking found in England in the late 1980s, and two of the countries in the 1998 study (Howson, 1998) were introducing or planning to introduce the teaching of probability into their primary schools. One wonders, though, exactly what mathematical purpose is served, or if time is well employed, if, as in one of these countries, the probability syllabus devotes six years of study to such colloquial terms as "never," "sometimes," "always," "likely," "unlikely," "probable," "expected," "certain," and "uncertain" before numerical probabilities are first introduced in Grade 6.

What has TIMSS to tell us about these attempts to teach probability and their effects on students' performances? At what stage does formal teaching begin to add anything to the students' knowledge of "street mathematics"?

There were six probability items in TIMSS, and the students' results on these items are shown in Table 2. (Here and elsewhere scores are printed in bold if the item was thought appropriate and in italics if this was not the case. The symbols /7 and /8 indicate the scores of Grade 7 and Grade 8 students, respectively. The "average" shown in this and later tables is the average of the relevant mean percent scores.)

Table 2. Mean Percent Scores on Six Probability Items

	1/7	1/8	2/7	2/8	3/7	3/8	4/7	4/8	5/7	5/8	6/7	6/8	Avg.
Belgium (Fl)	*83.1*	*84.8*	*77.4*	*78.6*	*89.7*	*85.8*	*70.9*	*78.2*	*73.1*	*68.4*	*63.5*	*70.7*	77.0
Canada	*68.7*	*76.2*	*58.3*	*62.3*	*85.5*	*89.8*	*66.0*	*73.7*	*49.3*	*56.9*	*62.9*	*67.2*	68.1
England	**61.7**	**71.0**	*50.0*	*50.4*	**80.9**	**86.0**	**61.6**	**75.3**	*35.6*	*39.1*	**76.8**	**78.3**	63.9
France	*60.9*	*72.0*	*46.4*	*55.2*	*81.7*	*81.5*	*54.6*	*68.2*	*43.5*	*54.4*	*56.3*	*65.3*	61.7
Hungary	**69.8**	**72.6**	*44.8*	*51.3*	**76.6**	**82.0**	**36.1**	**47.3**	*42.6*	*54.7*	**73.4**	**81.5**	61.1
Japan	*79.9*	*87.0*	*67.1*	*74.0*	*81.2*	*82.8*	*80.7*	*82.4*	*68.6*	*74.7*	*47.8*	*52.8*	73.3
Netherlands	*83.0*	*81.2*	*57.1*	*59.0*	*89.1*	*91.3*	*62.6*	*63.4*	*60.1*	*61.7*	*63.4*	*72.0*	70.3
Switzerland	*72.0*	*79.9*	*55.9*	*60.5*	*81.1*	*86.5*	*56.7*	*67.8*	*54.9*	*64.5*	*64.8*	*72.3*	68.1
United States	**65.9**	**67.6**	**46.8**	**53.4**	**82.0**	**85.8**	**66.3**	**73.9**	*37.0*	*47.2*	**55.6**	**60.2**	61.8

Items P1 (I9)[5] and P3 (M3) are linked and essentially can be tackled using "street probability": the probability of finding a needle in a haystack decreases with an increase in the size of the haystack. Item P1, however, asked students to identify which color would be most likely to be drawn from a set of cards having

$$\frac{1}{6}, \frac{1}{12}, \frac{1}{2}, \text{and} \frac{1}{4}$$

of green, yellow, white, and blue cards, respectively. This proved more difficult than Item P3 (e.g. 25 percent of English Grade 8 students opted for yellow, the fraction with the largest denominator).

Item P4 (N18) asked for the probability of drawing a chip with an even number on it from a bag containing nine chips numbered 1 to 9. It was answered very badly by Hungarian students[6]. Apart from this, having studied probability appeared to have improved student performance, although students in some of the other countries still answered the item exceptionally well.

Item P2 (K7) asked students how many blue pens there would be in a drawer containing 28 pens, if the probability of drawing a blue pen was

$$\frac{2}{7}.$$

We note that although English Grade 7 pupils were expected to be able to find the probability of, say, selecting a blue pen at random, even at Grade 8 they were not expected to be capable of working back from the probability to obtain the number of blue pens. The data would appear to justify the English panel's judgment on this situation.

Item P5 (O5) was similar to Item P2. When a cube is tossed the probability of its landing red face up is 2/3. How many faces of the cube are painted red?

Once again, the responses from those countries claiming to have taught probability are disappointing. More worryingly, roughly a third of the English and U. S. students Grade 7 students opted for 2 (the numerator) and slightly less for 3 (the denominator). Was it an ill-digested familiarity with a probability mantra that over-ruled the common sense of well over half the cohort?[7]

Item P6 (U) is valuable in revealing where "common sense" probability tends to break down. It concerns independent repeated trials: e.g., if red turns up 5 times running at roulette, what are the respective likelihoods of red or black turning up next time? Here, there is a tendency to believe either we are on a "streak" of reds which is likely to continue, or that black must turn up soon in order to even things out. Certainly the concept that the probability of obtaining a red will not be affected by what has recently occurred is something that has to be taught.[8] It was noticeable that, in those cases where students had to fall back on "common sense," then, in general, those from western countries performed better than those from elsewhere.

This range of items illustrates very forcibly where "common sense" probability begins to break down. It also illustrates the ability of many students to deal with straightforward probability items even if they had not been taught the topic. In retrospect it would have been advantageous to have included at least one open response item on probability that would have challenged the student to find an answer rather than to decide which of the options offered seemed most sensible. Nevertheless, questions must still be asked about the effectiveness of probability teaching in those countries that profess to teach it and especially about what benefits, if any, accrue from its early introduction. In particular, it would be good to know, for example, whether students have a sufficiently secure understanding of fractions, e.g., their ordering and equivalents, to benefit fully from a study of mathematical probability.

ALGEBRA

As in the case of probability, attempts have recently been made to introduce the subject "algebra" not only to all secondary school pupils, but also to those in primary schools. As was the case with probability, this has been largely brought about by re-defining what was meant by algebra, or, perhaps, to be fairer, by giving that title to what might better be called "pre-algebra." The latter term has come to be associated with the study of number patterns. Again, as in the case of probability, we then have somewhat bizarre consequences. Thus one country studied in Howson (1998), proposed to introduce algebra in Kindergarten, but not to use letters to denote variables or unknowns before Grade 6.

We shall comment on items not traditionally termed algebra first. The reader will then, in a later sub-section, be able to compare results obtained on the two types of algebra items.

Pre-Algebra: Number Sequences

Item A3 (L13) and the scores it produced stand apart from the others. In this item students were asked to replace one geometric symbol by another to retain a pattern. It might be dismissed as a trivial intelligence test item for Grade 7 and 8 students. Yet it does serve a useful purpose. The use of multiple choice items in tests such as TIMSS has been criticized on the grounds that students will not take such tests seriously: they will not read the questions but simply opt for one of the answers. After all, they have nothing to gain by replying correctly, or to lose by giving an incorrect answer. The scores obtained on this item suggest that students took the TIMSS tests seriously (But did the older students always take it as seriously as the younger ones?).

Item A1 (I4) was an open response item. It gave two sequences: one increasing by fives and one by sevens. The number 17 belonged to both and students were asked, "What is the next number to be in both sequences?" No algebraic solution could be expected at this level. All we should expect is counting on in fives and sevens or, from better pupils, the realization that the next common term will be 5×7 on from 17. We note how closely, the Netherlands and the United States apart, the scores cluster in Grade 8.

Item A2 (J18) was poorly worded, but not untypical of what can be found in texts and tests. We have a function table with one entry missing *and we are told nothing about the nature of the function*. Yet students are expected to supply the missing value by what we are informed are "routine procedures" (presumably, continue the pattern or assume the function is linear).

Table 3. *Mean Percent Scores on Pre-Algebra Items*

	1/7	1/8	2/7	2/8	3/7	3/8	4a/7	4a/8	4b/7	4b/8	Ave.
Belgium (Fl)	56.7	55.8	44.7	41.9	96.2	93.9	83.5	83.4	26.0	31.3	61.3
Canada	46.8	55.5	39.6	44.1	91.4	97.0	78.5	81.7	21.4	33.1	58.9
England	41.1	54.2	50.8	51.1	93.6	94.6	84.1	85.9	19.6	42.3	61.7
France	41.9	54.3	34.3	38.2	92.8	92.4	79.9	79.9	_12.2_	17.7	54.4
Hungary	42.0	56.2	43.9	49.2	92.6	92.7	83.9	91.0	_20.2_	_33.6_	60.5
Japan	51.6	59.9	48.2	56.0	97.3	96.2	88.6	94.1	43.5	52.2	68.8
Netherlands	36.5	38.1	45.9	54.9	87.3	90.5	82.4	84.0	28.7	38.3	58.7
Switzerland	50.4	55.1	37.1	45.3	95.1	94.8	79.9	86.0	26.6	38.1	60.8
United States	38.8	41.6	36.1	38.6	90.3	93.4	73.0	74.9	17.7	25.4	53.0

Item A4 (S01) was a two-part, open response item. The first part required the systematic counting of small triangles within a larger one. The second was more difficult (the underlined scores indicate that at least 25 percent of students omitted this part of the question), but essentially depended on spotting the number of small triangles that were added each time the size of the base of the larger triangle was increased by one unit. Students were asked how many smaller triangles there would be in a large triangle with base 8 units. The only method, then, available to the students was to write down the sequence up to its eighth term. In retrospect, a third, clearly algebraic part, say, "How many more small triangles are there in the nth figure than in the $(n\text{-}1)^{st}$?" would have been useful[9]. Once again, though, the item was well worth including (as the varying responses indicate), and the scores should give rise to concern in a number of countries.

In general, we see that there is relatively little difference to be observed between the averages of the nine countries studied on these pre-algebra items. Whether or not emphasis has been placed in the curriculum on such material does not appear to have been the most significant factor. However, once we look at individual items then some significant differences do appear.

Items Difficult to Classify

It is sometimes impossible to classify an item as being, say, algebra or geometry. The initial context may well be geometrical, but students may use a variety of methods to obtain the answer or, indeed, be constrained to use techniques and knowledge acquired when studying another area. Two such items would seem of particular interest.

Item A5 (I8), which was classified as geometry, gave the coordinates of two points on a straight line and asked students to identify, from its coordinates, another point on that line. Geometry was no doubt a reasonable classification although the item offers links with algebra and, in particular, with Item A2 above. If students have been taught co-ordinate geometry then they might be tempted to find the equation of the line and then to substitute the value $x = 5$. My guess, based on the results, is that this happened in some countries. This approach utilizes formal algebra. A second approach is to stay firmly within co-ordinate geometry and make use of the key fact concerning straight lines that equal increments in x always produce equal increments in y. Another way is to cast around for a number pattern. The two given points have coordinates 3 and 4 and so the next in the sequence would be 5. This restricts us to one possible distractor for which the second coordinates now give the sequence 2, 4, 6. Yet a further way is simply to sketch a graph. Clearly it is useful to know which students could, and which could not solve this problem, but it would have been considerably more informative to have known how students arrived at their solutions.

The second item to look at is Item A6 (T1) —"Fifty-four kilograms of apples in two boxes; 12 kg more in one box than the other. How many kilograms in each box?". This item clearly illustrates the great value to mathematics educators and curriculum designers of open response questions for which the data is well coded. TIMSS is to be congratulated for including such items, something its predecessors lacked. The problem, here classified as algebra, is solvable using several methods. It can be answered by formal algebraic means, by dividing 54 by 2 and then adding and subtracting 6, by subtracting 12 and then dividing by 2 to find the weight of the lighter box, or by trial and improvement.

Table 4. Mean Percent Scores on Difficult to Classify Items

	5/7	5/8	6a/7	6a/8	6b/7 alg	6b/8 alg	6c/7 sub	6c/8 sub
Belgium (Fl)	38.7	43.9	53.6	52.1	7.6	16.0	37.7	27.5
Canada	42.7	49.0	21.1	33.3	0.4	9.0	9.6	12.7
England	57.6	55.4	20.6	24.1	0.6	1.4	13.3	15.0
France	24.2	33.7	13.8	26.6	1.2	11.8	9.5	12.3
Hungary	47.0	50.6	28.7	45.5	21.0	46.8[10]	8.9	3.2
Japan	38.5	46.6	41.3	53.7	14.2	28.5	22.2	22.5
Netherlands	61.8	66.0	28.4	35.6	0.0	0.3	16.1	22.2
Switzerland	46.0	51.4	26.8	38.1	0.5	4.4	11.6	15.2
United States	37.3	40.9	19.8	24.7	2.4	10.2	9.8	6.8

The methods used varied greatly from country to country. Accordingly, the results for this item are tabulated in a slightly different manner. The columns headed 6a give the percentages of correct responses, those headed 6b show the percentage obtaining the correct answer by algebraic means, and those headed 6c show the percentage using the most favored of the methods listed above (i.e., the third: subtract 12 …).

This simplifies the data for this item, for, in particular, marks were also awarded for incorrect solutions that began correctly. The percentages given are those for students obtaining correct answers and therefore scoring full marks. There was no marked difference in the percentages of students being awarded one mark rather than the two awarded to those obtaining a fully correct answer.

The first question to be asked is what does it mean to say the item was appropriate? I can only believe (taking into account responses to similar questions on formal algebra) that the panel in England considered it to be an exercise in arithmetical problem solving. But clearly the panel in the Netherlands must have thought of the question as an algebraic one—for no one waits until Grade 8 to teach the required arithmetic. The Canadians, too, must have considered the item to be testing algebra. Yet we see that only in Hungary and in Grade 8 Japanese classes did the majority of successful students opt for an algebraic approach.

Now it could be argued that we want students to be able to solve such a question by whatever means they wish. Why teach algebra if there are simpler ways of solving the problem? In general, though, the percentage of students being able to solve this problem by any means was disappointingly small. Clearly something needs to be done to improve students' thought processes and techniques. It is noticeable that the countries with the best-performing students were those that taught formal algebra, even if the students did not choose to use their algebraic knowledge to crack this relatively small arithmetical and logical nut. (Although the U. S. data suggest that some students had been taught the algebraic method, it is not at all clear that the median student would have encountered it.) Of course, if one knows how to set the problem out algebraically, then the steps needed to obtain an arithmetic solution become obvious. Learning algebra could, it appears, increase a student's facility for "arithmetical" problem solving.

Another interesting feature of these results is the light they throw on the way complexity affects problem solving. If the item had concerned two girls who together had 12 dollars but one of whom had 2 dollars more than the other, then surely students would have found it relatively easy to calculate the amounts each one had. The level of complexity has not been raised greatly in arithmetical terms, but clearly sufficiently to deter and confuse. There is, I believe, a message for curriculum designers and, in particular, textbook writers. My feeling when studying the Grade 8 textbooks described in Howson (1995) was that in many countries the exercises tended to test whether or not a concept had been grasped and did so in a minimalist way, using only simple numbers. Yet, it is noticeable that once the arithmetical difficulty level is raised or the need arises to draw upon two or more distinct concepts then problems arise because of the students' lack of practice with this degree of complexity. If we look at pre-1960s textbooks, then the complexity of the exercises (intended, we recall, usually only for a selected elite) often seems excessively high. However, there would appear to be a danger that in trying to correct for this and, more significantly, to make mathematics accessible to all, we are failing to give many who could benefit from it the capacity to deal with complex mathematical problems. For although common sense might enable students to solve the problem of the two girls, once the complexity mounts then a stronger mathematical framework is needed for support.

Formal Algebra

The items on formal algebra throw some light on common sense and mathematics, but more importantly on the different weights attached to the teaching of formal algebra to average students in the various countries. Limitations on space prevent an item by item description and so we shall only draw attention to key data. Thirteen formal algebra items were studied and the results on these for the different countries are given below. The three numbers signify respectively, the number of items thought appropriate at Grade 7, the number thought appropriate at Grade 8, and the average mark attained by the students on the thirteen items.

Table 5. Formal Algebra Items

	No. of Items Appropriate for Grade 7	No. of Items Appropriate for Grade 8	Average on 13 Items
Belgium (Fl)	8	12	58.8
Canada	1	8	44.4
England	0	2	39.3
France	10	12	44.4
Hungary	12	13	57.7
Japan	8	11	69.0
Netherlands	0	2	42.6
Switzerland	0	6	38.7
United States	13	13	46.8

It will be seen that the different emphases placed on formal algebra mean that any attempt to interpret scores without secure data concerning the actual curricula covered by the students in the various countries would be fraught with danger. Attention has already been drawn to the unreliability of the U. S. curriculum data, but on this topic, experience gained from another study makes me doubt those concerning the French Grade 7.

The general impression gained from a study of the results on these items is that little of formal algebra is accessible to those who have not been specifically taught it. Exceptions to this were one item (U) asking students to identify the unknown in a question on ratio expressed entirely in words, and another (U) that was a simple extension of the use of "frames", familiar to many students even if the use of letters had not formally been met. However, items requiring familiarity with algebraic notation[11] usually proved inaccessible to students. Similarly, the multiplication of negative integers defies common sense.

In general, analysis of the results obtained on the other formal algebra items tells the same story: this subject must be taught, and, moreover, it takes time for students to come to grips with the ideas to which they have been introduced.

The low emphasis placed on formal algebra in England, the Netherlands and Switzerland is reflected in their students' attainments. Nevertheless, it is by no means obvious how much formal algebra should be taught to the average student at this level. However, the poor scores of England and Switzerland[12] on the item (Q7): $P = LW$. If $P = 12$ and $L = 3$, then W is equal to (a) 3/4, (b) 3, (c) 4, (d) 12, (e) 36 would seem a cause of particular concern. Such a lack of understanding seriously limits the amount of quantitative science that can be taught at this level.

GEOMETRY

Setting items on algebra is relatively easy compared with the task of setting them on geometry. For even if countries delay the teaching of formal algebra, there is fairly general acceptance of what formal school algebra should comprise. No such consensus exists about the teaching of geometry. Some countries (e.g., Japan and France) still teach geometry as a gateway to proof to Grade 8 students. (Relatively few countries do this, and as the tests were taken before students could develop a real understanding, it was impracticable to include items asking for geometrical proofs. This was a pity. A key question concerning the teaching of formal pure geometry is the percentage of pupils who can benefit from it. Any information about this would have been most valuable.) Elsewhere, the aims of geometry teaching are not so clear. Work is done on transformation geometry, but this rarely progresses far and often would seem to have no apparent long-term mathematical goal. Work on congruence and similarity is common but not universal and not always taught in a productive, successful way. Increasing emphasis is placed in some countries on co-ordinate geometry, a topic which forty years ago was usually seen as a senior high school topic. Clearly, the basic idea of co-ordinates is one that is both extremely valuable and can be taught at the Grade 7 and 8 levels. However, further progress may then be handicapped by a lack of algebraic facility. As a result of all these factors, the geometry items in TIMSS are not particularly imposing. This is not a criticism of those responsible for constructing them, but rather a consequence of the lack of generally accepted aims for geometry teaching in schools.

A typical geometry item (U) asked students to use knowledge of congruence and the angle sum of a triangle to calculate the measure of a named angle. The scores obtained varied greatly from country to country (in Grade 8, from the United States' 31.3 percent to Japan's 83.4). This was usually the case with any item having any degree of complexity, i.e., that required students to draw upon two or more items of factual knowledge. The high scores of Japan reflect what I have found to be the greater degree of complexity to be found in its textbook exercises.

The results on one item (J15) on similarity indicate that being taught the topic is not a necessary condition for being able to tackle the question. However, "similar" is a notoriously tricky word. Its everyday meaning frequently encompasses its strict geometrical one. Thus an English-speaking student might have chosen two non-obtuse, non-equilateral triangles as being "similar" in an obvious respect (and this is essentially what the item asked them to do). Moreover, countries that had not taught "similarity" may have substituted "the same shape" for this technical term, and so changed the nature of the item. One cannot, therefore, attach too much weight to the results obtained on it. The other item on similarity (P9) proved more straightforward to analyze, even if students found it much more difficult. In this instance, students had to calculate the length of one of the corresponding sides, i.e., it was a question involving ratio. Overall the scores on this item were very disappointing. Yet to what extent this was due to the students being deficient in geometrical knowledge or to arithmetical ability one does not know. The complexity of an item (K8) requiring students both to rotate and to reflect triangles mentally proved too much for many. A further item (U) also tested congruence and the ability to orientate triangles. On this occasion no arithmetical work was involved and the general level of student success was much higher with scores up to 81.7 percent (Japan, Grade 8). Only in two of the nine countries considered did fewer than 50 percent of Grade 8 students answer the question correctly: England (39.2 percent) and Hungary (48.8).

One aspect of the teaching of geometry is to encourage the development of spatial appreciation. This was tested in the only item (K3) to venture into 3-dimensional geometry which was answered well. It asked students to visualize what happened when a three-dimensional body (represented by a two-dimensional diagram) was rotated. Nowadays, with the constant exposure of students from the countries studied to television or computer-generated images of objects and bodies being turned, and with many constructional toys that necessitate the interpretation of diagrams representing solids, this item would appear to fall within the range of common sense. Some countries seem to devote more classroom time than others to encouraging the development of such facilities, but students' results do not appear to be very dependent upon classroom teaching.

Angle measure, of course, features in all curricula. However, there is still some divergence concerning when angle facts are taught and, for a number of reasons, it was difficult to interpret the results.

Another geometry question (J11) concerned the defining properties of a parallelogram. It was noticeable that some countries registered appreciable gains with maturity and also that, as with so many algebra items, the Netherlands students had clearly not been prepared to deal with a question of this nature.

Three items related to transformation geometry. One (O8) asked students to identify the centre of rotation given a triangle and its image and the second (M5) to identify the image of a pentagon after a half-turn about a given vertex. Unlike so many of the other items they do not rely on arithmetical or algebraic knowledge and could be classified as common sense items, although in both cases some benefits appear to have accrued from being taught the subject—possibly the vocabulary, e.g., half-turn, became clearer.

The third item (M2) asked students to identify the lines of symmetry of a rectangle. The item was claimed to be inappropriate only for Swiss Grade 7 students. Not unexpectedly, therefore, among Grade 7 students, the Swiss were least successful (58.3 percent), followed by those from the U. S. (66 percent). The other seven countries scored between 77.7 and 84.8 percent. The Grade 8 Swiss students scored 76.1 and those from the U. S. 69.9. However, the marks of the other seven countries now ranged between 71.8 and 82.5 percent. Several countries scored less well in Grade 8 than they did in Grade 7' the Netherlands showing a drop of 13 percentage points. One wonders then, what use, if any, was made of this geometrical fact in the higher grade. Where was this work on symmetry leading? How was it being developed?

FRACTIONS AND NUMBER SENSE

The importance of arithmetic and the stress traditionally placed upon it meant that many items were thought appropriate for all students. However, there were significant differences in the emphasis given to teaching fractions (both decimal and common) and operations on them. Thus the average scores on three items (J12, K09, L17) which tested, respectively, the division, addition and subtraction of common fractions ranged from England's 12.3 percent to Japan's 80.6.

Two items on operations on decimals illustrated similar enormous differences in curricular emphasis. The first (J14) tested an understanding of the magnitude of the result of the division ($24.56 \div 0.004$), an appreciation of which is particularly valuable for those using calculators. Again the range of percentage correct responses of Grade 8 students was great: from Japan's 71.0 and France's 70.0 to England's 18.6 and the Netherlands' 34.4. A second, open response item (M8) tested the technique of decimal multiplication (0.203×0.56). Once again, this had not been covered in England and was not well understood in the Netherlands. (In fact, even if we ignore the positioning of the decimal point, less than 9 percent of English Grade 8 students carried out the multiplication correctly; yet over 70 percent of their French and Hungarian counterparts obtained full marks.)

An important item (U) tested the ability to calculate an "average rate." Since an understanding of rates, like that of formulas (see above), would seem an essential item in the mathematical tool kit of any school leaver or any student of science, the scores obtained on this item should be a matter of concern in several countries. For less than 40 percent of Grade 8 students answered correctly in six countries, and less than 30 percent in England and the U. S. "Rates" is not an easy topic to come to grips with, as the results from those countries that thought the item appropriate (all except England at Grade 8) show: the learning process appears to be a relatively slow one. The wisdom of leaving the topic untaught until the last year or so of compulsory mathematics education is one, therefore, to be questioned.

As with formal algebra, then, we see that common sense offers little assistance when students are asked questions concerning aspects of number they are unlikely to meet in daily life. Again, too, important curricular decisions have to be taken by countries concerning the emphasis to give to such mathematics for students of average ability. Yet, it is important to stress that such decisions will not necessarily be the same at all times. The mathematical demands of everyday and scientific life, and the expectations of students, are constantly changing, as is the provision of technological aids: as a result, decisions will need to be continuously kept under review.

DATA REPRESENTATION

Most countries now seen to place considerable emphasis on helping students to exhibit data graphically and to interpret data represented in this way. The major problems to arise concerning curriculum design are when to introduce such work (and how subsequently to develop it) and to what extent students will simply pick up the rudiments of the topic either in other curricular subjects or from the media, in particular newspapers and TV. Two items were selected on the basis of the usual criteria. The first (U) was concerned with the conclusions which might be drawn for a whole population based on those obtained for a sample. The results appear to indicate that improvements come as much from increased maturity and an increased exposure to outside influences, as from teaching. The second item (O1) asked for the interpretation of a graph giving braking distances for a car traveling at different speeds. The results were somewhat equivocal: in three countries Grade 7 students outperformed their Grade 8 counterparts, in two of them the item being thought inappropriate at Grade 7 but appropriate at Grade 8.

PROPORTIONALITY

Proportionality is an important, but not an easy, topic and the majority of English and Swiss students had not done much on it even by Grade 8. Students from other countries that thought items were appropriate often failed to show any marked degree of success. A question (L14) asking students to supply missing numbers in a proportionality table was found extremely difficult by many English students (only 18.4 percent of Grade 8 students answered correctly). Significantly, 60 percent of English Grade 8 students (as against 55 percent at Grade 7) plumped for the distractors that gave an obvious number pattern (recall the emphasis on the teaching of "pre-algebra"): 3, 6, 10—a sequence of consecutive triangular numbers. The percentage of correct responses was, however, generally disappointing, for in the eight countries claiming the item as appropriate, only in Japan did more than one third of the students give the correct answer. It was difficult to draw firm conclusions from the attainment data on the other proportionality items.

CONCLUSIONS

Readers will not necessarily draw the same conclusions from the TIMSS data as I. My own are of two types: the first refers to common sense and the curriculum, the second to the place of comparative studies in informing the actions of educators and governments.

Certainly the data on the probability items cast doubt upon the benefits of an early introduction of that subject. Indeed as Item P5 suggests, this might even result in ill-digested rules and learning replacing sound common sense. Where intuition (or common sense) breaks down is clearly illustrated by Item P6. What is needed before the student reaches this point, then, would appear to be a brief structuring of intuition in mathematical terms, rather than an extended laboring of what is basically street probability.

Yet if probability is too difficult to be introduced into primary school mathematics, then there would appear to be strong arguments for devoting more time to the topic in the later years of secondary school. Currently, it seems that many students of probability will, even after some years' study, never reach the understanding of the topic that today's educated citizen requires and, indeed, of the need to correct certain intuitive ideas.

The emphasis now given to what I have termed "pre-algebra" yields somewhat different results. The data do suggest that if one emphasizes pattern spotting, then students will perform relatively better on such items than they do on mathematics in general. However, the data do not suggest that such a foundation is either necessary or sufficient when it comes to the learning of algebra proper. There is also the problem of the unthinking application of "pattern spotting", without any understanding of what is responsible for the generation of a particular type of pattern (e.g. the item on proportionality). It would seem that those who propose to spend time over many years on a protracted pre-algebra course should provide more solid evidence concerning its value to students.

I hope that this paper has also demonstrated that an international comparative study need not be merely a way of ranking countries by attainment, or of identifying social considerations that affect mathematical performance, but over which the educator can exercise no effective control. The data generated by such studies can also illumine key problems of mathematics education: the wisdom of certain curricular emphases, what is intuitive and what is not, ... However, if this is to be done successfully, then great care will have to be taken in the framing of items. Despite the criticisms contained in this paper of certain of its items, TIMSS made great steps forward in this direction. However, more could be done in future. Only in this way might it prove possible to throw more light on certain problems of curriculum construction and, in so doing, win over many of those who doubt the worth and validity of such large-scale, expensive studies.

NOTES

[1] The last, a report written for the English national Qualifications and Curriculum Authority (QCA), was never made publicly available. An anodyne summary of this report and of a similar one relating to primary textbooks and curricula from other countries was published as Howson et al, 1999.

[2] Formal education begins at different ages in different countries. It should be borne in mind, then, that Grade 7 is equivalent to Year 8 within the English system whereas for many Swiss students it will only be their sixth year of formal education.

[3] The Netherlands was the only one of these countries not to satisfy TIMSS' strict sampling conditions.

[4] The English students who took the TIMSS Population 2 tests would not necessarily have studied probability in the primary school although their results on, e.g., Item P6, strongly suggest that the vast majority had been introduced to the topic by the secondary level.

[5] The item labels used in *TIMSS Mathematics Items: Released Set for Population 2* are given in parentheses. "U" denotes an unreleased item. Items are multiple choice unless stated otherwise.

[6] 25 percent of Grade 7 students believed odd and even to be equally likely: a disappointing response from students who, so it was claimed, had been taught probability.

[7] In Grade 8, the English support for the denominator dropped, but that for the numerator rose.

[8] Of course, those who believe that red is more likely, those who believe that black is more likely, or those who know the "probability" answer, will all have the same chance of winning next time. Perhaps it is this fact that has prevented the mathematical belief failing to become part of "common sense".

[9] It must be stressed that all TIMSS items went through a long selection process and were field-tested. However, anyone with testing experience will know that it is only in retrospect that one realizes how items might have been improved.

[10] This indicates the number of students arriving at the right equation. That it is greater than the number in 7a/8 indicates that not all such students then went on to get the correct answer.

[11] The need to teach mathematical conventions was illustrated further by an item (J16) asking students to identify the coordinates of a point in the first quadrant. Swiss Grade 7 students, for whom the item was thought inappropriate, were often puzzled concerning the order in which the coordinates should be given.

[12] Even in Grade 8 only 52.6% of English students and 54.8% of Swiss ones selected the correct answer.

REFERENCES

Beaton, A. E., Mullis, I. V. S., Martin, M. O., Gonzalez, E. J., Kelly, D. L., & Smith, T. A. (1996). *Mathematics achievement in the middle school years: IEA's Third International Mathematics and Science Study (TIMSS)*. Chestnut Hill, MA: Center for the Study of Testing, Evaluation, and Educational Policy, Boston College.

Howson, A. G. (1991). *National curricula in Mathematics*. Leicester: Stanley Thorne for the Mathematical Association.

Howson, A. G. (1995). *Mathematics textbooks: A comparative study of Grade 8 texts*. Vancouver: Pacific Education Press.

Howson, A. G. (1998). *Primary school textbooks: An international study, 1998. A report commissioned by the Qualifications and Curriculum Authority.* London: Qualifications and Curriculum Authority.

Howson, A. G., Harries, T., & Sutherland, R. (1999). *Primary school mathematics textbooks: An international summary.* London: Qualifications and Curriculum Authority.

Chapter 8

ADVANCED MATHEMATICS: CURRICULA AND STUDENT PERFORMANCE

Geoffrey Howson

An oft-quoted remark by the comparative educator, Torsten Husén, is "comparing the outcomes of learning in different countries is in several respects an exercise in comparing the incomparable." In a strange way, the two aspects of the mathematics testing of TIMSS Population 3 students, i.e., students in their final year of upper secondary school, serve both to support and contradict this remark.

The first aspect of mathematics to be tested by TIMSS in Population 3 was the degree of mathematics literacy possessed by all students, not just those still studying mathematics.

The items were selected on the basis of what mathematics was thought essential to meet societal demands and those required for responsible citizenship. As a result, the items were not constructed with the curricula of the participating countries in mind, but, rather, the perceived future needs of students. All items were, for example, set in context and tested the ability to apply mathematics rather than to remember mathematical facts and techniques. The results can be found in Mullis et al. (1998), and it can be readily seen to what extent students from the countries concerned met the demands envisaged by those responsible for selecting the items. There would seem only one factor affecting this study that would result in our "comparing the incomparable," namely, the percentage of students who remain in education until the last year of secondary school and who, as a result, were included in the sample.[1] Otherwise, the mathematical demands likely to be encountered by these students do not differ substantially from country to country. The age of the students, the time they have spent studying mathematics, ..., are essentially irrelevant to the consideration "Do they, on leaving school, possess this necessary knowledge?".

The second aspect tested the attainments of students enrolled in specialist mathematics courses, and this proved much more complicated. The percentage of the age cohort enrolled in such courses differs widely from country to country, as does the time available for the study of mathematics. One result is that curricula vary considerably between countries both in the range and the depth of mathematical

D.F. Robitaille and A.E. Beaton (eds.), Secondary Analysis of the TIMSS Data, 113–123.
© 2002 *Kluwer Academic Publishers. Printed in the Netherlands.*

coverage. Much that is taught to some specialist students in some countries was, therefore, left untested. As a result, interpreting the data presented in a table, say, of the distribution of advanced mathematics achievement by country (see Mullis et al., 1998, Table 5.1) becomes far from straightforward. Indeed, it could be argued that to present results in this form, i.e. with countries arranged by "attainment ranking" rather than in alphabetical order, is an invitation to readers (and the media) to take an altogether too simplistic a view of the results. This aspect of TIMSS does attempt to "compare the incomparable." However, this does not mean that it was a worthless exercise: it simply means that the data must be interpreted with care. They still contain many messages for the participating countries of which heed should be taken. To decode these messages, however, it is essential to consider data at the level of individual items and usually to ask:

- Is this an item we should wish our students to be able to answer?
- If so, can a reasonable percentage of them do so?
- If not, is it the case that students everywhere have difficulty with this item, or do the results reveal specific problems presumably arising from the way the topic, or mathematics in general, is taught in our country?

Let us first consider three examples.

The first was a multiple-choice, unreleased item testing an understanding of averages (and might well have been set on the mathematical literacy test). It gave the (unequal) sizes of two disjoint sub-populations and an average for each and asked what the average for the union of these populations would be. (The item was set in context and did not make use of specifically mathematical vocabulary.) The correct answer was selected by 75.4 percent of Swiss students. On the other hand, over 52 percent of Greek and U. S. students opted for adding the two averages together and dividing by two. What does this tell us about the "specialist" mathematical students and the way in which they have been taught?

The second, open-response item was:

> In the $\triangle ABC$ the altitudes BN and CM intersect at point S. The measure of $\angle MSB$ is 40° and the measure of $\angle SBC$ is 20°. Write a PROOF of the following statement:
>
> $\triangle ABC$ is isosceles.
> Give geometric reasons for statements in your proof.
> [A diagram with angle measures marked in was supplied.]

Full marks on this item were awarded to 64.9 percent of Greek and 9.7 percent of U. S. students (see Lo1 (K18)[2] in Table 2).

A third, unreleased example was a "common sense" multiple choice item asking which of four statements could be inferred, using the "pigeon hole" principle, from given data. (Again, the question would not have been out of place in a "logical literacy" test.) The correct answer was identified by 81 percent of Israeli, 70 percent of U. S. and 33.8 percent of Greek students (see Lo2 in Table 2).

These three examples should suffice to indicate the amount of information that is lost when one only takes averages: information that clearly identifies weaknesses and strengths in a country's mathematics teaching. Moreover, the messages in these cases are not confused by "comparability" issues.

The purpose of this paper, then, is to look in detail at selected items. By considering how student performance varied on and between items (with the emphasis on the mathematics tested rather than on the countries from which the students came), we hope to throw light on key pedagogical issues. Moreover, we shall see what, if any, marked differences could be observed in the curricula of the countries concerned.

Clearly, if one is designing a curriculum for 2 percent of the age cohort (Russian Federation), then one would expect it to be more challenging than one for 75 percent (Slovenia). One would also expect the former to score more highly than the latter. So as to get rid of such obvious anomalies, attention will be concentrated on those countries which indicated that they had enrolled between 10 and 20 percent of the age cohort in the specialist mathematics course[3]: namely, Canada, the Czech Republic, France, Greece, Sweden, and Switzerland, plus Australia, Italy and the United States which met the "percentage" condition but which failed to meet the TIMSS guidelines concerning sample percentage rates (see Mullis et al., 1988).

SOME PRELIMINARY DATA

In Table 1 we give data extracted from a variety of tables in Mullis et al. (1988) intended to present an overall picture of the performance and the background of the students from the countries on which attention is focused.

Table 1. Advanced Mathematics performance of selected countries

Country	MTCI (%)[1]	Mean Age	Mean score[2]	Mean top 5 %	Mean top 10 %[3]	Pct. male	Gender differ- ence[4]	TCMA (82)[5]
Australia	16	17.8	525	643	589	55	14	71
Canada	16	18.5	509	620	567	53	39	70
Czech Rep.	11	18.1	469	558	485	41	92	80
France	20	18.2	557	645	612	63	23	80
Greece	10	17.7	513	592	513	69	11	-
Italy	14	19.1	474	569	520	61	24	-
Sweden	16t	18.9	512	608	564	69	23	62
Switzerland	14	19.5	533	629	575	54	56	72
United States	14t	18.0	442	543	485	51	31	82

* The notes for Table 1 may be found on page 123.

SOME QUESTIONS CONCERNING AIMS, CURRICULA, AND STUDENT ATTAINMENT

What has TIMSS to tell us about students' ability to prove and reason?

Logical reasoning and proof should pervade all mathematics, and five items would seem specifically to test students' ability to prove and reason. We have already described two: Lo1 (K18) asked students to provide a proof in Euclidean-style geometry, while Lo2 (U), which might also be classed as combinatorics, was a simple example of the pigeon hole principle. The third, Lo3 (L11), tested, within context, students' ability to comprehend the relation between a statement and its contrapositive. Both the last two items could be classified as "common sense" (as is borne out by the students' performances). It will be noticeable here, as elsewhere, that panels sometimes classified as "inappropriate" items that required students to think rather than to recall what had been taught in lessons. (Note that in the following and subsequent tables, bold type indicates scores on "appropriate" items; italics indicate "inappropriate"; and the scores for Greece and Italy indicate that no data on appropriateness were supplied.)

Item Lo4 (U) was a more specialized, open response item and asked for the logical steps needed to complete a proof by mathematical induction. This would have meant nothing to a student who had not been taught it. Finally, Lo5 (U), an open response item, asked students to provide a proof in the context of coordinate geometry. This could be done either by using straightforward analytic methods for calculating the midpoint of a line segment, or by recalling the properties of parallelograms. (Since the item has not been released, its essential simplicity cannot be adequately described to, or appreciated by, readers.)

Table 2. Performance of selected countries on Logic and Reasoning items

Country	Lo1	Lo2	Lo3	Lo4	Lo5
Australia	**40.8**	**65.2**	*78.3*	**36.1**	**57.7**
Canada	**34.3**	**66.4**	*70.4*	**8.4**	**29.4**
Czech Rep.	**33.9**	**57.0**	*84.1*	**15.8**	**27.6**
France	**53.4**	**61.8**	*77.3*	-	**64.7**
Greece	64.9	33.8	73.2	79.5	49.5
Italy	35.1	47.0	76.9	0.4	31.3
Sweden	**40.7**	**78.5**	**66.6**	*0.2*	**12.0**
Switzerland	**46.3**	**68.0**	*83.0*	*14.4*	**29.9**
United States	**9.7**	**70.0**	**69.8**	**0.9**	**21.3**

Clearly, Lo4 stands apart from the other items. The performance of the Greek students is particularly noteworthy (87 percent of Greek females obtained full marks). The data also call into question the understanding of the principle of induction that students in those other countries that taught it have gained. The results on Lo1 and Lo5, which call for simple proofs, should ring alarm bells in some countries. But what is one to make of the lack of transferability between formal proving and common sense reasoning exhibited in their different ways by, say, Greek and U. S. students? We are often told that Euclid provides a strong foundation for logic and proof: but to what extent is the knowledge gained in its study readily transferable (see also Item G2 below)?

Have foundations been securely laid?

Several items tested students' understanding of the function concept, functional notation, and graphing of functions: of course, many other items also made use of such knowledge. All the items tested basic knowledge upon which further mathematics could be developed and accordingly all were thought "appropriate." It is valuable to see to what extent countries were successful in laying firm foundations with all students.

Item F1 (U) tested understanding of the notation for the composition of functions; F2 (U), the ability to graph a piecewise linear function given its value in different intervals; F3 (U), (set in context) the ability to judge the interval in which one function was greater than another; while F4 (U) asked about points with integer coordinates on the graph of a given function. Item F5 (U) was technically trickier and asked students to identify the interval within which a rational function took negative values; F6 (U) sought to find the equation of the graph resulting from the translation of the graph of a given function. Item F7 (U) was hard to classify since it had a degree of complexity absent in other function items. Had it been an open response item, it would have demanded knowledge of calculus and maxima and minima. As a multiple choice item it could be solved by algebra and substituting various values for x into a mensuration formula (given in the formula sheet). Item F8 (K1) tested the (rather obvious) relationship between two variables in an equation.

Table 3. Performance of selected countries on knowledge of functions items

Country	F1	F2	F3	F4	F5	F6	F7	F8
Australia	59.9	67.7	65.4	50.4	53.5	50.3	34.9	78.5
Canada	69.8	55.0	67.1	47.8	54.2	50.2	34.4	88.0
Czech Rep.	20.5	50.7	64.8	66.9	56.2	27.8	36.3	81.4
France	87.3	84.3	49.7	59.5	83.5	20.0	30.7	94.9
Greece	85.8	68.5	51.0	67.4	64.6	51.6	34.4	83.6
Italy	48.4	63.3	54.1	63.4	65.9	34.1	35.9	82.4
Sweden	59.6	67.0	70.2	59.8	67.7	58.0	41.7	90.2
Switzerland	58.7	65.9	70.4	74.3	64.2	26.8	43.1	95.7
United States	54.7	43.7	57.4	32.1	42.4	44.3	25.3	67.1

The scores on these items are of interest. In general these tend to cluster, but we note the strong performance of Sweden (which has a restricted curriculum and time allocation for its specialist mathematics course) and the weakness of American students (who, so it is claimed, have an all-embracing curriculum). In particular, we note the high performance of Swedish students on F6 and F7: items which students from other countries tended to find difficult. The variation in scores of French students also suggests questions concerning curricular coverage and emphases.

Is an understanding of basic concepts sacrificed to the acquisition of techniques?

The items in TIMSS appear to have been selected more to ensure content coverage rather than to answer specific pedagogical questions. Nevertheless, usually one can find a small number of items that cast light on such issues. Here we select four: two referring to exponential growth and decay and two concerned with the differential calculus.

Item Ex1 (K13) was an open response item that asked, "The number of bacteria in a colony was growing exponentially. At 1 p.m. yesterday the number of bacteria was 1000 and at 3 p.m. yesterday it was 4000. How many bacteria were there in the colony at 6 p.m. yesterday?" It demanded nothing more than the knowledge that in equal time intervals the number of bacteria would increase by a constant multiplicative factor, plus some very simple arithmetic. The second, Ex2 (L3), a multiple choice item on exponential decay, gave the general formula for calculating the mass, y, after t days, including two unknown constants for the initial value and for the rate of decay. Students were then asked to identify the value of the latter constant for an element whose half-life (term defined) was 4 days. The item was not particularly demanding, but did require facility in dealing with logarithmic and exponential functions. Even allowing for the fact that one item was open response and the other multiple choice, the fact that the (overall) international average on the first was 27 percent and on the second 44 percent is both surprising and disturbing.

On differentiation, one item, Diff1 (U), was concerned with what I would call "basic understanding" and several items with technique, of which we shall consider a typical one, Diff2 (U). The former, open response item provided students with the graph of a piecewise linear function defined on the interval $-3 \leq x \leq 3$ and asked them to say for what values of x it was (a) not continuous and (b) not differentiable. Item Diff2 was a typical question on technique and asked students to identify the derivative of a function of the form

$$\frac{a}{\overline{)bx+c}} .$$

Once again our eyes are drawn to disparities between the scores on different cognitive aspects. Consider, for example, the differences in the Greek scores on Ex1 and Ex2, or on the items on differentiation. Of course, the word "continuous" carries meaning by itself and so the high scores on Diff1(a) are not unexpected. But, the low scores on Diff1(b) compared to those obtained by showing technical proficiency (not overwhelming in some cases!) on Diff2 should give rise to concern.

Table 4. Performance of selected countries on Exponent and Differentiation items

	Ex1	*Ex2*	*Diff1(a)*	*Diff1(b)*	*Diff2*
Australia	**33.9**	**55.9**	**53.8**	**7.2**	**63.1**
Canada	**35.2**	**46.4**	**60.8**	**9.7**	**51.7**
Czech Rep.	**24.2**	**35.2**	**47.7**	**5.8**	**30.6**
France	**24.7**	**47.3**	**83.1**	**7.3**	**56.0**
Greece	8.4	64.7	67.4	7.4	64.1
Italy	21.2	29.6	56.7	11.4	59.2
Sweden	**51.2**	**65.8**	**36.9**	**1.6**	**34.2**
Switzerland	**26.3**	**37.6**	**55.1**	**16.3**	**50.9**
United States	**24.4**	**39.2**	**45.9**	**7.9**	**35.4**

Can students solve problems and deal with complexity?

What distinguishes a "problem" from an "exercise" is a matter that has been debated at length elsewhere. Let us assume here that, in the case of a "problem," it is not immediately obvious exactly what mathematics is being tested. This definition clearly limits the opportunities for testing problem-solving abilities within a short test or examination intended to cover a wide curriculum: for problem solving requires time for contemplation of the problem, analysis of the situation and the identification of the mathematics that will allow one to solve the problem[4]. Indeed, only one item set to Population 3 students clearly qualifies as such. This item, PS1 (K14), was: "A string is wound symmetrically around a circular rod [this was illustrated in a diagram]. The string goes exactly four times around the rod. The circumference of the rod is 4 cm and its length is 12 cm. Find the length of the string." Students had to visualize the surface of the rod being cut and then flattened to form a rectangle, after which the problem became an easy application of Pythagoras' Theorem. As will be seen, the results should give rise to great alarm in several countries.

Here, we shall study only a small selection of items involving "complexity." The first, in fact, hardly qualifies for that description: it is an extremely basic exercise in probability.

This item, Pr1 (K11), tested the calculation of probability as the number of favorable cases divided by the (given) number of possible cases (i.e., the same concept as for Population 2). The second item, Pr2 (L10), asked for the probability that, given two independent alarms with differing probabilities of operating, "at least one alarm would sound." It is disturbing that scores on what might be seen as one of the first parts of mathematical probability impossible to classify as "common sense" should be so low—and this a key concept tested in a relatively straightforward manner. What was particularly alarming was the high percentage of students who opted for distractors that indicated that safety was not improved when a second alarm was installed. "Common sense" was, in fact, ignored.

Table 5: Performance of selected countries on Problem Solving and Complexity items

	PS1	Pr1	Pr2	G1a	G1b	G2
Australia	13.9	76.1	45.5	62.6	26.4	16.7
Canada	12.2	60.9	27.5	64.6	25.1	15.1
Czech Rep.	8.3	42.9	22.1	79.6	23.4	33.3
France	4.1	61.9	33.8	87.7	38.3	34.7
Greece	5.0	29.1	17.6	60.8	11.7	12.5
Italy	6.1	33.7	11.6	72.0	27.8	19.5
Sweden	23.6	64.4	52.9	41.4	11.9	25.5
Switzerland	17.2	62.5	40.3	82.0	45.6	33.7
United States	3.8	62.4	21.3	49.8	10.7	8.6

Item G1 (U) was on transformation geometry. It required students to draw the image on a given grid of a given triangle under (a) a reflection in an axis and (b) under a rotation of 90° in an anticlockwise direction about an exterior point of the triangle. One might have imagined that at this level students would have been asked to deal with reflection in a line other than an axis or a line parallel to one. The last part, which was a little more complex, was surprisingly poorly answered.

Finally, G2 (K10) was an interesting item set on pure geometry: "AB is the diameter of a semicircle k, C is an arbitrary point on the semicircle (other than A or B), and S is the centre of the circle inscribed into $\triangle ABC$. [A diagram was supplied.] Then the measure of $\angle ASB$ (a) changes as C moves on k, (b) is constant for all positions of C but cannot be determined without knowing the radius, (c) equals 135° for all C, (d) equals 150° for all C." Again, carrying out a multi-step problem appears to have proved too much for many students.

CONCLUSIONS

In the introduction to this paper stress was put on the difficulty of making comparisons at this level —they did not disappear in the course of this study. Some things did, however, become clearer. First, the items fitted the countries' curricula somewhat better than was suggested by the TCMA. Nevertheless, certain countries were disadvantaged if one concentrates only on overall rankings. As was well known, certain topics, e.g., the principle of mathematical induction, complex numbers and vectors, remain on the fringe of the specialist mathematics curriculum. In addition, Sweden, for example, does not appear to deal with the co-ordinate equation of the circle, let alone other conic sections; neither does it put emphasis on combinatorics, probability or statistics (although clearly many students had done some work on these topics). Australia, Switzerland, and Canada also have noticeable gaps. Here, however, we get back to the nub of the problem of comparisons: not all countries share the same aims for senior high school education. In particular, views differ markedly on the degree of specialization that should be encouraged. If two-thirds of students study mathematics for less than three hours a week, then the curriculum will have to be considerably sparer than that of a country in which almost all students study the subject for five or more hours a week—and spend a proportionate amount of time on homework. In that respect it is no wonder that the French and the Greek students show a much higher level of technical facility than their peers elsewhere. Once one departs from the standard type of question, however, the gap between these countries and the others changes and in some cases is reversed—and one wonders to what extent students are being trained "to jump through hoops" rather than to develop sound mathematical understanding and powers. Here, though, one meets another imponderable: do some of the good performances of, say, the Swedes and the Swiss spring as much from extra maturity (look at the average ages) as from less technique-centred mathematics teaching?

Similar problems arise when one considers the results of the mathematics literacy tests, for in this Sweden and Switzerland came out on top of those countries that satisfied the sampling criteria[5]. Here, another basic question arises: how should weightings be allocated between the two aims of equipping all students with a high degree of mathematical literacy, and on ensuring that national requirements for a mathematical elite are met?

Yet, not all the extra mathematics instructional time enjoyed by, say, Greek and French students was spent polishing up the techniques tested in TIMSS. A good number of the items set to specialist students might have been expected to have been covered pre-16: for example, all those on pure geometry, two of the three probability items, and all except one on statistics. However, trigonometry was virtually untested, and there was little emphasis on algebra. 3-D work was largely absent. Indeed, vast swaths of, for example, the contemporary French and Greek syllabuses were left unexamined in TIMSS (see, e.g., Howson, 1991). Of course it would be impossible to test all these additional topics in a short TIMSS test, since the more advanced the mathematics becomes the more difficult (and potentially misleading) it is to set multiple choice or even open response items that can be answered quickly. Moreover, many of the topics would not appear in the curricula of other countries. What TIMSS did was to design a test on basic material that was expected to be familiar to all specialist mathematics students and to show how students from different countries fared on these. That alone was worthwhile and the resulting data will, as was pointed out earlier, provide most, if not all, countries with food for thought. As it was, facility levels were decreasing as one began to reach the frontiers of the students' knowledge. Pressing on much further could have proved altogether too discouraging!

Another well known factor is again made remarkably evident in this study: the marked difference between scores on open response items and those on multiple choice ones. It may well be that there is a strong correlation between rankings on both types of question, but often the results are remarkably disparate, as is the case, for example, on the two items on exponential growth quoted above. One still wonders, then, to what extent students really understand a topic. Again, one observes marked differences between students' employing common sense to solve problems (see, for example, the items on logic), and on their ability to supply formal proofs for relatively straightforward results. These differences must, surely, derive from different teaching priorities. Again, the TIMSS study is useful in focusing our attention on such problems: for showing that they can be overcome, but that, to a large degree, they are independent.

The main conclusion from this study, therefore, is that little is to be gained from studying the ranking lists of countries to be found in Mullis et al. (1988). Indeed, their simplicity may well prove a tempting trap into which politicians might fall. There are some clear warnings in such tables, of which heed should be taken; but, they, by themselves, do not provide value for use. A country wishing to use TIMSS data to improve its mathematics teaching must carry out a careful study of responses to individual items; to the validity and importance they assign to these; and to the way that students' successes or failures can be linked, not only to specific topic areas given the country's curriculum content and pedagogy, but, perhaps more importantly, to the varying cognitive demands of the individual items.

NOTES

[1] There was considerable variation in these percentages even among the Western European countries, e.g. in Italy, 52percent of the age cohort were sampled, in Norway 84 percent. Such differences can, however, be taken into account by readers when interpreting the student attainment data.

[2] The item label given in parentheses is that used in *TIMSS Released Item Set for the Final Year of Secondary School.* "U" denotes an unreleased item.

[3] The structures of upper secondary education in the participating countries and characteristics of the students tested (e.g. definitions of what comprised an "advanced mathematics course") can be found in Appendix A of Mullis et al., 1998.

[4] Moreover, a true "problem" raises questions of reliable marking, for students might make mark-worthy efforts while pursuing different paths and failing to reach a solution. They pose questions for tests/examinations with a large number of entries and/or examiners. That problem solving could not be more thoroughly tested in TIMSS was perhaps a matter for regret, but understandable. That in many countries it would now seem to be insufficiently tested in national external examinations is more serious. It would appear that high-attaining young mathematicians (and this means very many more than participate in International Mathematical Olympiads) are being given insufficient training, and opportunities to develop and display their abilities in problem-solving.

[5] The Netherlands (which did not satisfy these criteria) came out on top overall.

NOTES FOR TABLE 1

[1] The MTCI (Mathematics TIMSS Coverage Index) is a measure of the percentage of the age-cohort taking the specialist mathematics course.

[2] The mean scores were obtained after a statistical weighting adjustment to percentage marks obtained on individual items. As a result it is not possible to give a "total mark" out of which the given scores were obtained.

[3] The marks shown in the 5 percent and 10 percent columns are the hypothetical means of the marks scored by these percentages of the age cohort, based on the assumption that those not classed as "specialist mathematicians" would have obtained fewer marks than those who were.

[4] The gender difference is obtained by subtracting the mean score for females from that for males.

[5] A panel in each of the participating countries was asked to indicate which items were considered appropriate for their students (i.e. the mathematical content would have been expected to have been taught). We reproduce these opinions as an indication of what mathematics was included in that country's curriculum for specialist mathematicians. Although there are obvious anomalies, the data for the "specialist mathematicians" give a more accurate impression than do those for Populations 1 and 2. In all, 82 points could have been obtained from all the items set and the numbers of points arising from "appropriate items" are shown in the column headed TCMA (Test-Curriculum Matching Analysis).

REFERENCES

Howson, A. G. (1991). *National Curricula in Mathematics*. Leicester: Stanley Thorne for the Mathematical Association.

Mullis, I. V. S., Martin, M. O., Beaton, A. E., Gonzalez, E. J., Kelly, D. L., & Smith, T. A. (1998). *Mathematics and science achievement in the final years of secondary school: IEA's Third International Mathematics and Science Study (TIMSS)*. Chestnut Hill, MA: Center for the Study of Testing, Evaluation, and Educational Policy, Boston College.

3 FOCUS ON SCIENCE

Chapter 9

EXPLORING POPULATION 2 STUDENTS' IDEAS ABOUT SCIENCE

Marit Kjærnsli, Carl Angell, and Svein Lie

The aim of this chapter is twofold. One purpose is to illuminate Population 2 students' conceptual understanding of some central phenomena and principles within the natural sciences. The TIMSS international database is a rich source for various secondary analyses, and, in our view, particularly for investigating the conceptual understanding that students around the world have. In addition to examining how much students know and understand within science, we will pay attention to what understandings and what conceptions they have.

The second aim of this chapter is to look for consistent similarities and differences among countries concerning patterns of students' responses to selected items. In a given country, these response patterns depend on curricular antecedents in addition to general cultural and natural factors, such as language, landscape and climate. We will investigate to what extent similar response patterns are found in countries with similar sets of curricular antecedents and cultural factors. The idea is first to group countries according to cultural and geographical factors, and then to investigate the similarity of the response patterns within each of these groups compared to other groups. Differences between countries may originate from cultural, geographical, or curricular factors. We cannot empirically distinguish between these factors when it comes to explaining differences in response patterns. However, in a few cases we will comment on the possible role of individual factors.

The present analyses were conducted mainly on the basis of the free response items, as these usually allow a deeper understanding of students' ideas than do multiple choice items. Elsewhere (Angell, Kjærnsli, & Lie, 2000) we have demonstrated that the TIMSS two-digit coding of responses to these items provides a potent tool for the analysis of students' conceptual understanding in international achievement tests. But multiple choice items can also provide valuable information about students' ideas, in particular if the distractors have been carefully constructed for that purpose.

Since the fundamental paper by Driver and Easley (1978) there have been numerous research studies on students' conceptions across a range of science topics.

D.F. Robitaille and A.E. Beaton (eds.), Secondary Analysis of the TIMSS Data, 127–144.
© 2002 *Kluwer Academic Publishers. Printed in the Netherlands.*

Overviews (e.g., Wandersee, Mintzes, & Novak, 1993) and bibliographies (e.g., Pfundt & Duit, 1994) have been published. The theoretical paradigm for this research activity is the constructivist view of learning. The core of this theory is that students learn by constructing their own knowledge. When outer stimuli are treated in the mind together with earlier knowledge and experience of the issue at hand, new insights may be formed. Obviously, within such a framework, it is of crucial importance for science educators and teachers to be aware of students' conceptions within a topic prior to instruction in order to accomplish successful learning.

Selection of science topics for analysis

As was said above, this chapter aims at exploring 13-year-old students' conceptions concerning some central topics in classical science. The science topics were chosen using three criteria. Firstly, there had to be an adequate and released Population 2 item that addressed the key point of the phenomenon or principle at hand. Secondly, the topic had to be regarded as one of the cornerstones of classical science, without being too advanced for 13-year olds. Thirdly, taken together the topics had to represent some breadth of coverage the science framework (Robitaille et al., 1993) and allow some general statements to be made about students' understanding of basic scientific concepts. The selected topics and items are listed below. (Item labels in parentheses are the original TIMSS Population 2 labels.)

- the constancy of mass in physical processes: When melting, ice is converted into an amount of water of exactly the same mass (Q18);
- constancy of temperature during a process of change of state: viz., melting (Y2);
- the dependency of the process of photosynthesis on light in maintaining an ecosystem (X2b);
- gravity as a force acting on all bodies regardless of their movement and position (K17);
- the crucial role of water balance and temperature regulation for the human body (O16);
- the Earth as a planet with a rotation around its own axis that causes day and night (Q11);
- the ever-changing surface of the Earth, due to building and breaking forces (J1).

Coding rubrics for free-response items

A necessary tool for making diagnostic analyses of responses to free response items is a coding system that encompasses both the correctness dimension and the diagnostic aspect. In TIMSS this was provided by a two-digit system, originally proposed by the Norwegian TIMSS team. The process of development and the scope and principles of the coding rubrics have been described by Lie et al. (1996). The idea behind the two-digit system is to apply one two-digit variable to take into account correctness as the first digit (2, 1, or 0 points) and codes for method/error/type of explanation as second digit. The second digit taken alone has no separate meaning across items.

An important feature of the codes is their empirical basis, as all codes are based on authentic responses from many countries in the field trial of the items. The set of actual codes for a particular item therefore reflects the types of responses actually given by students. Furthermore, score points are allocated from a judgment of quality based on what can be expected from students at age 13, not on requirements of "correctness" from an *a priori* subject matter point of view.

Clustering of countries

We wanted to cluster countries into groups that shared common characteristics. These common characteristics should be well defined from a cultural or geographical perspective, and furthermore, the grouping of countries should also somehow be based on TIMSS achievement data from Population 2. Following an approach proposed by Zabulionis (1997), country clusters were constructed based on patterns of differential item functioning in the form of p-value residuals (after subtracting over-all p-values by items and by country) for all science items. These residuals act as measures of how much more or less difficult an item was, compared to what was "expected" on the basis of the country's general international overall results and the international difficulty of the item.

In Table 1 we have listed the actual country groups. For these groups we have both meaningful cultural or geographical similarities (validity) and reasonably high achievement similarities (reliability). As a measure of this reliability, Cronbach's alpha is given for each group. The following countries did not fit well into any of the groups and were therefore not included: Austria and Germany (the two were closely linked), France and French Belgium (linked together), Israel, Kuwait, Colombia, the Islamic Republic of Iran, and South Africa. Reference will be made to some of these countries individually. The North European group may seem a bit odd, in particular because of the presence of Switzerland. However, this country as well as Flemish Belgium and the Netherlands, "wanted" to belong to this group according to the empirical clustering of the data. Therefore, we "allowed" them to do so in the present chapter. It is interesting to note that in the SMSO study (Survey of Mathematics and Science Opportunities) linked to TIMSS, Switzerland was often compared to Norway with regard to both teaching style and curricular emphases (Schmidt et al., 1996).

Table 1. Groups of countries

Group	Countries	Cronbach's alpha
East Asia	Hong Kong, Japan, Korea, Singapore and Thailand	0.52
East Europe	Bulgaria, Czech Republic, Hungary, Latvia (LSS)*, Lithuania, Romania, Russia, Slovak Republic, Slovenia	0.68
English-speaking	Australia, Canada, England, Ireland, New Zealand, Scotland, United States	0.88
North Europe	Denmark, Iceland, Norway, Sweden, Belgium (Fl)*, Netherlands, Switzerland	0.73
South Europe	Cyprus, Greece, Portugal, Spain	0.44

* The Flemish (Fl) and French (Fr) educational systems in Belgium participated separately in TIMSS. Latvia is annotated LSS for Latvian Speaking Schools only.

ITEM-SPECIFIC RESULTS

In the following sections, results will be presented and discussed, item by item. The focus will be on the nature of students' conceptual understanding as indicated by the occurrence of particular responses within country groups and some individual countries. We have tried to avoid drawing simplistic conclusions about students "having" or "not having" a particular idea, misconception or "conceptual framework," based on responses to stand-alone items. Different aspects of the cognitive networks will be activated depending on the context created by the item.

We do not aim at very precise quantitative statements. Nevertheless, the order of magnitude for percentages of code occurrences is relevant when discussing the occurrence of and background for typical patterns of student thinking. Typically, standard sampling errors for percentages for a country are of the order of a few percent (for the international average, less than one percent). Furthermore, the within- and between-country inter-marker reliability for free response items has been reported by Mullis and Smith (1996). Based on the reported reliabilities, it seems that valid comparisons between frequencies for groups of countries can be made within a few percent. In our presentation of results, frequency distributions will be reported to the nearest percent without any further comments. Results are given for the upper of the two grades included in Population 2. In most countries this means Grade 8 (Beaton et al., 1996). It should also be mentioned that even if response distributions are given for the actual codes for students' responses to free-response items, the descriptions themselves have been simplified. In particular, examples given in the original coding guide (TIMSS, 1995) have been omitted. Furthermore, in a few cases response categories are combined.

For the discussion of the effect of country clustering we need some measure of the proportion of the variability in response patterns that is "explained" by the grouping of countries. For each item we have therefore calculated the between-group variance to total variance ratio (both sum of squares pooled over all response categories).

Constancy of mass (Q18)

One may argue that the history of physics, to a large extent, has been a search for fundamental conservation laws. It has been important to define basic quantities that are conserved under certain conditions. The theme for the first item is a typical example, the concept of mass. Mass is the fundamental ontological concept, a measure of the "amount of matter," a quantity which is conserved during change of state, which is the issue of the item at hand.

> Q18. A glass of water with ice cubes in it has a mass of 300 grams. What will the mass be immediately after the ice has melted? Explain your answer.

For full credit a statement that implies that the mass has not changed is required, and, in addition, an adequate explanation (see below). Partial credit is given for a correct response with inadequate or no explanation. Table 2 shows the coding guide and the response distributions given in percent.

Table 2. Coding guide and results for item Q18

Code	Response	East Asia	East Europe	English-speaking	North Europe	South Europe
Correct Response						
20	300g with a good explanation	33	26	34	31	25
Partial Response						
10	300g. Explanation is inadequate	6	4	4	6	4
11	300g. No explanation	2	9	1	7	4
Incorrect Response						
70	More than 300g. With explanation	20	10	20	15	15
71	More than 300g. No explanation	1	4	1	2	2
72	Less than 300g. With explanation	16	5	14	12	8
73	Less than 300g. No explanation	1	2	1	2	2
79	Other incorrect	6	4	7	5	8
Blank		15	35	18	20	33

The country clustering "explains" 42 percent of the variability in response patterns on this item. However, the main between-group differences are largely found among the patterns of incorrect responses and non-responses, whereas clusters of countries account for very little of the variation in correct responses. If we concentrate on the correct aspect of no change of mass (Codes 20, 10, and 11 taken together), discarding for the moment the explanation part, we see that around 40 percent of the students in all country groups responded correctly. Given the clear and direct question in this item, it seems to be a straightforward interpretation that about 40 percent of students "believe" in conservation of mass in the case of melting ice.

The rather large non-response rates show large variations between country groups, with particularly high rates in the East and South European groups. The incorrect responses are divided between increased mass and decreased mass, with the former being somewhat more frequent. During the marking process in Norway we looked into the explanations given to these responses, which all are necessarily "incorrect." Some interesting sources for the misconceptions were evident. Most typically, a common way of reasoning starts from the fact that "ice is lighter than water, since it floats." As is often the case, this daily life expression is in conflict with correct scientific terminology, which states that ice has lower density but is not lighter. Since the process of melting transforms ice into water (which is regarded as "heavier"), the conclusion for these students is that we will end up with more mass.

Explanations of the correct responses could be very different and still receive full credit. The short and easy direct reference to the general principle of constant mass was not given by many. It may well be that some students know this principle, but still do not regard a statement of this "law" as appropriate. After all, referring to a general principle, albeit perfectly correct, really does not explain anything ("Why is it so?"), beyond restating the correct response ("that"). As soon as they try to give an explanation in terms of volume and density, or even microscopic particles, students tend to get confused.

Change of state (Y2)

The next item, Y2, also focuses on a quantity which, under certain conditions, is constant. Again the topic is the melting process . The phenomenon focused on here is that during this process the temperature remains constant at the melting point.

> Y2.   One day when the temperature was just below 0°C, Peter and Ann made snowballs. They put a thermometer into one of the snowballs and it showed 0°C. They tried to make the snowball warmer by holding it in their hands. What do you think the thermometer showed after two minutes?

The concepts of heat and temperature are fundamental to physics, and freezing, melting, boiling, heat transfer, evaporation, and condensation are all phenomena very closely connected to our everyday life. If the notion of everyday physics should have any meaning, the area of heat and temperature must be important. This theme is taught at different levels in school around the world, and a number of common misconceptions have been reported (e.g., Erickson & Tiberghien, 1985; Thomaz, Valente, & Antunes, 1995). Furthermore, a proper discrimination between heat and temperature is very demanding at this age level. Many students have the conception that transferring "heat" to a body always means increasing its temperature.

The response distribution, given in percent, is shown in Table 3. The international mean on this item is only 14 percent for a correct response with an acceptable explanation. Another 30 percent have a partially correct response. This indicates that a substantial part of 13-year-old students do not have adequate knowledge of the fact that snow cannot get warmer than 0 degrees. Another reasonable correct response could have been that snow is a very good insulator and therefore the temperature inside would not change. However, very few students responded in this way.

Table 3. Coding guide and results for item Y2

Code	Response	East Asia	East Europe	English-speaking	North Europe	South Europe
Correct Response						
20	Reports 0 degrees or mentions "the same temperature." Explanation includes: "Snow cannot be warmer than 0 degrees" or similar.	11	10	13	15	6
29	Other correct.	3	3	4	2	3
Partial Response						
10	0 degrees or "the same temperature". No or incorrect explanation.	22	16	15	16	17
19	Other partial correct.	15	11	15	19	9
Incorrect Response						
70	Above 0 degrees, because the hands are warm.	23	12	25	12	20
71	Above 0 degrees, because the snow melts.	2	6	5	4	6
72	Above 0 degrees. No explanation.	2	4	2	4	2
79	Other incorrect.	11	12	13	14	17
Blank		12	26	10	14	24

The effect of country clusters is very similar to that for the previous item. The between-group variance is about 40 percent; but, just as for the previous item, most of this is due to large between-group differences in the non-response category. East Europe and South Europe have significantly higher frequencies of non-response, and this may indicate some cultural difference when it comes to willingness (motivation) to answer. There are, however, some large differences between countries within the groups. For example in East Asia, Hong Kong has 34 percent correct or partially correct while Japan has 74 percent; and, in North Europe, Denmark has 41 percent correct or partially correct, while Iceland has 62 percent.

The most common incorrect response is "above 0 degrees because the hands are warm" (Code 70). North and East Europe have fewer responses in this category, a feature that may be explained by more experience with snow and ice in these countries. Also individual countries seem to support that pattern. The two "winter countries" Japan and Korea both have relatively low percentages on Code 70 (7 percent and 12 percent, respectively), whereas Singapore, with its very good overall results, has a percentage for Code 70 as high as 41. If we combine the three codes— 70, 71, and 72—we find that, internationally, 24 percent of students seem to believe that the temperature will increase to above 0 degrees.

The role of light in maintaining the ecosystem (X2b)

The process of photosynthesis is fundamental for life on Earth. Much research has focused on students' understanding of how plants grow and where their "food" comes from (e.g., Driver et al., 1984; Wandersee, 1983). There is also some research showing that light is not regarded as necessary for photosynthesis (Bell, 1985; Stavy, 1987). In TIMSS there are three items that, in different ways, focus on photosynthesis. Here we want to discuss the second part of item X2 (i.e., Item X2(b) below) where the students were asked to explain the role of light in maintaining the ecosystem in the aquarium. The "photosynthesis" was not mentioned in the stem, and it was not required that they explicitly mention that concept to get the response correct.

X2. In the picture of an aquarium six items are labeled.

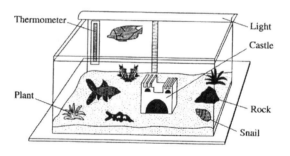

Explain why each of the following is important in maintaining the ecosystem in the aquarium.

(a) the plant

(b) the light

From the results in Table 4 we can see that there are large differences between the groups; and, for this item, the country clustering can "explain" more than in any of the other items we discuss here: almost 60 percent of the variability of response patterns. The between-group component of the variance is pronounced in every response category except the unimportant Code 11. East Asia stands out as the best group. More than half of the students in East Asia responded correctly as opposed to the English-speaking group where only one-fifth of the students' responses were correct. In spite of the large between-group effect, there are large differences within the East Asia group, correct responses varying from 78 percent in Singapore to 26 percent in Hong Kong.

It is interesting to see which concepts the students use in their responses. From Table 4 we can see that almost all of those who have correct answers in the East Asia countries referred to photosynthesis explicitly: in Singapore as many as 70 percent. Examples of student answers in this category are "to help the plant to make photosynthesis," and "the light provides energy for the plants to make food." In English-speaking countries and in the North Europe group we see the opposite. In New Zealand and Norway, for example, photosynthesis is referred to in only one fourth of the correct responses. Many students in these countries gave answers such as "the light makes the plants grow" (Code 19).

Table 4. Coding guide and results for item X2B

Code	Response	East Asia	East Europe	English-speaking	North Europe	South Europe
Correct Response						
10	Refers to photosynthesis	44	22	9	10	18
11	States that the light provides energy without any further detail	3	3	2	3	2
19	Other correct	6	14	11	18	14
Incorrect Response						
70	Says the fish need the light to see	8	9	22	12	16
71	Says we need light to see	4	1	5	2	5
72	Says light provides warmth for the fish	6	6	20	12	6
79	Other incorrect	18	17	21	27	19
Blank		11	27	12	18	22

Some of the incorrect responses (e.g., "So we can see the fish," Code 71) could arguably have been correct if the question had been why we have light in an aquarium. There are, however, very few responses in this category. Many students state that the light is needed so the fish will have "a good time," "so the fish can see," Code 70, or "the light provides warmth to the fish," Code 72. Both these categories say something about conditions for the fish, and almost half of the students in English-speaking countries answer in this way. It seems to be a pronounced cultural effect that caring for pets plays a particularly important role in these countries.

In Part A of this item the students are asked to explain why the plant is important in maintaining the ecosystem in the aquarium. The results show that this aspect of photosynthesis is better known. Internationally, 43 percent answered that the plant produces oxygen, and, for example in countries like New Zealand and Norway, where very few students referred to photosynthesis in Part b, half of the students answered that plants produce oxygen. It seems that students are much more informed about the substances taking part in photosynthesis than in the role of light as the energy source for the process.

Gravity (K17)

The next item (Item K17) is about an apple falling to the ground due to gravity, a concept that is far from easy. Gravity acts on the apple whether it is falling or not, regardless of its position and movement. However, when the apple is resting on the ground there is also a force from the ground pointing upwards. The force of gravity and this force from the ground have the same size, and the apple is at rest because the sum of forces is zero. If we neglect the air resistance, only gravity acts on the apple when it is falling.

Students' understanding of mechanics is probably the domain within physics that is most frequently explored by science educators. Many such studies show that students in all age groups all over the world have an unsatisfactory understanding of simple but crucial topics of mechanics. And often, the students' ideas and concepts have been shown to be surprisingly similar in many countries (Duit & Treagust, 1995). Looking at Table 5 we see that this item shows remarkably good results. Perhaps all the attention connected to this research has had a positive influence on teaching and learning around the world.

K17. The drawing shows an apple falling to the ground. In which of the three positions does gravity act on the apple?

A. 2 only
B. 1 and 2 only
C. 1 and 3 only
D. 1, 2, and 3

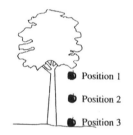

The response patterns are quite similar across groups of countries, the between-group part of the pooled variance amounting to only about 25 percent. More than 60 percent correct responses in East Asia represents a very satisfactory result, and all the countries within this group score above the international average. However, there are large differences between some countries within groups, East and South Europe in particular. For example in East Europe, the Czech Republic has 81 percent correct while Russia has only 42 percent correct and 41 percent for distractor B. Furthermore, in South Europe Spain has 55 percent correct compared to Greece with 30 percent correct and 49 percent for distractor B.

Table 5. Results for item K17

	East Asia	East Europe	English-speaking	North Europe	South Europe
A	4	7	6	10	8
B	25	25	30	27	36
C	6	7	8	8	11
D*	64	57	56	53	44
Blank	0	3	1	1	1

The most common incorrect response is B, which indicates the belief that gravity acts only when the apple is falling. Distractor A most likely indicates the same belief. (It is, however, not entirely clear if the apple in Position 1 is moving or not). The international mean response frequency is 34 percent for these two alternatives together. The conception that gravity acts only when the effect is observable, is known to be a common misconception. As diSessa (1993) says, according to many students' views, only the movement needs to be explained, and gravity provides this explanation in the form of a "moving force." Forces are therefore comprehended as cause for movement and not as cause for changes in the movement (Sjøberg & Lie, 1981).

On the other hand, those students who choose Distractor C may think that forces only act when the apple is at rest and no forces act on the falling apple in Position 2. Something similar appears when we are talking about weightlessness. When an astronaut is "weightless," gravity is acting on both the spaceship and the astronaut. There are, however, no forces acting between them, and therefore the astronaut has this feeling of weightlessness. It is when we are at rest on the ground that we "feel" our weight. This may explain some students' response to this item. In some countries almost 20 percent chose C: e.g., Iceland, Latvia, and South Africa.

Water balance and temperature regulation (O16)

Regulation of the human body's temperature and balance for total body water are vital processes in human physiology. Water is crucial for the internal environment. The two sources of body water are metabolically produced water and ingestion of water. Perspiration is controlled by a mechanism directed toward temperature regulation. The mechanism of thirst is certainly of great importance, since water for evaporation from the skin by the production of sweat, can be made up only by ingestion of water. It is, however, also true that our fluid intake is often influenced more by habit and sociological factors than by the need to regulate the body water. Item O16 assesses students' understanding of these processes in human physiology.

O16.   Write down the reason why we get thirsty on a hot day and have
to drink a lot.

The country clustering can "explain" about 45 percent of the variability of
response patterns for this item, with group effects in all response categories. From
the results in Table 6 we see that there are large differences between East Asia (with
above 80 percent correct responses) and the other groups of countries. Correct
responses are divided in three categories. To get Code 10 or 12 the students had to
refer to perspiration and also its cooling effect. In all groups of countries there are
few students that explicitly refer to the cooling effect. The very frequent simple
reference to the perspiration of water was coded 11 or 13. The combination of these
pairs of categories has been done here due to the difficulty and unimportance of
distinguishing between whether replacement of water is implicitly mentioned (Codes
10 or 11) or not (Codes 12 or 13).

Table 6. Coding guide and results for item O16

Code	Response	East Asia	East Europe	English-speaking	North Europe	South Europe
Correct Response						
10+12	Refers to perspiration and its cooling effect.	10	12	8	3	4
11+13	Refers to perspiration.	68	40	50	62	51
19	Other acceptable explanation.	5	3	3	3	2
Incorrect Response						
70	Refers to body temperature (being too hot) but does not answer why we get thirsty.	2	4	5	3	5
71	Refers only to drying of the body.	4	11	22	11	8
72	Refers to getting more energy by drinking more water.	2	4	1	3	6
79	Other incorrect.	5	12	6	9	14
Blank		4	16	4	5	12

It seems that the importance of perspiration for the cooling effect is not a central
part of teaching in many countries, except in Hungary where 42 percent of the
students referred to this effect. In Kuwait and Korea they also had above 20 percent.
It was discussed whether Code 10 represented a better answer than the other
responses and therefore "deserved" 2 points. From a psychometric point of view this
can be supported by an analysis of the Norwegian data that has shown that students
in this category have higher overall scores than the two other categories.
Furthermore, a similar pattern appeared in the results for those in Population 3 (the

item was a "link item": i.e., given in all three populations). The references to the cooling effect dramatically increase in frequency from Population 2 to 3 and from the vocational to the academic branch of upper secondary school (Angell, Kjærnsli, & Lie, 2000). However, a closer look at the item itself reveals that the students are asked to write down "the reason why" we get thirsty, thus implicitly asking for "the one" reason. A response that also refers to the "function" of perspiration, namely temperature regulation of the human body, is definitely a more advanced response, but it cannot reasonably be given a higher score. However, this example also gives another demonstration of the power and flexibility of the coding system. When performing a diagnostic analysis, the codes can be compared and combined according to the main issues under consideration.

From Table 6 we see that most of the incorrect responses have been coded as 71, responses such as, "Your throat gets dry." Among individual countries, France stands out with as many as 41 percent in this category, and of the next seven countries, six of them are English-speaking.

Code 72 represents an interesting misconception. These students seem to believe that you have to drink because you get exhausted, and that you get more energy by drinking water. This misconception is more common than seen from the response distribution, simply because a number of responses included sweating in addition to this incorrect statement and therefore were scored as correct.

Day and night (Q11)

This item , Item Q11, deals with why we have day and night on Earth. The correct answer is of course that the Earth rotates around its own axis (A).

Q11. Which statement explains why daylight and darkness occur on Earth?

   A. The Earth rotates on its axis.

   B. The Sun rotates on its axis.

   C. The Earth's axis is tilted.

   D. The Earth revolves around the Sun.

It is common to think of the Sun at rest in the center of the solar system, and all the planets including the Earth orbiting around the Sun. In addition, each planet is rotating around its own axis, so the whole movement is quite complicated. Maybe this complex structure to some extent could explain why as many as 42 percent of the students choose D. There are relatively few students from the various countries who chose the two other distractors (see response distribution in Table 7). Looking at the response alternatives, we see that all of them as isolated statements are true, even if they do not represent correct answers to the question at hand. This fact may well offer a particular challenge to the students.

Table 7. Results for item Q11

	East Asia	East Europe	English-speaking	North Europe	South Europe
A*	59	47	36	46	45
B	4	4	5	7	5
C	4	4	8	4	4
D	32	40	49	41	42
Blank	1	6	2	2	5

East Asia clearly stands out as the best group on this item with an average of 59 percent correct responses, and with all of those countries appearing above the international average (44 percent). The English-speaking countries are at the bottom with 36 percent correct answers. In this group the U. S. is on top with 55 percent correct responses and Scotland at the bottom with 21 percent, which means large differences between countries within this group. Also, in North Europe there are large differences between countries (Netherlands 62 percent, Iceland 21 percent). The between-group part of the total variability is 30 percent for this item, but the major part of this is provided by the difference between East Asia and the English-speaking countries. The response patterns for the remaining groups of countries are strikingly similar.

Astronomy appears in science curricula in many countries around the world, often at the primary level to capitalize on children's interest and enthusiasm in this area and to enrich their experiences of science (Sharp, Bowker, Mooney, Grace, & Jeans, 1999). In spite of this, it is remarkable that so few students know why it is day or night, which definitely is an everyday (and night) phenomenon! It is worthwhile comparing this finding with the curriculum in the countries participating in TIMSS 2003. A large majority of countries indicated in a questionnaire that the explanation of day and night is taught even at Grade 4. It is easy to underestimate the difficulty for children of understanding the models we have developed for the solar system and which we use to explain day and night, and also the seasons. Most of the models bear no easy relation to any concrete observation the children can make. For example, the two-dimensional figures in many textbooks are difficult to understand if students are not shown a three-dimensional model as well.

Surface of the Earth (J1)

The last item, Item J1, addresses the fundamental aspect of what we may call the Earth's ever-changing surface. In the 19th century there were active debates before the "old Earth" geological paradigm was settled: i.e., that the Earth is at least many millions of years old. Only by accepting this large time scale could the observed slow building and breaking geological processes be argued as providing an explanation for the surface features. The fact that the 20th century stretched the time span even further to some billions of years, mainly by astronomical evidence, is of minor importance from our point of view here. The crucial point is that what we observe as very insignificant geological processes in daily life or during a human life span, in a geological time scale can (and do so continuously) lift mountains and break them down.

The results are exhibited in Table 8. The four response alternatives may be characterized as the "building" (A), the "breaking" (B), the correct "building and breaking" (C), and the "static" (D) responses. There is a striking similarity between responses from the country groups, and in particular there are about 40 percent correct responses in all of them. The percent of the variability of response patterns that can be accounted for by the grouping of countries is only about 20 percent, the lowest for all items under consideration.

J1.   Which BEST describes the surface of the Earth over billions of years?

A. A flat surface is gradually pushed up into higher and higher mountains until the Earth is covered with mountains.

B. High mountains gradually wear down until most of the Earth is at sea level.

C. High mountains gradually wear down as new mountains are continuously being formed, over and over again.

D. High mountains and flat plains stay side by side for billions of years with little change.

What mainly concerns us here is the large portion of students responding D. This response strongly indicates that the student really does have the view of a static Earth. The format of this item requires each student to read through all response alternatives and select the "BEST" explanation. Most likely, since D comes last, all three "active Earth" responses have "actively" been rejected.

Table 8. Results for item J1

	East Asia	East Europe	English- speaking	North Europe	South Europe
A	18	9	10	10	8
B	17	13	18	14	20
C *	49	41	42	43	38
D	16	33	28	30	31
Blank	1	4	1	3	2

It should be a concern among science educators in many countries that the basic geological paradigm seems to have been understood by so few students by the end of compulsory schooling. Even if differences between country groups are small, there are some interesting and very surprising results for individual countries. One might suspect that countries with earthquakes, volcanoes, or even high mountains with glaciers would be more likely to focus more, in the media and in school, on geological processes. However, the results are not better, and in particular there are not fewer "static Earth believers," in countries such as Switzerland (31 percent) or Iceland (44 percent). On the other hand, there are particularly few students in this category in Japan (8 percent) and Korea (3 percent). Korea also stands out as having a very high percentage of correct responses (76 percent). We conclude that neither cultural nor geological factors can explain much of the variability between countries in understanding the phenomenon discussed here.

CONCLUSION

The first aim of the item-by-item analyses in this chapter has been to explore students' conceptions of some fundamental ideas in science. The overall results for each selected item show striking similarities in the general response patterns. As a rough guide we may say that something like 25 to 40 percent of the students seem to have demonstrated an adequate understanding of the phenomena discussed here. Furthermore, the analyses have revealed some very distinct misconceptions that are common around the world. We can describe these findings by summing up the misconceptions of some international "typical student," representing a considerable percentage of students at age 13 around the world:

- Day and night occur because the Earth orbits the Sun.

- Earth's surface remains the same over billions of years.

- During the process of melting, ice increases its temperature above 0°C, and the mass is increasing because water is "heavier" than ice.

- Gravity acts only when we can observe the effect: i.e., when an apple is falling and not when it is at rest.

- Even if the need to drink to replace water lost due to perspiration is well known, the cooling effect on the body by perspiration is not recognized.

- The main role of light in an aquarium as an ecosystem is to care for the fish, for visibility, or to maintain the temperature.

The second aim has been to study differences and similarities in response patterns between countries. Country groups were formed according to cultural/geographical regions and similarities of patterns of correct responses for all science items for Population 2. When we consider the total patterns of all response categories for individual items, we find that the grouping of countries can "explain" between 20 percent and 60 percent of the total variability. For some items (particularly J1-Surface of the Earth, and K17-Gravity) the between-group differences are rather small, whereas for other items (X2b-The role of light, and O16-Water balance) there seems to be a pronounced cluster effect. It is difficult to see any clear reason for this different behavior among items, partly because the role and number of response categories vary from item to item. Obviously, there are curricular impacts within some areas, accounting for significant differences between countries that cannot be understood by cultural or geographical factors alone. In order to investigate the differences in student response patterns across countries further, it would be necessary to carry out detailed curricular comparisons between countries for the topics discussed here.

REFERENCES

Angell, C., Kjærnsli, M., & Lie, S. (2000). Exploring students' responses on free-response science items in TIMSS. In D. Shorrocks-Taylor & E. W. Jenkins (Eds.), *Learning from others. International comparisons in education. Science & Technology Education Library volume 8*. Dordrecht: Kluwer Academic Publishers.

Beaton, A., Martin, M. O., Mullis I. V. S., Gonzales, E. J., Smith, T. A., & Kelly, D. A. (1996). *Science achievement in the middle school years. IEA's Third International Mathematics and Science Study (TIMSS)*. Chestnut Hill, MA: Boston College.

Bell, B. (1985). Students' ideas about plant nutrition: What are they? *Journal of Biological Education, 19(3)*, 213 – 218.

diSessa, A. A. (1993). Toward an epistemology of Physics. *Cognition and Instruction, 10 (2 & 3)*, 105-225.

Driver, R., Child, D., Gott, R., Head, J., Johnson, S., Worsley, C., & Wylie, F. (1984). *Science in Schools. Age 15: Report No.2. Report on the 1981 APU survey in England, Wales and Nothern Ireland*. London: Assessment of Performance Unit, Department of Education and Science, HMSO.

Driver, R., & Easley, J. (1978). Pupils and paradigms: A review of literature related to concept development in adolescent Science students. *Studies in Science Education, 5,* 61-83.

Duit, R., & Treagust, D. F. (1995). Students' conceptions and constructivist teaching approaches. In B. J. Fraser and H. J. Walberg (Eds.), *Improving Science education*. Chicago, IL:The University of Chicago Press.

Erickson, G. & Tiberghien, A. (1985). Heat and temperature. In R. Driver, E. Guesne, & A. Tiberghien. (Eds.), *Children's ideas in Science*. UK: Open University Press.

Lie, S., Taylor, A., & Harmon, M. (1996). Scoring techniques and criteria. In M. O. Martin & D. Kelly (Eds.), *Third International Mathematics and Science Study technical report. Volume 1: Design and development*. Chestnut Hill, MA: Boston College.

Mullis, I. V. S. & Smith, T. A. (1996). Quality control steps for free-response scoring. In M. O. Martin & I. V. S. Mullis (Eds.), *Third International Mathematics and Science Study: Quality assurance in data collection*. Chestnut Hill, MA: Boston College.

Pfundt, H., & Duit, R. (1994). *Bibliography: Students' alternative frameworks and Science education* (4th ed.). Germany: IPN at the University of Kiel.

Robitaille, D. F., Schmidt, W. H., Raizen, S., Mc Knight, C., Britton, E., & Nicol, C. (1993). *Curriculum frameworks for Mathematics and Science. TIMSS monograph no. 1*. Vancouver, Canada: Pacific Educational Press.

Schmidt, W. H., & Jorde, D., et al. (1996). *Characterizing pedagogical flow. An investigation of Mathematics and Science teaching in six countries*. Dordrecht: Kluwer Academic Publishers.

Sharp, J. G., Bowker, R., Mooney, C. M., Grace, M., & Jeans, R. (1999). Teaching and learning astronomy in primary school. *School Science Review, 80*(292), 75 – 86.

Sjøberg, S., & Lie, S. (1981). *Ideas about force and movement among Norwegian pupils and students. Report 81-11*. Oslo: University of Oslo.

Stavy, R. (1987). How students' aged 13 – 15 understand photosynthesis. *International Journal of Science Education, 9*(1), 105 – 115.

Third International Mathematics and Science Study. (1995). *Coding guide for free-response items, Populations 1 and 2.* TIMSS Doc. Ref.: ICC897/NRC433. Chestnut Hill, MA: Boston College.

Thomaz, M. F., Valente, M. C., & Antunes, M. J. (1995). An attempt to overcome alternative conceptions related to heat and temperature. *Physics Education, 30*(1), 19 – 26.

Wandersee, J. H. (1983). Students' misconceptions about photosynthesis: A cross-age study. In H. Helm & D. Novak (Chairs), *Proceedings of the International Seminar about Misconceptions in Science and Mathematics.* Ithaca, NY: Cornell University.

Wandersee, J. H., Mintzes J. J., & Novak, J. D. (1993). Research on alternative conceptions in science. In D. Gabel (Ed.), *Handbook of research on science teaching and learning.* New York: Macmillan Publishing.

Zabulionis, A. (1997). *The formal similarity of the mathematics and science achievement of the countries. Vilnius University pre-print No 97-19.* Vilnius: Vilnius University.

Chapter 10

SCIENCE ACHIEVEMENT: A CZECH PERSPECTIVE

Jana Paleckova and Jana Strakova

In 1991, following the collapse of communist rule, the Czech Republic became a member of IEA and decided to participate in the Third International Mathematics and Science Study. TIMSS was the first opportunity for the Czech Republic and many other post-communist countries to compare their achievement with other countries in a large-scale educational survey. The results were much anticipated in educational research and policy-making circles. Czech students in Population 2, aged 14 at the time of testing, achieved very high results in both mathematics and science. However, the results of students in their final year of secondary school were disappointing. After the release of the TIMSS results, educators from all over the world expressed interest in the Czech educational system and teaching practices, seeking explanations for different aspects of its performance.

The aim of this chapter is to present contextual information that will allow the reader to better understand the Czech Republic's results on the science portion of the TIMSS test. Some additional findings on various aspects of student achievement on the TIMSS science assessment, gained through national analyses of the TIMSS data, are also presented. At the end of the chapter, similarities and differences in achievement among the post-communist countries are discussed.

This chapter was written in the autumn of 2000, more than five years after the TIMSS data collection. During this period, many changes occurred in the educational system and in Czech society as a whole. The description and explanations presented in this chapter are based on the situation as it existed in 1995. To what extent these explanations remain valid and useful for interpreting student achievement in TIMSS-R in 1999 is a subject for further exploration.

SCIENCE EDUCATION IN THE CZECH REPUBLIC

Structure of the Education System

In 1995, eight years of compulsory education were provided at two levels of basic school: Grades 1 to 4 and Grades 5 to 8 for 6- to 14-year olds. During the 1995–96 school year, one year after the TIMSS testing, basic school was extended to

D.F. Robitaille and A.E. Beaton (eds.), Secondary Analysis of the TIMSS Data, 145–155.
© 2002 *Kluwer Academic Publishers. Printed in the Netherlands.*

a ninth year. The end of basic school is a crucial moment in the educational career of every Czech student. At this point, students are supposed to have acquired the basics of general education and may apply for entry to secondary school: academic, technical, or vocational. Academic schools provide demanding academic training for higher education, while technical schools combine general and vocational education. Both offer four years of study leading to a leaving examination in four subjects: the Czech language, a foreign language, and two subjects of the student's choice. Upon successful completion of these examinations, students may apply for university entrance. Vocational schools provide programs with a practical orientation, most of which may be completed in three years.

The number of students admitted to the various tracks in individual secondary schools is regulated. It is considered important to gain admission to a highly reputed school or track, since it is not easy or common to change after admission. When applying for secondary school, students must present report cards from the last two grades of basic school and pass an entrance examination. These examinations, some of which are very demanding, usually consist of mathematics and the Czech language. During the last two grades of basic school, classroom teachers systematically prepare students for the examinations. It is not uncommon for students to be more diligent in their study of all subjects in order to obtain the high grades needed for admission to the secondary school of their choice.

The Role of Science in Basic and Secondary School Curricula

Mathematics and sciences have always played an important role in compulsory education. For the first four years of basic school, science is taught as an integrated subject called "first science." Beginning in Grade 5, biology, physics, and chemistry are taught separately. Earth science is taught as part of geography. The time allocated to the science subjects, more than 25 percent of the total in the last grades of basic school, is one of the highest among the European Union as well as central and eastern European countries. Furthermore, the ministry of education requires students to master a large number of topics within the science subjects.

At most secondary schools, mathematics and science, as well as other general curriculum subjects (languages, history, and civics, for example), are given a much lower priority than at the basic school level. Academic schools are the only ones that continue to provide students with a full general education. In both technical and vocational schools, the emphasis is on technical subjects and the general subjects are limited according to the needs of the particular specialization. In technical schools, the ratio of general subjects to vocational ones is 2:3, meaning that students spend 40 percent of their time on mathematics, languages, and social and natural sciences while the remainder of the curriculum is devoted to specialized subjects such as technical drawing or mechanics. In vocational schools, the average ratio of general subjects, vocational subjects, and practical training is 1:1:2, with only 25 percent of school time allocated to mathematics, languages, and sciences. In 1995, the academic program was completed by less than 20 percent of the grade cohort, while more than 50 percent of the grade cohort completed vocational education.

Teaching Practices

In Czech schools, science is taught in a traditional manner with a similar lesson structure for all science subjects. At the beginning of the lesson, teachers usually question a few children at the blackboard or give the whole class a short test. They review the subject matter covered in the last lesson by asking questions. In the second half of the lesson, teachers introduce a new topic by lecturing from the blackboard or, occasionally, by performing an experiment to demonstrate a science phenomenon. Students are required to be quiet while the lesson is being taught, paying attention and writing notes in their exercise books. They rarely perform science experiments.

Students are expected to be well prepared for science classes. They are tested regularly during the school year, with an emphasis on oral examinations. Each student undergoes a 10-minute oral examination in front of the whole class twice per semester, usually without advance warning. In some classes, students also write tests comprised of open-ended questions, such as describing a phenomenon, drawing and describing a biological object, solving a physics problem, or balancing a chemical equation.

Teacher Education

All teachers in basic and secondary schools are required to complete four to five years of university education. Teachers in the first level of basic school are trained in all subjects that are part of the elementary school curriculum. Teachers at the second level of basic school and secondary school teachers are required to specialize in two subjects. During their university studies, theoretical education in their field of study is emphasized. Although the teacher education curriculum for general subjects includes some courses in psychology and pedagogy, the greatest emphasis is on subject matter knowledge. Teaching methods and practices and alternative approaches are covered only marginally. As an example, the syllabus for the education of physics and mathematics teachers at the second level of basic school is shown in Table 1 on the next page.

TIMSS IN THE CZECH REPUBLIC

Because there was no tradition of participation in comparative international studies in the Czech Republic, the TIMSS testing was in many respects an unusual event. It was the first international study to be carried out in the Czech Republic and the first to use standardized tests. It gave educators their first opportunity to compare student results with those of their counterparts in other countries, as well as their first opportunity to compare results at the school level.

Table 1. Syllabus for physics and mathematics teacher education (5-year program): number of lessons allocated to compulsory subjects

Subject	Number of Years	Lessons Weekly
Calculus	2	4
Linear algebra	1	4
Geometry	2	3
Algebra and theoretical arithmetic	1	3
Probability and statistics	0.5	4
Problem-solving methods in mathematics	1	2
Didactics of mathematics	1.5	3
History of mathematics	0.5	2
Mathematical methods in physics	1	4
Physics	4	5
Introduction to measurement in physics	0.5	1
Physical experiments	2.5	4
Physical experiments in school	1.5	2
Didactics of physics	1	3.5
Informatics	1	3
Psychology	1	3
Pedagogy	1	3
Foreign language	1	2
Physical education	1	2
School-based practice teaching	-	6 weeks

Czech educators were acutely aware of the lack of comparative information about education and their lack of experience in conducting large-scale assessments. It was to address these concerns that the Czech Republic opted to participate in all components of TIMSS. The Population 1 tests were administered to students in Grades 3 and 4 of basic school and the Population 2 tests, including the performance assessment, to Grades 7 and 8. Students in the final grade of all three types of secondary schools participated in Population 3.

The format of the test, particularly the use of multiple choice items, was unfamiliar to most Czech students. The performance assessment component was also unfamiliar since Czech schools have no tradition of evaluating experiments or activities. In general, neither students nor teachers seemed to have trouble with the format of the tests.

STUDENT ACHIEVEMENT

Results for Populations 1 and 2

Czech students performed very well on both the mathematics and science components of TIMSS at both population levels. In the science component of the Population 1 test, only three countries achieved significantly higher results than Czech students in Grade 3, and only two countries had higher results in Grade 4. In Population 2, no country had a score that was significantly higher than that of the Czech Republic for Grade 7. For Grade 8, only students from Singapore had results that were significantly higher.

The results for the performance assessment component were not as outstanding as the results for the written test—they were only slightly above the international average. Czech students exhibited the greatest difference in science achievement between the written test and the performance assessment among countries that participated in both tests.

It is probable that the strong results at the Populations 1 and 2 levels can be attributed to the emphasis placed on science education in basic school and to the close match between the TIMSS tests and the Czech curriculum. In addition, the testing was well timed in the Czech Republic, coinciding with a period during which Grade 7 and 8 students were studying for secondary school entrance examinations. Teachers of all subjects, including the sciences, were conducting extensive reviews of the subject matter covered in basic school in order to prepare the students for secondary study.

Strengths and Weaknesses of the Czech Performance in Population 2

In order to understand the strengths and weaknesses of Czech students in Population 2, we examined 30 items according to the performance expectation categories assigned by the TIMSS science framework (Robitaille & Garden, 1993), and interpreted them in the context of the curriculum and teaching practices. The items examined consisted of the 15 with the highest positive differences between the national and international percent correct scores and the15 with the highest negative differences.

The content of the "best" 15 items was covered in the intended curriculum. Twelve of the 15 fell into two performance expectation categories: understanding simple information and understanding complex information. The remaining three items belonged to three performance expectation categories: applying scientific principles to develop explanations, gathering data, and making decisions. From the perspective of the Czech curriculum, however, these particular items could more appropriately be classified as understanding simple or complex information, since all the facts needed to respond correctly to the items were taught in science classes.

Among the 15 items with the highest negative differences, seven fell into the performance expectation categories understanding simple or complex information. These items required students to apply knowledge in a context that was not familiar to them from their science classes, thus representing a greater challenge than other tasks from the same category. The students faced the same kind of difficulties when responding to the other eight items that were classified as applying scientific principles to develop explanations (four items), applying scientific principles to solve quantitative problems (one item), abstracting and deducing scientific principles (one item), and designing investigations (two items).

Despite the fact that the relevant topics were covered in the Czech curriculum, Grade 8 students did not do as well on some of the items as might have been expected. They often had difficulty with items that were presented in ways or contexts different from those used during their lessons. For example, one item asked students to draw a diagram to explain how water from one place can fall in another place. Czech students were taught how to draw a diagram of a water cycle but many failed to respond correctly because the phrase "water cycle" was not mentioned explicitly.

The fact that the sciences were taught separately and that the students were not accustomed to applying knowledge gained in one science subject to another one also contributed to their relatively poor performance on some items. For example, one item asked students to explain what happens to an animal's atoms when the animal dies. In Grade 6 physics and chemistry students learn about atoms, molecules, and the structure of matter. Despite having learned a lot of facts about the topic, students were not able to use their knowledge to respond to the item and their results were surprisingly poor. They were not able to think about an animal, a topic studied in biology, as an object composed of atoms and molecules, concepts learned in physics and chemistry.

Results for Population 3: Scientific Literacy

Czech students in their final year of secondary school did not perform as well in international terms as Czech students in Population 1 and 2 on the TIMSS test. In the science literacy component of the Population 3 test, students from 9 countries (out of a total of 21) had significantly higher scores than students from the Czech Republic. The Czech Republic, the Russian Federation, and Hungary had the largest gaps between the relative performance of Grade 8 students and students in the final year of secondary school. On some of the link items used in both populations, senior secondary students achieved results lower than their younger colleagues. For example, an item that asked students to explain why it is necessary to paint steel bridges was answered correctly by 85 percent of Grade 8 students but only by 76 percent of students in the final year of secondary school.

Policymakers in the Czech Republic were alarmed by the magnitude of the difference between the achievement of students in academic and vocational schools at the Population 3 level, the greatest among participating countries. The academic students performed very well compared to their international counterparts, but their proportion within Population 3 was small.[1] At the same time vocational students' achievement was quite low and the proportion of these students within Population 3 was much higher than in other countries.

On an item that asked whether the amount of light energy produced by a lamp is more than, less than, or the same as the amount of electrical energy used, for example, a correct response was given by 49 percent of academic school students, 41 percent of technical school students, and only 8 percent of vocational students. While students from academic and technical tracks performed well on this item compared to their counterparts in Grade 8 (20 percent correct), the results for vocational students were disappointing.

It may be helpful to consider both the curriculum content and the structure of secondary education when analyzing achievement on the science literacy test and the huge differences in scores between tracks. A high proportion of Population 3 students, those enrolled in vocational and technical streams, received only a limited amount of general education during their secondary studies. They were no longer required to study the sciences or to demonstrate competence in these subjects. Academic students and those technical students who intended to pursue university study, on the other hand, were preparing for university entrance examinations at the time of the TIMSS testing and working at achieving high grades in all subjects to improve their chances of admission. These students were better able to compete with their international counterparts on the literacy test.

A second factor impacting the achievement of Population 3 students is the way science is taught in basic school. A traditional teaching approach coupled with an overloaded science curriculum and an emphasis on knowledge in a broad range of subjects leads to rote memorization without deeper understanding. Subject matter that is memorized in this way is easily forgotten, particularly in secondary schools where science is not emphasized.

Results for Population 3: Physics

In the Czech Republic, the physics component of TIMSS was administered to students in the last year of the *gymnasium* or academic track of secondary education. Their achievement was disappointing for many experts who had touted the high quality of our academic education. Students in 9 of the 16 participating countries achieved results significantly higher then Czech students.

The sample selection for the physics assessment was widely discussed in the Czech Republic and is viewed as the main cause of the low performance. Physics was a compulsory part of the curriculum for all *gymnasium* students, with at least two physics lessons per week for three years. In fact, many of these students were not interested in physics. When they apply to the *gymnasium*, students choose among several streams, the most common being streams specializing in the natural sciences or humanities. There are also streams offering enriched education in mathematics, foreign languages, and computer programming.[2] Lesson plans for all streams are very similar, but the curricula and the teachers tend to be more demanding in the specialty subjects. Students usually choose these subjects for their leaving examinations and are expected to demonstrate better knowledge of these subjects than students in the other streams. It was argued that only students who chose physics for their leaving examination should have been included in the specialist sample instead of all students in the final grade of gymnasia. As a more in-depth analysis of the TIMSS data will show, these students could indeed compete successfully with their contemporaries from other countries. However, the percent of these students in Population 3 in the Czech Republic was extremely low, around one percent.

Gender Differences

Statistically significant gender differences favoring boys were found in the science component of both Populations 1 and 2. There was also a statistically significant gender difference in the achievement of physics specialists in Population 3. In Population 2, the Czech Republic was among the group of countries reporting the greatest gender differences. But an interesting discrepancy was found at this level: despite the fact that Grade 8 boys performed better than girls in all the science content areas (earth science, life science, physics, chemistry, environmental issues, and the nature of science), girls' school results in physics, chemistry, and biology were higher than boys'. Table 2 contains the average marks for boys and girls on their final report for the first term of Grade 8. The scale ranges from 1 to 5, with 1 being the highest mark.

Table 2. Average marks of boys and girls for the first term of grade 8

	Marks (1–5; 1 is highest mark)	
	Girls	Boys
Mathematics	2.5	2.8
Physics	2.3	2.4
Chemistry	2.3	2.7
Biology	2.0	2.3
Czech	2.4	2.9
Foreign language	2.0	2.5

In performing evaluations teachers subconsciously include criteria not directly related to academic performance; many of these, such as diligence and attentiveness during classes tend to favor girls. Some of the assessment requirements typical of Czech classrooms include learning facts, which also tends to favor girls.

The difference in the achievement of boys and girls in their last year of secondary school on the science literacy component was among the greatest of all the participating countries, with boys achieving higher marks than girls. This was also the case with the physics test, on which the Czech Republic ranked among three countries (the Czech Republic, Switzerland, and Slovenia) with the greatest statistically significant gender difference.

REACTIONS OF THE CZECH SCIENCE EDUCATION COMMUNITY TO TIMSS

The TIMSS test was well regarded by basic school science teachers because it required students to use concepts and facts that were taught in the classroom. They considered the unfamiliar questions, such as asking for the application of acquired knowledge, to be appropriately challenging. The multiple choice format and the integration of the different science subjects into one test were regarded as unusual. Nevertheless, these aspects of the test did not seem unsuitable to either teachers or students.

The tests for Populations 1 and 2 matched the Czech curricula rather closely, allowing students to demonstrate the knowledge they had acquired at school. In Population 1, only 10 items out of 97 were designated by the experts as inappropriate for fourth grade students. In Population 2, 10 items out of 135 were considered inappropriate for eighth graders. In both populations, the Czech Republic was among a group of countries with the highest test-to-curriculum match. The items regarded as inappropriate belonged to different fields of science, that is, there was no particular subject area that was not covered in the Czech curriculum. The items that were unfamiliar to Czech students usually required some specific knowledge that was not part of the curriculum. For example, Czech students do not learn how to test for carbon dioxide using lime water, nor about how to work with food webs. They do not study the composition of the electromagnetic spectrum, the velocity of light, or the principle of the transmission of sound.

Most Grade 8 teachers wanted to use the TIMSS items to assess their students at a later date. Out of the 32 items they evaluated, only 4 were found unacceptable by 75 percent or more of the teachers. These were deemed too easy for students in Grade 8.

In comparison with the traditional tests or examinations taken by Grade 8 students, the TIMSS test had a much narrower curriculum coverage. It did not require students to solve complicated physical problems, demonstrate knowledge of chemical nomenclature, enumerate chemical equations, demonstrate detailed knowledge of the periodic table, describe in detail the organs of various animals and plants, or classify them within the system. All of these are typical of what Grade 8 students are supposed to study during science classes in the Czech Republic.

The situation is very different for Population 3, with many teachers and educational professionals in the Czech Republic stating that it was inappropriate to test students in mathematics and science at this level. They argued that the aim of secondary education is to give students a professional education and that they should only be assessed in areas relevant to their profession. A second group, however, who disapproved of the science curriculum's focus on knowledge to the neglect of skills and applications, asserted that the TIMSS assessment was not the appropriate tool to highlight the insufficiencies of the educational system. Although they valued the experience of administering an international assessment, they would have preferred a test that focused on real-life applications and processes rather than on facts and routine procedures.

STUDENT ACHIEVEMENT IN CENTRAL AND EASTERN EUROPEAN COUNTRIES

If we compare the average achievement of Grade 8 students from the central and eastern European countries that participated in TIMSS (Bulgaria, Hungary, Latvia, Lithuania, Romania, the Russian Federation, and the Slovak Republic) with that of Czech students on select science items, we can see that although the average percent correct scores of these countries are different, they follow very similar patterns of achievement. In other words, students from all of these countries performed relatively better or worse than the international average on the same items. Although the countries differed significantly in terms of the amount of time allocated to science in the compulsory portion of the education system, there were some similar features. The teaching approach, for example, tended to be traditional, with a greater emphasis on learning facts and a lesser emphasis on acquiring experimental skills. The science subjects were taught separately as physics, chemistry, biology, and geography. These common features led to similarities in students' strengths and weaknesses with respect to both performance expectations and content categories in all eight countries. For example, all eight were relatively stronger in life science and weaker in the area of environment and the nature of science. This may be because life science is the first science subject to be taught separately in the eight countries and it receives a great deal of emphasis. Environmental issues have always been neglected in central and eastern European societies, and this area receives relatively little coverage in school curricula.

NOTES

1 In the Czech Republic the proportion of academic students was 14 percent, the lowest proportion among 21 participating countries. The proportion of vocational students was 57 percent. In only two countries, Germany and Switzerland, was the proportion of vocational students higher.

2 This practice of streaming gymnasium classes is less and less common. Most classes are currently regarded as general. In the final grades of secondary school, students may specialize in elective subjects of their choice.

REFERENCES

Robitaille, D. F. and Garden, R. A. (1993). *TIMSS Monograph No. 1: Curriculum frameworks for Mathematics and Science*. Vancouver, British Columbia: Pacific Educational Press.

Chapter 11

TIMSS SCIENCE RESULTS FOR HONG KONG: AN AILING DRAGON WITH A BRITISH LEGACY

Nancy Law

The TIMSS science results are worrying for Hong Kong, and not only because her students' achievement ranks among the lowest for developed countries and falls far short of that of the other "Asian Dragons" participating in the study: Singapore, Japan, and Korea. Further examination of the achievement data reveals that the performance profile of Hong Kong students was very different from the international norm. Items that tested non-contextualized knowledge were found to be difficult internationally but easy for Hong Kong students, while items testing the application of knowledge and reasoning in familiar contexts were found to be easy internationally but turned out to be otherwise for Hong Kong students. Hong Kong students were found to be good at items that tested simple information and routine procedures, but weak on complex reasoning and application of knowledge.

In fact, the TIMSS test design helps to reveal serious problems in science education for Hong Kong. Besides ensuring that there is a balance in test items for different science content areas, the TIMSS test design also categorized items into four major performance expectation categories: (1) understanding; (2) theorizing, understanding, and solving problems; (3) using tools and routine procedures; and, (4) investigating the natural world, with sub-categories within each major category. For the TIMSS achievement test, there were two different kinds of test item formats: multiple choice and free response items requiring either short answers or extended responses. An examination of the achievement data reveals that Hong Kong students were particularly good at multiple choice items and items that tested simple information and routine procedures but weak on complex reasoning and items demanding verbal explanations.

Relative changes over time provide another productive dimension for international comparison. In SISS, one of the results that attracted a lot of attention was the fact that Hong Kong, Singapore, and England had very similar performance profiles: all three systems performed very poorly at both Populations 1 and 2 while ranking top in Population 3 (Holbrook, 1990a). One of the popular explanations put forward at that time was that both the Singapore and Hong Kong science curricula

D.F. Robitaille and A.E. Beaton (eds.), Secondary Analysis of the TIMSS Data, 157–175.
© 2002 *Kluwer Academic Publishers. Printed in the Netherlands.*

were very much modeled on the English curriculum. A decade later, the student achievement ranking in TIMSS changed dramatically from that found in SISS. During the decade that elapsed between SISS and TIMSS, both England and Singapore undertook major revisions of their science syllabuses while Hong Kong did not made any change to the primary and secondary science syllabuses. In TIMSS, both Singapore and England showed remarkable progress in their students' science performance, ranking first and sixth respectively at Population 2. This paper explores these peculiarities in greater detail through in-depth comparisons of the Hong Kong data with the achievement data of three other countries—Korea, Singapore, and England—for the different item categories and further discusses possible factors that might have led to such peculiarities.

Another intriguing finding from the TIMSS science data was that the impact of the students' socio-economic background on achievement was much smaller for Hong Kong compared to most other countries. This paper will explore this finding by comparing the patterns of performance for different socio-economic groups across different item formats and item categories in order to provide a better understanding for this result.

THE PERFORMANCE PROFILE OF HONG KONG STUDENTS: A PICTURE OF MEDIOCRITY

As mentioned earlier, the TIMSS science results for Hong Kong ranked among the lowest for the developed countries and fell far short of the three other Asian countries (Singapore, Japan, and Korea) in both Populations 1 and 2. This ranking was made on the basis of the mean scores of the participating countries. Hong Kong's Population 2, upper grade students achieved a mean score of 522 in science, compared to a mean score of 516 internationally (Beaton et. al., 1996). The Hong Kong Population 1 students scored an average of 533 compared to a mean of 524 internationally. Thus, for both populations, the mean scores for Hong Kong were slightly above the international mean. However, the score distribution for Hong Kong was very different from the international population distribution and was skewed to concentrate around the mid-lower scores so that most students' performance ranked as mediocre while very few students were either very high achievers or extremely low achievers. Specifically, while Hong Kong's mean scores are slightly better than average for both Populations 1 and 2, only 7 percent reached the 90th percentile on the international scale, and only 22 percent reached the top quartile of the international scale for the upper grade in Population 2. For Population 1 (Grade 4), the equivalent statistics were 4 percent and 17 percent, respectively, suggesting an even smaller proportion of high achievers and a narrower spread of lower scores than that for Population 2. This paper reports on explorations of patterns of performance based on the TIMSS Population 2 upper grade data.

Patterns of Performance

An important strength of the TIMSS design is the comprehensive curriculum framework used in the data collection and analysis. The written achievement test comprised two different formats: multiple choice (MC) and free response (FR) items (the results of the optional performance assessment test will not be discussed here). Careful exploration of the patterns of performance across item formats and item categories in the written achievement test reveals that Hong Kong students perform very differently from their international counterparts.

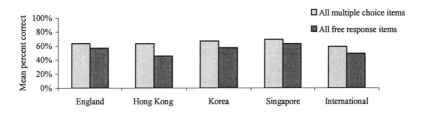

Figure 1. Mean percent correct for different item formats in different countries.

As is evident from Figure 1, while the overall percentage correct for multiple choice items was higher than for free response items, the difference was relatively small and below 10 percent for England, Korea, and Singapore, except for Hong Kong where the difference was as much as 18 percent. One interpretation of this result is that Hong Kong students had a relatively lower ability to express their ideas in prose form as compared to answering multiple choice questions. However, the different item formats may also be testing different abilities other than language. This will be explored in the following section.

The Performance Differential across Performance Expectations

In the TIMSS curriculum framework, the curriculum is characterized by the concepts, processes, and attitudes of school mathematics and science that are intended for, implemented in, or attained during the students' schooling experiences. All curricular items were analyzed in terms of three dimensions: subject matter content, performance expectations, and perspectives (Robitaille et al., 1993). The performance expectations dimension was considered central to a country's curricular visions and intentions as it reflects what students are expected to be able to do with science at particular points in their schooling. In the TIMSS achievement test design, there was an attempt to use the same framework to categorize assessment items so as to reflect the greater complexity of item differences beyond that of the subject matter content being tested. The performance expectations aspect is a reconceptualization of the cognitive behavior dimension and aims to describe, in a non-hierarchical scheme, the main kinds of performances or behaviors that a given test item or block of content might be expected to elicit from students. In the TIMSS Science curriculum framework (Robitaille et al., 1993), there were four performance expectations categories:

- understanding
- theorizing, analyzing, solving problems
- using tools and routine procedures
- investigating the natural world

Each of the above categories had further subcategories. For example, understanding can be subcategorized into understanding of simple information or of complex information. The distribution of the different categories of science test items is summarized in Table 1.

Table 1. Distribution of score points* across performance expectation categories and item formats in the Population 2 science achievement test

	Understanding Simple Information	Understanding Complex Information	Theorizing, Analyzing & Problem Solving	Using Tools & Procedures	Investigating
Multiple Choice score points	53	30	8	8	3
Free Response score points	2	12	36	0	2

*   The number of score points is used in preference to the number of items as some free response items had a full score greater than 1, and the number of score points reflects more closely the emphasis in terms of the test composition.

As can be seen in Table 1, the TIMSS test design in fact has a strong emphasis on testing the understanding of information, especially the understanding of simple information via multiple choice questions. From results presented in Figure 2, the four systems examined do not show the same relative patterns of performance across the five performance categories, though all found questions on understanding simple information to be the easiest. For Hong Kong, the greatest performance difference occurred between understanding simple information and theorizing and explaining, a phenomenon that is not shared by the other three systems. In fact, because of the TIMSS test design, the low performance of Hong Kong students in free response items cannot be separated from their low performance in items related to theorizing, analyzing, and solving problems.

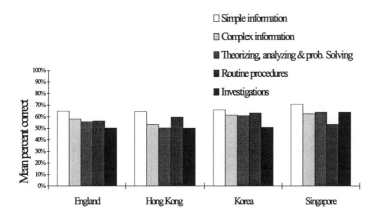

Figure 2. Mean percent correct for items categorized by performance expectations in four education systems.

PATTERNS OF PERFORMANCE: A CLOSER LOOK

The examination of patterns of performance across different item formats and performance expectations categories helps to reveal some intriguing patterns of performance. In order to explore further the relative strengths and weaknesses of Hong Kong students in comparison with the three other countries, the results for some items testing similar content, but set in different problem contexts, are discussed below.

Photosynthesis and Chloroplasts

There were two test items related to photosynthesis, one of which was a multiple-choice item belonging to the performance category "understanding simple information," while the other was a free-response item belonging to the category "theorizing, analyzing, and solving problems." It is interesting to note from Figure 3 that three of the four systems whose student achievement data were listed as well as the international mean achievement indicate that students found Item X2b more difficult than Item K18, and this difference in performance is much more prominent for Hong Kong than the other countries. Both questions require the knowledge that plants produce food under sunlight. However, if we use Li et al.'s (2002) (see Chapter 15) categorization of TIMSS science items into two categories, *everyday knowledge* and *scientific knowledge*, then K18 would be testing scientific knowledge since it involved the knowledge of a scientific term, "chloroplast," which students would be unlikely to know except through formal science studies while X2 could arguably be answered on the basis of everyday knowledge. Unlike students in the three other systems in this comparison, Singaporean students performed much better on Item X2b, indicating that they probably have a much better grasp of the concepts of an ecological system and photosynthesis though they may not remember the exact scientific terms. In fact, 23 percent of students in Hong Kong thought that plants were needed in aquariums to improve the aesthetic environment of the aquarium or for the enjoyment of the fish while more than 30 percent thought that light was important to enable vision for humans or fish or to provide warmth to the fish. Such misconceptions are relatively rare among the comparable group of Singaporean students.

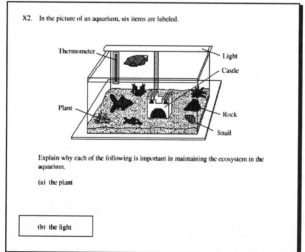

Figure 3. TIMSS items related to photosynthesis.

Atmosphere and Life on Planets

This section analyses a set of three items related to the function of different gaseous components of the earth's atmosphere.

	Mean % correct
O12. Air is made up of many gases. Which gas is found in the greatest amount? A. Nitrogen B. Oxygen C. Carbon dioxide D. Hydrogen	England 17 Hong Kong 50 Korea 41 Singapore 58 International 27

	Mean % correct
R4. Write down one reason why the ozone layer is important for all living things on Earth.	England 38 Hong Kong 56 Korea 57 Singapore 79 International 54

P3. Jane and Mario were discussing what it might be like to live on other planets. Their science teacher gave them data about the Earth and an imaginary planet, Athena. The table shows these data.

	Earth	Athena
Atmospheric Conditions	21% oxygen	10% oxygen
	0.03% carbon dioxide	80% carbon dioxide
	78% nitrogen	5% nitrogen
	ozone layer	no ozone layer
Distance from a Star Like the Sun	148,640,000 km	103,600,000 km
Rotation on Axis	1 day	200 days
Revolution Around Sun	365 ¼ days	200 days

Mean % correct

England 89
Hong Kong 50
Korea 90
Singapore 96
International 80

Write down one important reason why it would be difficult for humans to live on Athena if it existed.

Figure 4. Items related to atmospheric gases and their impact on life.

Item O12 tests simple understanding of knowledge about the most abundant gas in the atmosphere. Item R4, asking for the importance of ozone in the atmosphere, was categorized as testing understanding of complex information. Item P3, which asked for one reason why the fictitious planet Athena is not suitable for human habitation, was categorized under "theorizing, analyzing, and solving problems." While it may be argued that, of the three performance expectation categories, "theorizing, analyzing, and solving problems" is the most demanding, the international data indicates that O12 is the most difficult while P3 was an extremely well answered item. Using Li et al.'s (see Chapter 15) categorization, O12 would be testing scientific knowledge while both R4 and P3 could be answered on the basis of everyday knowledge. Both R4 and P3 require the respondent to understand how knowledge about properties of gases and the atmospheric composition could be used to determine whether certain situations would be fit for human habitation.

Here again, Hong Kong students' performance stood apart. Students were able to answer simple knowledge items but did not find it easy to apply that knowledge in specific contexts. Thus while 56 percent of Hong Kong students were able to point out that ozone helps to screen out dangerous UV rays to protect animals, less than 17 percent were able to suggest that Athena is not suitable for human habitation because it does not have ozone in its atmosphere.

Starting and Extinguishing Fires

This same sort of disparate achievement patterns is found repeatedly across domains that have different kinds of items in the same content area included in the test. As a last example, two items on combustion are presented for further examination.

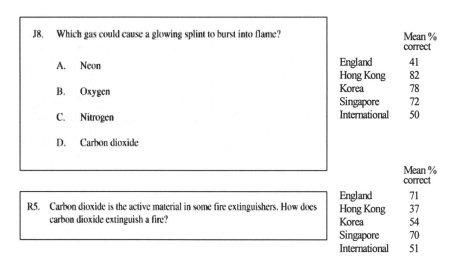

Figure 5. Items related to combustion property of common gases.

It is apparent from the item difficulties quoted for the two items above that, even though the international item difficulties for these items were very similar, students in different countries did not find them to be of similar difficulty at all. Item J8 is actually a standard test for presence of oxygen found in common science textbooks in Hong Kong. Hong Kong students found no problem in producing an answer for it and were in fact the top performers on this item. However, Item R5 which asked about how carbon dioxide extinguishes a fire, and required as a major part of the answer that carbon dioxide prevents oxygen from reaching the fire, was answered very poorly by Hong Kong students. In fact, for both Singapore and Korea, their students' performance on R5 was also somewhat below that of J8, though much less dramatically than Hong Kong. England's results, on the other hand, exhibited very different patterns. English students found J8, which tested simple chemistry information, to be rather difficult (41 percent correct) while R5, which was categorized as a "theorizing, analyzing, and solving problems" item to be quite easy (71 percent). In fact, it can be seen that a number of Western countries exhibited a pattern of achievement similar to that of England: for example, Australia, Austria, Canada, New Zealand, and the United States.

HONG KONG HAS A HIGH CONCENTRATION OF HARDWORKING, ROTE LEARNERS

So far, the discussion has accepted the TIMSS performance expectation categorization of items uncritically. It could be argued that some of the "theorizing" items might be successfully completed by rote learning too, such as Item R5 on fire extinguishers, and the difference in performance across education systems might just reflect a difference in the contexts within which the science content was presented. One of the contributing factors to Hong Kong students' poor performance on "theorizing" items may be an overly academic science curriculum that gave little attention to developing students' understanding of science in everyday contexts.

On the other hand, Hong Kong students' remarkable ability to excel on items testing understanding of simple information demonstrated their strength on "academic" items that involve memory of scientific facts and information. Hong Kong students' high performance on multiple choice items as opposed to very poor performance in free response items (Law, 1996) is not just a reflection of the poor communication skills of Hong Kong students. Hong Kong students appeared to have difficulty applying known facts and scientific principles to develop explanations and in the analysis and solution of problems.

The above findings seem to reveal that Hong Kong students are very good at memorization. This is consistent with work conducted in recent years on the conceptions of learning and the learning behavior of Chinese learners in Hong Kong. Watkins and Biggs (1996) put together a collection of papers to examine how Chinese students and their teachers see the context and content of their learning. A major theme to be addressed there was, "How can Chinese learners be so successful academically when their teaching and learning appears to be so focused on rote memorization?"

Watkins (1996) explored how secondary school students perceived the changes in their approach to learning and its causes as they progressed through primary and secondary schooling. He found that they all seem to have reached or passed through three clear cut stages. Students were found to report an initial stage (at primary and sometimes through to junior secondary level) where the goal was to achieve through reproduction, and the strategy was to rote learn everything. Later, as they progressed through secondary school, the focus was still on achievement through rote learning, but the strategy changed to rote learning important things since the increased memory load required them to be selective about what to memorize. As the students moved further along the academic ladder, they realized that their memorizing capacity would be enhanced by understanding the material. Senior secondary students would typically be at the third stage where the focus was still on achievement through reproduction, but the strategy used was to understand first. So, they learned to appreciate understanding, but that was only to serve their objective of reproducing answers to achieve good grades.

Tang (1996) attributed, as a major contributor to the heavy emphasis on memorization, the backwash effect from the overwhelming use of tests that test low-level outcomes in teacher-generated, school-based assessment. Marton et al. (1996) argued that memorization does not preclude understanding and reported two forms of memorization with understanding: memorizing what is understood and understanding through memorization. However, the results from the TIMSS test items discussed above indicate that Hong Kong students are very good at items that test memory of scientific knowledge, but are much less capable of understanding how the memorized knowledge could be applied in everyday contexts, be they concerned about the aquarium, the atmosphere, or the fire extinguisher.

WHAT IS A FAIR TEST? WHAT IS A GOOD SCIENCE CURRICULUM?

The differential achievement profile across different performance expectation dimensions for the different educational systems discussed above highlights the importance of examining the foci and objectives in science curricula beyond that of subject matter content. The results also point strongly to the impact of test design on the performance outcome in studies of student achievement. Given that the level of achievement of students can be so very different for items on essentially the same content, rankings in student achievement can be an artifact of the instrument even when curriculum coverage is fully taken into account.

Such considerations pose two serious challenges. One is a challenge to international comparative studies of student achievement: what composition of items should a "fair" test have? The second is a challenge to curriculum developers and science educators: the assumption that theorizing, analyzing, and problem solving is a more advanced outcome of learning building on lower levels of knowledge and comprehension (Bloom et al., 1971; Klopfer, 1971) does not seem to hold. The TIMSS results in fact point to the possibility that different curriculum processes can lead to qualitatively different learning outcomes, which have been assumed to be hierarchically related to each other before. If this is indeed the case, should science education focus on the development of particular kinds of performance outcomes, or should we aim to target a "balanced" profile of performance outcomes? Obviously, another important question to explore in this regard is whether there are identifiable links between the curriculum and the very different performance profiles observed in

TIMSS across the systems studied. This question will be explored in a later section of this paper.

FURTHER PATTERNS OF PERFORMANCE: SOCIO-CULTURAL AND SOCIO-ECONOMIC DIMENSIONS

So far, this paper has presented a general picture of mediocrity for the population of Hong Kong students as a whole in terms of tackling items requiring the application of knowledge to everyday contexts, but rather successful in tackling those requiring memorization of scientific knowledge. Is this performance pattern different for students from different socio-economic backgrounds? Education research has consistently found higher achievement to be associated with students of higher socio-economic status (SES). Education sociologists such as Bernstein have also found that middle class families tend to exercise different power structures at home compared to their working class counterparts, thereby encouraging their children to develop their own abilities to critically evaluate situations based on their own knowledge and understanding, and to express their own ideas. For example, he suggests that, "Despite clear indications of improvements in working class/race/gender educational chances, social class is a major regulator of the distribution of students to privileging discourse and institutions" (Bernstein, 1996). Thus students from higher SES backgrounds may excel more on items requiring critical evaluation and application of ideas than their counterparts from lower SES backgrounds. Consequently, students from different SES backgrounds might have different performance patterns for items in different performance expectation categories and formats. Specifically, while Hong Kong students performed very poorly on the free response items testing theorizing, analyzing, and solving problems, would students from higher SES backgrounds exhibit markedly better performance on those items, because of the kind of home influence that is expected based on general sociological findings?

For Hong Kong, student results at Population 2 upper grade showed a difference in means of only 42 points (with the population mean at 522) between the highest socio-economic group (as measured by the number of books at home) and the lowest. The variance across different SES groups in terms of their overall scores was very small, and, as shown in Table 2, this difference is comparatively small compared to other countries as well. From this we can see that there is indeed very little variation between different socio-economic groups in terms of their overall performance in science. This can be contrasted with several other countries, where the mean difference in achievement across different socio-economic groups was much larger.

Besides the very small difference in performance across different SES groups in Hong Kong, another intriguing observation is that, for the mean achievement score of the lowest socio-economic group (students with 10 or fewer books at home), Hong Kong students ranked fourth in the 18 education systems that satisfied all the sampling and participation requirements and had sufficient data to produce a statistic to that effect, but ranked twelfth out of 18 in the group with the most books. This may be interpreted as a positive indicator of the strong concern of Hong Kong families for their children's education, as Asian families influenced by the Confucian tradition are known to be very concerned about the education of their children (Watkins & Biggs, 1996). However, this does not explain the much larger differences across SES groups in Korea and Singapore, two countries that have been referred to as part of the "Confucian-heritage culture" by Ho (1991). Biggs (1996) also reported similar approaches to learning and classroom environments in these cultures.

Table 2. Difference in scores between the group with more than 200 books in the home and the group with 0-10 books for various countries for Population 2 (upper grade)

	Mean score	Difference in scaled score*
England	552	124 (23%)
Hong Kong	522	36 (07%)
Korea	565	87 (15%)
Singapore	607	81 (13%)

* Figures in brackets represent score difference as a percentage of population mean score.

Research on the impact of SES on student academic achievement has found that family process variables (what parents do to support learning) have a stronger influence on achievement than family status variables (who the families are) (Christenson, 2001; Kellaghan et al., 1993). Walberg (1984) reported that social class or family configuration predicted only up to 25 percent of the variance in achievement while family support for learning or interaction style predicted up to 60 percent of the variance in achievement. The small difference in performance across different SES groups in Hong Kong is probably a reflection of similar family processes across SES groups.

Performance Variation across Item Types for Different SES Groups

A more detailed analysis of the performance of students in different SES groups for the four performance expectations categories with the most number of items (namely multiple choice simple information, multiple choice complex information, free response complex information, and free response theorizing and explaining) for the four selected systems is presented in Figure 7. It is evident from these results that the performance profile across different performance expectations for the different SES groups shows rather complex variations for the four systems and that the results from Hong Kong stood apart from the other systems in several important aspects.

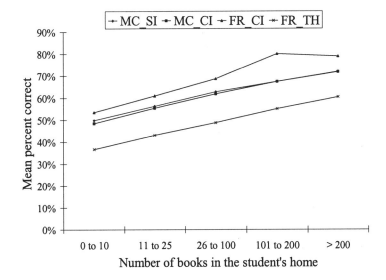

Figure 7a. Student performance on science items by performance expectation categories: England.

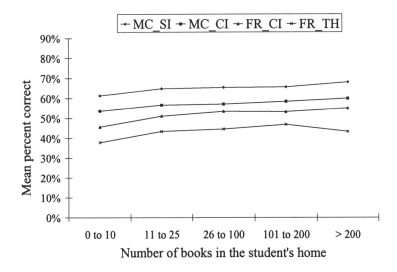

Figure 7b. Student performance on science items by performance expectation categories: Hong Kong.

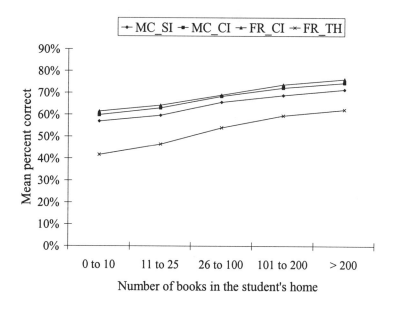

Figure 7c. Student performance on science items by performance expectation categories: Korea.

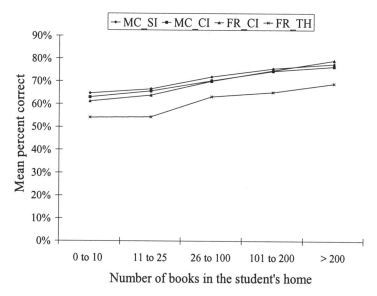

Figure 7d. Student performance on science items by performance expectation categories: Singapore.

Note: From the Figure 7 legends, *MC* represents "multiple choice"; *FR* represents "free response", *SI* represents "simple information"; and *CI* represents "complex information."

First of all, the results indicate that the performance profile for the different item types differ across the different systems. While free response items testing theorizing, analyzing, and problem solving (FR-TH) proved to be most difficult for students in all systems irrespective of their SES background, they excel on different item types. Hong Kong students performed best on multiple choice items testing understanding of simple information (MC-SI). English and Korean students performed best on free response items testing knowledge of complex information (FR-CI); while Singaporean students performed equally well on all item types except free response items on theorizing, analyzing, and problem solving. Another noteworthy observation is that the relative performance profiles across item types remain the same across SES groups in all four systems. This seems to indicate that the differences in relative performance profile across item types within a system are essentially a result of the differences in school curriculum emphasis rather than differences in family background.

Secondly, differences in performance profile across item types are most noticeable for Hong Kong, followed by England, and smallest for Singapore. The magnitudes of the differences in performance across item types are again relatively stable across different SES groups within the same system, giving indication that such differences arise from system level variations.

CURRICULUM MATTERS

As mentioned earlier in this paper, Hong Kong had a very similar achievement profile to that of Singapore and England in SISS at all three populations. One speculation at the time was that these three countries followed a "British style educational system."

> ... Hong Kong obtained a low mean score [for both Populations 1 and 2] compared with the other countries. However, it is difficult to clearly conclude that the low score is solely related to curriculum coverage as other countries offering British style educational systems, e.g. England and Singapore also achieved low scores, even though test validities were higher.
>
> (Holbrook, 1989)

> The countries offering a British style education (England, Hong Kong and Singapore) show the greatest gains between Populations 1 and 3 (and 2 and 3) and the data illustrates their emphasis on reaching a high level of science achievement for the selected few at the expense of the general population.
>
> (Holbrook, 1990b)

In the space of 11 years between SISS and TIMSS, the science achievement in these three systems diverged. At the Population 2 level, Singapore ranked top and England ranked sixth, while Hong Kong remained low at the sixteenth position. What can we attribute these changes to? During the period between the studies, Singapore changed its science curriculum at both the primary and the lower secondary levels, and England launched its first national science curriculum. On the other hand, there was no change at all made in the science curriculum at the lower secondary level in Hong Kong during this period. In fact the junior secondary science curriculum in force during the 1990s in Hong Kong was based on the Scottish integrated science curriculum developed in the 1960s and first implemented in Hong Kong in the mid-1970s. Thus Hong Kong has been using a legacy science curriculum for a quarter of a century.

It is not possible in this chapter to get into a deep cross-national analysis of curriculum differences. Some indication of the curriculum differences can be obtained from examining the findings from the TIMSS science curriculum analysis (Schmidt et. al., 1997). Unfortunately, England did not participate in the curriculum analysis study. Comparisons of the science textbooks from Hong Kong with those from Korea and Singapore for Population 2 upper grade (Schmidt et al., 1997), show that Hong Kong textbooks had no coverage at all of investigating the natural world, while Korea had up to 10 percent and Singapore had up to 20 percent in this performance area. On the other hand, Hong Kong textbooks put a great deal of emphasis (up to 30 percent coverage) on using tools, routine procedures, and processes, while the textbooks from the other two countries both devoted 10 percent or less to this performance expectations category.

From a more detailed breakdown of the performance expectations under each broad category, differences in patterns of emphases that reflect differences in students' learning outcomes in these three systems emerge. While much of the space in Hong Kong science textbooks was concerned with understanding, most of it was devoted to the understanding of simple information, with less than 10 percent given to complex or thematic understanding. For Korea, only 50 percent of the textbook space was devoted to understanding simple information while the understanding of complex information constituted up to 30 percent of the textbook space. Singapore science textbooks at this level devoted up to 20 percent of the space to complex information. For the performance category theorizing, analyzing, and problem solving, Hong Kong textbooks did not have any coverage of abstracting and deducing scientific principles or developing explanations, while the textbooks in Korea and Singapore covered all five subcategories.

These reflect, to a large extent, the emphasis in science curriculum guides in Hong Kong. Most of the performance expectation codes found from the analysis of the science curriculum guides were also associated with understanding content and the use of tools and procedures. These rather low-level learning outcomes expectations dominated even the science syllabuses at the senior secondary level. Variations in curriculum intentions in performance expectations across different school levels for the category of "theorizing and problem solving" were found to be rather idiosyncratic. In the primary science curriculum guide, this performance category was totally absent, increasing to 5 percent for the junior secondary integrated science curriculum guide and an average of about 13 percent in the science subjects. While one might expect this pattern to persist through to the advanced level, the senior biology syllabus has only 1 percent of its content on theorizing and problem solving and another 1 percent on investigating. The senior physics syllabus has only 6 percent of its blocks on theorizing and problem solving and investigating is completely absent.

The ability to communicate science effectively is generally recognized as an important learning objective in science education, but communicating does not figure as an important objective in any of the syllabuses except at the primary level. Moving from a comparison of the performance expectation categorization of textbooks and curriculum guides to an inspection of the content categorization of the science syllabuses at different school levels in Hong Kong, one is immediately struck by the near complete absence of content other than the traditional basic science disciplines. Other than earth science, life science, and the physical sciences, there were only two content areas that were being covered: that of science, technology, and mathematics in junior secondary science and Grades 10 and 11 physics, most of which would be concerned with the application of mathematics to science in terms of numerical problem solving. Environmental and resource issues were only mentioned to any significant extent in the primary level syllabuses and were completely absent from the junior secondary science and both physics syllabuses. Other, more modern content areas such as the history of science and technology, nature of science, and science and other disciplines were basically absent from all the eight syllabuses. This is a rather revealing, though disappointing, finding. It is thus apparent that the science performance profile of the Hong Kong students in TIMSS has strong links to the relative emphases and idiosyncrasies of the science curriculum being enforced.

SUMMARY

There are four main findings from this cross-national analysis of the TIMSS science results from Hong Kong, England, Korea, and Singapore. First, the format of the assessment items matters very much in international comparative achievement tests. Hong Kong students would out-perform many other countries if most of the test items were objective-type items.

Second, and this is linked to the first finding, students' performance may vary dramatically across items testing the same subject matter domain knowledge but focusing on different performance expectations. Moreover, there are cross-national differences in the profile of achievement across different performance expectations. Hong Kong students were found to perform best on multiple choice items testing understanding of simple information. English and Korean students performed best on free response items testing knowledge of complex information, while Singaporean students performed equally well on all item types except free response theorizing items that all four systems found to be the most difficult.

Third, it was observed that the relative achievement profiles across item types remained the same across SES groups in all the four systems studied. This seems to indicate that the differences in relative performance profile across item types is essentially a result of the differences in school curriculum emphasis rather than differences in family background. An initial exploration of the differences in science curriculum as revealed by the curriculum analysis results provided some evidence for this.

Fourth, the performance profile is also one of "mediocrity" that has a relatively small dispersion with few really low performers and a much smaller proportion of high achievers. This same profile of mediocrity is found again in the TIMSS–R results in 1999 (Martin et al., 2000). Hong Kong students' science achievement profile reveals that they perform best on items requiring memorization of simple knowledge and are least able to cope with items that require the application of scientific knowledge to everyday contexts. This same performance profile is found for students from different SES backgrounds.

These findings from the TIMSS science results reveal specific weaknesses in Hong Kong students' science performance, but they also provide indications that such weaknesses may be related to the science curriculum. It is hoped that such findings will help science educators in their efforts to improve the quality of science education in Hong Kong.

REFERENCES

Beaton, A. E., Martin, M. O., Mullis, I., Gonzalez, E. J., Smith, T. A., & Kelly, D. L. (1996). *Science achievement in the middle school years: IEA's Third International Mathematics and Science Study*. Chestnut Hill, MA: TIMSS International Study Center, Boston College.

Bernstein, B. (1996). *Pedagogy, symbolic control and identity: Theory, research and critique*. London: Taylor and Francis.

Biggs, J. (1996). Western misperceptions of the Confucian-heritage learning culture. In D. A. B. Watkins, & J. B. Biggs. (Eds.), *The Chinese learner: Cultural, psychological and contextual influences*. Hong Kong: CERC & ACER.

Bloom, B. S., Hastings, J. T., & Madaus, G. F. (Eds.). (1971). *Handbook on formative and summative evaluation of student learning*. New York: McGraw Hill.

Christenson, S. L., & Sheridan, S. M. (2001). *Schools and families: Creating essential connections for learning*. New York: The Guilford Press.

Ho, D. Y. F. (1991). *Cognitive socialization in Confuscian heritage cultures*. Washington, DC: Workshop on Continuities and Discontinuities in the Cognitive Socialization of Minority Children.

Holbrook, J. (1989). *Science education in Hong Kong: Primary and junior secondary science, Vol. 1*. Hong Kong: Hong Kong IEA Centre, University of Hong Kong.

Holbrook, J. (1990a). *Science education in Hong Kong: Achievements and determinants*. Hong Kong: Faculty of Education, University of Hong Kong.

Holbrook, J. (1990b). *Science education in Hong Kong: Pre-University science, Vol. 2*. Hong Kong: Hong Kong National IEA Centre, University of Hong Kong.

Kellaghan, T., Sloane, K., Alvarez, B., & Bloom, B. S. (1993). *The home environment and school learning: Promoting parental involvement in the education of children*. San Francisco: Jossey-Bass Publishers.

Law, N. (1996). Hong Kong students' science achievement in the international comparison. In N. Law (Ed.), *Science and mathematics achievements at the junior secondary level in Hong Kong*. Hong Kong: TIMSS Hong Kong Study Centre, University of Hong Kong.

Klopfer, L. E. (1971). Evaluation of learning in science. In B. S. Bloom, J. T. Hastings, & G.F. Madaus (Eds.), *Handbook on formative and summative evaluation of student learning*. New York: McGraw Hill.

Martin, M. O., Mullis, I. V. S., Gonzalez, E. J., Gregory, K. D., Smith, T. A., Chrostowski, S. J., Garden, R. A., & O'Connor, K. M. (2000). *TIMSS 1999 International Science report: Findings from IEA's repeat of the Third International Mathematics and Science Study at the eighth grade*. Chestnut Hill, MA: International Study Center, Boston College.

Marton, F., Dall 'Alba, G., & Tse, L. K. (1996). Memorizing and understanding: The keys to the paradox? In D. A. B. Watkins & J. B. Biggs. (Eds.), *The Chinese learner: Cultural, psychological and contextual influences*. Hong Kong: CERC & ACER.

Robitaille, D., McKnight, C., Schmidt, W. H., Britton, E., Raizen, S., & Nicol, C. (1993). *TIMSS Monograph No. 1: Curriculum frameworks for mathematics and science*. Vancouver: Pacific Educational Press.

Schmidt, W. H., Raizen, S. A., Britton, E. D., Bianchi, L. J., & Wolfe, R. (1997). *Many visions, many aims: A cross-national investigation of curricular intentions in school science, Vol. 2*. Dordrecht: Kluwer Academic Publishers.

Tang, C. B. (1996). How Hong Kong students cope with assessment. In D. A. B. Watkins, & J. B. Biggs. (Eds.), *The Chinese learner: Cultural, psychological and contextual influences*. Hong Kong: CERC & ACER.

Walberg, H. J. (1984). Families as partners in educational productivity. *Phi Delta Kappan, 65*, 397-400.

Watkins, D. A. (1996). Hong Kong secondary school learners: A developmental perspective. In D. A. Watkins, & J. B. Biggs. (Eds.), *The Chinese learner: Cultural, psychological and contextual influences*. Hong Kong: CERC & ACER.

Watkins, D. A., & J. B. Biggs (Ed.). (1996). *The Chinese learner: Cultural, psychological and contextual influences*. Hong Kong: CERC & ACER.

Chapter 12

SCIENCE ACHIEVEMENT: A RUSSIAN PERSPECTIVE

Galina Kovalyova and Natalia Naidenova

The principal goal of this paper is to present an analysis of the TIMSS results for the Russian Federation and other countries of interest from the Russian perspective of teaching science in secondary school. A second goal is to contribute to a better understanding the state of science education in the Russian Federation.

International comparative studies provide opportunities for the evaluation of educational systems in national and international contexts using uniform measurement instruments. When these instruments are developed, they are based on international criteria and priorities in education. We should thus be aware that international studies may not be able to evaluate educational systems in relation to their national goals. However, international studies do provide valuable opportunities for revealing the strong and weak points of educational systems and for contributing to modernization and reform. An international comparative study such as TIMSS gives countries an opportunity to view their own educational systems from the outside, but also allows countries to share complex information about the functioning and outcomes of education in their own countries with the international community. It is important to know to what extent a country's students are mastering the content of the national curricula and how this content reflects international trends and developments in education.

In analyzing the TIMSS results we sought answers a number of questions. What are the main differences between science education in the Russian Federation and other countries? How do the TIMSS results illustrate these differences? What should be done in order to improve science achievement in the Russian Federation? How do the TIMSS results contribute to a new era of educational reform in the Russian Federation?

Some attempts at analysis have already made in publications related to TIMSS (Kovalyova et al., 1996, 1998, 2000). In this article, we provide additional discussion of the results and present further viewpoints to aid readers in interpreting the results and their context in the Russian education system.

D.F. Robitaille and A.E. Beaton (eds.), Secondary Analysis of the TIMSS Data, 177–191.
© 2002 *Kluwer Academic Publishers. Printed in the Netherlands.*

THE TIMSS-95 RESULTS FOR THE RUSSIAN FEDERATION

The current state of science education in the Russian Federation may be better understood in the context of the TIMSS-95 results. Secondary school students studying advanced physics (Population 3 specialists) achieved high scores relative to their international counterparts, placing the Russian Federation among the top three countries in this category. These results support the notion that the Russian secondary physics course provides a solid foundation for success in responding to the majority of TIMSS items. Russian physics students demonstrated a mastery of almost all physics topics included in the test and skill in solving basic problems in physics, especially those that required knowledge of mathematics.

Attention should be given to two points in the interpretation of the Russian results. In 1995, only a small proportion of the age cohort, less than five percent, studied advanced physics, a figure that has not changed significantly since the 1960s. Furthermore, the average age of the Population 3 students in the Russian Federation who participated in TIMSS was 16.9, making them the youngest among all participating countries. By way of comparison, the highest mean age was 19.5 years, for students from Switzerland (Mullis et al., 1998).

At the Population 2 level, Russian students demonstrated high levels of science achievement with a mean score that was significantly higher than the international average. In terms of mean achievement, their scores in life science were higher than in the other science content areas (earth science, physics, and chemistry). However, their results in chemistry were higher than the international mean by a greater margin than in the other science content areas (Beaton et al., 1996).

On half the Population 2 science items, Russian students' results were higher than the international mean. The items were drawn from a variety of topics and included specific subject knowledge and skills. The items on which the Russian mean was 15 to 30 percent higher than the international mean were related to features of the Earth, the atomic structure of matter, and the structure and functions of plants and animals.

Russian students experienced the greatest increase in achievement between Grades 7 and 8 among all participating countries, 54 units on the international scale. A particular feature of science education in the Russian Federation may account for the increase, namely, the intensive, systematic study of science subjects in middle school. In Grade 8, students study all science subjects, including chemistry, as separate courses.

Not all of the Russian Federation's results were strong. At the Population 3 level, generalist students achieved results that were lower than the international average in science literacy. They demonstrated weakness in applying knowledge to real-life situations and in explaining phenomena. They experienced difficulty responding to interdisciplinary science items from the nature of science content area. They also had difficulty interpreting quantitative information about real processes presented in the form of diagrams, tables, or graphs. These results were quite unexpected given the intensive nature of secondary science instruction in the Russian Federation.

To illustrate these findings consider a free-response item from the science literacy Population 3 test, Item A7 show below. The item assesses the application of the physical concept of pressure in explaining a situation from every-day life. The two-digit coding system used in TIMSS permitted a deeper analysis of student responses and made it possible to identify different patterns of responses in different countries.

> A7. Some high-heeled shoes are claimed to damage floors. The base diameter of these very high heels is about 0.5 cm. Briefly explain why very high heels may cause damage to floors.

Table 1 shows the achievement data and coding information for Item A7 from selected countries. For the purpose of the analysis some coding categories have been combined. Sampling errors for the country percentages are small, and less than one percent for the international average.

Table 1. Distribution of student responses among coding categories for Population 3, Item A7 (selected countries)

	Distribution of student responses among coding categories (percent)						
	Fully correct		*Partially correct*		*Incorrect*		*No response*
Country	*20*	*21*	*10*	*11-13, 19*	*70*	*76, 79*	*90, 99*
Canada	21.6	28.4	4.4	14.0	8.5	15.2	7.4
Czech Republic	14.1	13.0	1.0	21.0	16.8	24.5	8.6
Russian Federation	24.1	6.6	5.5	16.6	4.4	21.7	20.8
Sweden	22.9	22.8	3.5	21.5	10.4	12.5	5.3
International Average	18.6	21.7	2.9	17.0	11.2	15.5	12.7

Only about 30 percent of Russian students gave a fully correct response that used the appropriate scientific terminology (Codes 20 and 21). The international average was 40 percent. Twenty-two percent of Russian of students (the sum of Codes 10 to 13 and 19) gave a partially correct response or a correct response without an explanation, for example one that demonstrated an understanding of the problem but used physics terminology incorrectly. The comparable international average was 17 percent. These results are troubling. Virtually all secondary students in the Russian Federation study physics for five years, including the final year of secondary school. The majority of them are aware of the concept of pressure and ought to be able to solve simple problems calculating pressure using force and area. In some countries with higher results, by contrast, not all students study physics or science in their final school year. In Canada, for example, only 55 percent study science or physics, and the comparable figure in Sweden is 43 percent.

The differences between boys' and girls' responses to this item were statistically significant in almost all of the participating countries. Only in Cyprus did girls perform better than boys on this item.

Two salient points that are illustrative of the way physics is taught in the Russian Federation emerge from an analysis of students' responses to this item. There were two patterns evident among students who responded correctly to this item. The majority of Russian students used the term "pressure" explicitly and demonstrated an understanding of the concept (Code 20). In other countries an almost equal number of students responded similarly, or showed an understanding of the concept of pressure without naming it (Code 21). Only a very small proportion of Russian students, 4 percent, referred to the sharpness of the heels in their response (Code 70), providing this answer without any scientific explanation. In other countries the number of students giving this response was greater: 16.8 percent in the Czech Republic, for example. These observations were also made in the analysis performed by Angell et al. (2000).

A large number of Russian students, more than 20 percent, omitted this item. Omission was typical for many free-response items in the Populations 2 and 3 tests, perhaps because students preferred to omit items rather than provide an obviously unscientific response.

Structure and Content of Science Education

Analyses of the structure of school science among the TIMSS-95 countries, of the content of the TIMSS tests, and of the match of the tests to Russian science curricula reveal two important differences in the structure and content of general and advanced science education. The TIMSS-95 countries were split into two almost equal groups with respect to the organization of their science curricula at the Population 2 level: those teaching integrated science (21 countries) and those teaching separate science courses (18 countries). The Russian Federation belongs to the latter group along with most European countries. Based on student achievement data, it is not possible to draw any conclusions about the advantages of one system over the other.

The Test-to-Curriculum-Matching Analysis (Beaton et al., 1996) showed that the content of the Population 2 test only partially matched the Russian science curriculum. About 40 percent of the Population 2 items were not covered in the majority of science classes in the Russian Federation. The rest of the TIMSS test matched the science curriculum in the Russian Federation but did not closely conform with the basic material studied. The bulk of the knowledge and skills attained by Russian students in science, therefore, was not covered by the test (Aksakalova et al., 1995). On the Population 3 physics test, the majority of items corresponded to the general and not the advanced physics curriculum. All secondary school students had been exposed to the majority of the physics test content.

To further this analysis, we paid special attention to the science curricula of central and eastern European countries which, during the period from 1970 to 1990, had had many common features. The Test-to-Curriculum-Matching Analysis for Population 2 showed that for Grade 7 there were only 4 items among 135 that corresponded to the curricula of all countries. They were Items D6 (seeds come from which part of a plant), I13 (measuring temperature), J9 (tree rings), and N6 (most basic unit of living thing). At the Grade 8 level there were only 34 items (about 25 percent) for which the experts indicated a match to the country's curriculum. These findings illustrate the differences between the countries and the limitations of potential comparisons. In spite of these limitations the information was widely used by science subject matter specialists as the basis for making decision about curricular change.

Equity Issues in Science Education

There are two areas of concern relating to equality of educational opportunity in science education in the Russian Federation. There are great differences in mean student achievement in relation to school location and student gender. It is important to note that these differences do not occur in relation to mathematics education. The school location factor, in particular, has a significant impact on science achievement in the Russian Federation. The farther a school was from the center of a region, the lower the mean science achievement on TIMSS. The lowest-achieving schools were located in rural areas. Table 2 presents the TIMSS-95 data for Populations 2 and 3 for urban and rural schools. All three differences are statistically significant and favored urban schools over rural schools.

Table 2. Mean achievement of students in urban and rural schools (TIMSS-95)

	Mean achievement		
	Science *Population 2*	*Science literacy* *Population 3*	*Physics* *Population 3*
Urban schools	541	490	551
Rural schools	517	449	484

In addition, the intraclass correlation was calculated using hierarchical linear modeling. The percent of variation in score achievement between schools, or the so-called intraclass correlation, is given in Table 3. The analysis also shows that the dependence of the results on school variables is higher for advanced physics students, with the intraclass correlation increasing from Population 2 to Population 3.

These results were expected. While middle schools offer general education without streaming, upper secondary schools offer two streams: general secondary education and specialized secondary education. (The Population 3 science literacy test was administered to these two streams.) The differences among upper secondary schools, therefore, are greater than the differences among middle schools. It should also be noted that, as a rule, upper secondary schools enroll higher-achieving students than middle schools, since lower achieving students generally attend vocational schools rather than upper secondary. The specialized secondary schools are very different from general secondary schools, and tend to enroll higher-achieving students. Some of these schools are affiliated with universities and act as feeder schools.

Table 3. Intraclass correlation (x100) for TIMSS-95

Population 2 (Grade 8)	Population 3 (Non-specialists)	Population 3 (Specialists)
30.2 %	49.3 %	58.8 %

Science subject matter specialists have traditionally not considered gender issues a problem. For many years, all Russian students, regardless of gender, interest, or ability studied the same science courses. In the majority of TIMSS countries, including the Russian Federation, no significant gender differences in mathematics were found. In science, gender differences were noted in many countries, although the findings varied from content area to content area. In the majority of countries, boys' results in physics were higher than girls'. In the Russian Federation, Grade 7 boys performed better on all content areas of the Population 2 science test, on which there were large proportions of items that did not match the Russian curriculum. Grade 8 boys also scored higher than girls in the earth science, chemistry, and physics content areas of the Population 2 test, as did boys on the Population 3 physics test. On the advanced physics test, boys' results were significantly higher than girls' on the majority of items. It appeared also that boys' results were higher than girls' on items not taught in class and on items with practical or technical applications.

Figure 1 presents information about the proportion of Russian girls and boys in Populations 2 and 3 in rural and urban schools. Figure 2 shows the distribution of mean student achievement by gender and school location for both populations. Differences in the achievement of boys and girls increased from Population 2 to Population 3, independent of the location of the school. Boys are more likely than girls to attend vocational schools after graduating from middle school. In the general education stream of upper secondary school, 37 percent are boys and 63 percent, girls.

Among Grade 8 students the difference in mean achievement between rural girls and rural boys was not statistically significant. However, the difference between urban girls and urban boys was significant, with boys scoring significantly higher than girls. Among Population 3 physics specialists, the situation was similar. Among Population 3 non-specialists who took the science literacy test, the mean

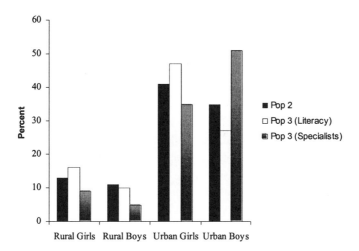

Figure 1. Distribution of female and male students in rural and urban schools.

achievement of rural girls was significantly lower than the results for all other groups of students. Rural boys attained significantly higher scores than rural girls, although their scores were lower than urban boys'. The mean science literacy score of urban boys was significantly higher than all other groups. Urban girls had higher scores than rural girls, although they were lower that those of urban boys. These differences in student achievement raise serious questions about general science instruction and student motivation that must be addressed.

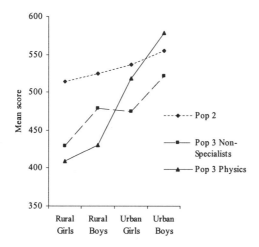

Figure 2. Mean science achievement by gender and school location.

ACHIEVEMENT OF RUSSIAN STUDENTS IN TIMSS-99

The TIMSS-99 results yielded only one new message of importance: that nothing had changed in science education in the intervening years. For a country in transition, the fact that educational outcomes had not declined was good news in that it might reflect the essential stability of the educational system in spite of negative economic and social contexts.

The mean results attained by Russian Grade 8 students were higher than the international average, as they had been in 1995. Achievement in earth science, life science, physics, and chemistry was also above the international average. In environmental and scientific inquiry Russian achievement was approximately equal to the international mean. In relation to the national mean score, Russian students had higher scores for earth science, physics, and chemistry, roughly equal scores in life science, and scores below the mean for environmental and scientific inquiry. Multiple comparisons of the results did not show any change from 1995 to 1999.

TIMSS-99 collected more detailed information about school location. Table 4 shows this information along with student achievement data. If we combine data from rural schools with data from schools located in settlements with populations of less than 3000 inhabitants, as was done in 1995, and label them as rural, then combine all urban data (for all urban subcategories) into one category and label them as urban, the results in 1999 were very similar to those from 1995.

Table 4. Mean student achievement, by content area and by school location (TIMSS-99)

School location	Science	Physics	Chem- istry	Earth Science	Life Science	Environ - ment	Nature of Science
Rural area	523	519	532	539	511	486	498
Settlement	501	488	512	494	462	482	478
Small city	539	536	545	550	513	504	494
Middle city	520	507	515	522	496	506	510
Large city	542	519	535	536	531	469	505
Regional Center	545	545	538	554	526	506	515

Mean achievement, Population 2

There were differences between the results of rural (rural area and settlements) and urban schools (all subcategories) in both the overall science score and in the physics score. The lowest results in the Russian Federation were attained by students from schools located in settlements, and the highest from students in regional centers. Statistically significant differences in achievement were found only in content areas where the associated knowledge was learned at school (physics, environment, and nature of science). Because the TIMSS test included many practical items about nature in the earth science and life science content areas, rural students may have used their own life experience to respond to these items. Rural students' practical experience in these areas may be greater than that of urban students. These findings may indicate that the lower achievement of rural compared to urban students in the Russian Federation may be due to lower opportunities for them to learn science at school, since urban schools tend to be better equipped and to employ more highly qualified teachers.

COMPARISONS BETWEEN TIMSS AND IAEP-II RESULTS

To evaluate the extent to which TIMSS reflects the state of science education in the Russian Federation, it is important to examine several questions. Is TIMSS the only study that provides an international perspective on science education in the Russian Federation? If not, what could other studies teach us, and how would their results correlate to TIMSS?

An international project, IAEP-II (International Assessment of Educational Progress), was administered by the Educational Testing Service in 20 countries in 1991. IAEP-II was the first international comparative study in which Russian schools participated, and the Russian Federation took part as a part of the USSR. The USSR sample included only Russian-speaking students aged 13, the same age cohort as TIMSS Population 2. More than two-thirds of the USSR sample came from the Russian Federation.

On the IAEP-II test, Soviet 13-year-old students achieved rather high results in mathematics and science, placing among the five top-scoring countries. However, the results for different science content areas and for items with different performance categories were not uniform (Lapointe et al., 1992; Kovalyova, 1992).

According to the 1991 IAEP Assessment Objectives for Mathematics, Science, and Geography, the items on the science test were divided into three categories corresponding to three different levels of cognitive processes required to deal with science content at different levels of complexity. These categories were knows facts, concepts, and principles; uses knowledge to solve simple problems; and integrates knowledge to solve more complex problems. Figure 3 shows the science results for selected countries participating in IAEP-II for these three categories. National results for the three categories of items are shown in relation to the international mean for each category. The *y*-axis shows the difference between the country result and the international mean for each group of items. The *x*-axis presents three columns for the three groups of items. For each country the points corresponding to the results for the three groups of items are connected by a line illustrating the country profile in relation to the international mean.

Two groups of countries can be clearly identified from the figure. The first group, consisting of Hungary, Slovenia, and the USSR, had better results in knowledge than in application or integration of knowledge. The second group, consisting of Switzerland, Israel, Canada, France, and England had better results for the integration category.

If we disregard the absolute values, we observe two categories of results that differ by the inclination of the line. The results for the first group of countries may indicate that their science curriculum strongly emphasizes the mastery of factual knowledge and skills in their application to situations, skill in the application of known procedures, and a lower emphasis on higher-order intellectual skills. The USSR belonged to this group. Soviet students attained much lower results on the nature of science portion of the test, e.g., in using scientific methods of observation, classifying and inferring, interpreting data, formulating a hypothesis, designing experiments, and conducting investigations (Kovalyova et al., 1992; Firsov et al., 1994).

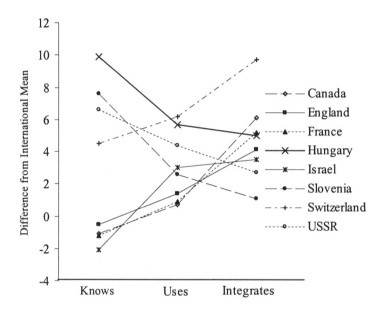

Figure.3. IAEP-II results for science.

Are the IAEP-II results similar to those of TIMSS-95 and TIMSS-99? Unfortunately, TIMSS did not publish comparisons of achievement in different cognitive behavior categories. All conclusions about student achievement in TIMSS were made on the basis of content areas. In order to make some observations and draw some conclusions from this perspective, we would have to perform an item-by-item analysis. Although the measurement instruments used in IAEP-II and TIMSS were different, the TIMSS results in 1995 and 1999 confirmed the main findings of IAEP-II, even though many changes have occurred in the education system in the intervening years.

EDUCATIONAL INNOVATIONS AND THEIR IMPACT ON THE TIMSS RESULTS

A reader examining the chapter on the Russian Federation in the TIMSS encyclopedia (Robitaille, 1996), will find a description of the new goals for science education in the Russian Federation, major changes and current issues in the science curriculum, information about the new science textbooks, and recent trends in pedagogy. The chapter was written in 1994-95, during a period when the major changes in school education were influenced by fundamental social transformations in the Russian Federation. School education gained a new emphasis, with the general intellectual and cultural development of students becoming one of the principal objectives for teaching. The focus was on finding a balance between academic, traditional, and practice-oriented individual approaches in teaching.

Because only a short time had elapsed between the implementation of these innovations and the administration of the test in 1995, the TIMSS results were not expected to reveal any great changes in achievement patterns. The new goals for science education had been articulated but were not yet implemented in textbooks or in the teaching and learning process. In essence, Russian achievement in TIMSS-95 was very similar to that in IAEP-II. It is not difficult to explain the apparent lack of change. Science education in the Russian Federation had essentially not changed since the end of the Soviet period, in spite of efforts stemming from *perestroika* in the 1980s and early 1990s.

Explaining the lack of change from IAEP-II to TIMSS-99 is more difficult. New textbooks containing the revised goals had been developed. Many of them used new approaches in structuring science education, and most were designed to meet the needs of students of different interests and abilities. However, the majority of schools continued to use the old textbooks, and the majority of teachers still employed the traditional teaching methods of the old Soviet schools.

Russian schools have reacted very slowly to change. The reasons go beyond budgetary concerns (for example, no funds to buy new textbooks), and seem to be connected to Soviet educational policy, school tradition, curriculum, and organization of the teaching-learning process. Stereotypes about the significant advantages of Soviet schools, supported by results from international Olympiads in various school subjects as well as by the results of international studies like IAEP-II and TIMSS, did not promote a national understanding of the need for serious educational reform.

A Quick Look At Science Education in the USSR

During the Soviet period educational goals were established in documents issued by the Central Committee of the Communist Party of the Soviet Union (CPSU). The Program of the CPSU (1969) suggested the following as the goals of general education:

... a sound knowledge of the basis of the sciences, mastering the principles of the communist outlook, labor and polytechnic grounding according to the growing development of science and technology, with due regard for the requirements of society, the abilities and interests of students as well as the esthetical and physical education of the next generation.

These general goals of education determined the priorities in formulating the learning objectives for all subjects. So, for science education, the priority objective was "sound knowledge of the basis of the sciences."

The proportion of time spent studying mathematics and science in the USSR occupied about half the school curriculum. The science subjects—geography, biology, physics, and chemistry—were studied as separate disciplines. All students took the same science courses, but only the most able had a real chance to learn advanced mathematics and physics. In 1959, the first experimental secondary school promoting advanced mathematics education was established in Moscow. Several years later, two advanced mathematics and physics boarding schools for gifted children affiliated with universities in Novosibirsk and Moscow were established. In 1972, there were no more than 2000 classes providing advanced education in all of the USSR, with a coverage of about two percent of the population of secondary school leavers.

The amount of information taught was accepted as the most important indicator of the quality of education. During the reform of science education that took place in the 1960s and 1970s under slogans such as "raising the scientific level of education," and "approaching the level of the modern sciences," science courses were overloaded with high-level, theoretical material. Teachers were not permitted to deviate from the curriculum. Reproducing and applying knowledge in familiar situations and performing according to known rules or instructions were strongly emphasized.

Even in those subjects that were impossible to teach without experiments and problem solving, practical work was conducted according to instructions. Problems were solved using procedural knowledge. For example, the physics curriculum (1986) used in secondary schools until the 1990s included no laboratory work or free investigation that could be done without detailed procedural instructions. Typical problems were presented for all subtopics. Students were trained intensively in the solving of these exact problems during the lessons. University entrance examinations for admission to many professional courses included examinations in physics that mainly involved the solving of these typical problems. In preparing for the entrance examinations, one had to learn to solve these kinds of problems. A few elite universities, such as Moscow University, selected the most talented applicants from special schools or boarding schools for gifted students. But even in specialist schools, the mastery of basic facts and the solving of typical problems with intensive use of mathematics prevailed.

An overestimation of students' ability to master the basic sciences led to a significant gap between curricular requirements and realistic expectations for student achievement. Reports prepared during the Soviet years by researchers at the Institute of Content and Methods of Education of USSR Academy of Pedagogical Sciences (now the Institute of General Secondary Education of the Russian Academy of Education) indicated satisfactory results in formal knowledge and basic skills, but low results on problem solving. For example, the report on the assessments of general school students' physics achievement in 1986-87 stated that only 20 percent of students could solve problems requiring the application of knowledge to new situations.

The difficulties in changing science education in the Russian Federation may have roots in an ideology that was deeply established when the country was part of the USSR. The paradigm of the ideal Soviet citizen was to be a cog in the big state machine that was formed during the Stalinist period. The quality of the citizen in this framework was determined by his or her correspondence to the place prescribed to him or her. Cogs were naturally interchangeable. It does not matter to a cog which part of the machine it was to be used for; its valiant labor would be realized in fulfilling its function. A cog did not have its own interests.

Naturally enough, cogs were the products of mass production from Soviet schools. Many features of the Soviet system of education described above were related to this concept. For example, the interchangeability of cogs was provided by a uniform production process. From this followed a lack of variability in Soviet schools, including uniformity in study plans and curricula. Cogs were designed to fulfill well-defined functions and not to have their own interests. From this followed the technocratic orientation of schools and the dogmatic character of education: coaching for mastery of technical routine procedures instead of the development of creative abilities, and suppression of individual initiative and student independence.

The idea of children as cogs completely contradicts democratic reform in society. There is no place for cogs in the new millennium because that is not the characteristic of a free, enterprising, and creative personality. We should halt the production of cogs if we really wish to advance the goal of developing a democratic society in conjunction with a market economy. For this reason, the slogan and goal of post-communist school reform is humanization (making education more "human"), meaning shifting the center of gravity from state requirements toward the need of the developing personality of a child (Firsov et al., 1995).

It is impossible to solve this problem without reconstructing the systemic elements that formed the foundations of the previous system. This reconstruction should not lead to the loss of past achievements, in particular, Russia's traditionally strong achievement in science and mathematics. Nevertheless, high achievement must be attained in conjunction with developing the personality.

Modernization of education at the beginning of the new millennium

It is more and more clear that the role of education in modern society is increasing, that real societal change cannot take place without the modernization of education. In 2000, the Russian government began to develop a new educational reform program. It declared that the education system should shift from a regime of survival to a regime of development. The main directions of reform were articulated in the document "Main directions of social-economic policy of the Government of the Russian Federation from a long-term perspective," and the associated plan of activities was adopted in 2000. The priorities in the modernization of the structure and content of education were raising the quality of education, providing equal access to education, developing effective mechanisms for the transmission of social requests to the educational system, and broadening public participation in managing education.

The 2001-02 school year marked the start of a nation-wide experiment in modernizing the content of school education. This experiment has been in the planning stage for several years, with approximately 2000 schools slated to participate. These schools, which are located around the country, will in time become centers for disseminating the new school model and new educational technologies to the other 60,000 schools.

Reducing the content of education through the elimination of unnecessary terminology that is no longer used in educational, professional, or personal circles was brought to the forefront of the reform. Also at the forefront was a change in teaching methods and a reorientation to individual, independent work, and self-learning. This may promote a shift away from mastering large volumes of knowledge, to the development of competencies needed for students' personal and professional development. In addition, a reduction in content will allow more time for active, independent work individually and in small groups, in classrooms and out of school. Less emphasis should be placed on terminology and memorization of facts and formulas, and more on conducting observations, experiments, investigations, and fieldwork, and on finding solutions to practical and creative problems.

In the new documents on education in the Russian Federation we can see that many of the problems raised in this paper, including equality issues in education, are being addressed. Ideas about the need for reform are not new, but to get effective results it is necessary for these notions to be shared by the whole nation. In a recent interview, the Minister of Education said that education has attracted the attention of the entire country, adding that, for the first time in the past hundred years, the budget for education in 2002 would exceed all other budgets, including that of defense. This is a good start for the new reform. Only the future will tell what its effectiveness will be.

REFERENCES

Aksakalova, S. E., Kirillova, I. G., Kovalyova G. S., Korotchenko, A. S., Miagkova, A. N., Reznikova, V. Z., & Tarasov, J. P. (1995). *The relationship between the TIMSS assessment content and what students in the Russian Federation have studied in science by the Population 2 and Population 3 ages.* Moscow: Russian Academy of Education.

Angell, C., Kjaernsli, M., & Lee, S. (2000). Exploring students responses on free-response science items in TIMSS. E.W.Jenkins and D. Shorrocks-Taylor (Eds.), *Learning from Others: International Comparisons in Education.* Dordrecht, Netherlands: Kluwer Academic Publishers.

Beaton, A. E., Martin, M. O., Mullis, I. V. S., Gonzalez, E. J., Smith, T. A., & Kelly, D. L. (1996). *Science achievement in the middle school years: IEA's Third International Mathematics and Science Study (TIMSS).* Chestnut Hill, MA: Boston College.

Educational Testing Service. (1990). *The 1991 IAEP assessment. Objectives for Mathematics, Science, and Geography.* Princeton, NJ: Author.

Firsov, V. V., Kovalyova, G. S., & Loginova, O. B. (1995). Transition to a market economy: Applications for curriculum and teaching in post-communist society. Paper for the Wold Bank Report *Russian Education in Transition.*

Kovalyova G. S. (Ed.). (1992). *Comparative assessment of students' achievements in mathematics and science (according to the results of international study IAEP)* (in Russian). Moscow: Institute of General Education, Russian Academy of Education.

Kovalyova G. S. (Ed.). (1996). Comparative analysis of mathematics and science achievement of students of the basic schools in Russia. *Third International Mathematics and Science Study - TIMSS. Volume 2.* (Denitseva L. O., Djukova S. E., Kovalyova G. S., Korotchenko A. S., Krasnianskaia K. A., Mjagkova A. S., Naidenova N. N., Reznikova V. Z., Suravegina I. T.) (in Russian). Moscow: Russian Academy of Education.

Kovalyova G. S. (Ed.). (1998). Comparative evaluation of mathematics and science achievement of secondary school graduates in Russia. *Third International Mathematics and Science Study - TIMSS. Volume 4.* (in Russian). (Denitseva L. O., Kovalyova G. S., Koshelenko N. G., Krasnianskaia K. A., Loshakov A. A., Nurminsky I. I., Naidenova N. N.). Moscow: Russian Academy of Education.

Kovalyova G. S. (2000). Russian Federation. In D. F. Robitaille, A. E. Beaton and T. Plomp. (Eds.). *The Impact of TIMSS on the teaching and learning of Mathematics and Science.* Vancouver, BC: Pacific Educational Press.

Lapointe, A. E., Askew, J. M., & Mead, N. A. (1992). *Learning Science.* Princeton, NJ: IAEP, ETS.

Mullis, I. V. S., Martin, M. O., Beaton, A. E., Gonzalez, E. J., Kelly, D. L. & Smith, T. A. (1998). *Mathematics and Science achievement in the final year of secondary school: IEA's Third International Mathematics and Science Study (TIMSS).* Chestnut Hill, MA: Boston College.

Program of the Communist Party of the Soviet Union. (1969). Moscow, "Politizdat" (in Russian).

Physics curriculum for general secondary school. (1986). Moscow, "Prosvetschenie" (in Russian).

Robitaille D. F. (Ed.). (1996). *National contexts for mathematics and science education: An encyclopedia of the educational systems participating in TIMSS.* Vancouver, BC: Pacific Educational Press.

Chapter 13

TIMSS SCIENCE RESULTS SEEN FROM A NORDIC PERSPECTIVE

Marit Kjærnsli and Svein Lie

In this chapter we describe some characteristic features of the TIMSS science results as seen from a Nordic perspective, in particular; from a Norwegian one. Four of the five Nordic countries participated in TIMSS: Iceland and the three Scandinavian countries, Denmark, Norway, and Sweden. (Finland did not take part, but participated as the only Nordic country in the repeat of TIMSS in 1999.) These four Nordic countries have much in common historically, culturally, and politically. In the Scandinavian countries almost the same language is spoken, whereas Icelandic is different, but quite similar to the old common Nordic language. During the last six or seven hundred years Sweden and Denmark have rivaled each other for hegemony in the area, while Norway, Iceland, and Finland have been the smaller brothers in union with one or the other most of the time.

Today all the Nordic countries are independent democracies with a strong social democratic tradition. They are relatively homogeneous societies with highly developed social welfare systems. There are also strong and friendly cultural links among the Nordic countries, even if the European Union has split these countries into members (Denmark, Finland, and Sweden) and non-members (Iceland and Norway). Sweden has the largest population, eight million, whereas Denmark, Finland, and Norway all have between four and five million. Iceland has a small population, about 200,000. It should also be mentioned that the climate is similar in all of these countries. Denmark stands out from the other Nordic countries in two respects: firstly, due to its position closer to the other North European countries it appears more "continental" both in climate and culture; and secondly, its relatively high population density makes Denmark somewhat different from the typical Nordic countries which have wilderness within easy reach of even the largest cities.

Nordic Education Policy and Science Education

Concerning education, there has been a common and strong commitment in all of the Nordic countries toward equal opportunity for all. There have been frequent

D.F. Robitaille and A.E. Beaton (eds.), Secondary Analysis of the TIMSS Data, 193–208.
© 2002 *Kluwer Academic Publishers. Printed in the Netherlands.*

educational reforms in the last decades, which have mutually influenced each other, so that we may speak of a "Nordic" educational policy. Characteristic features are compulsory schooling until the age of 16 with no ability-based streaming, free upper secondary schools covering both academic and more vocational lines of study, and a relatively late start to formal schooling at age 7. (The only exception has, until recently, been Iceland where schooling starts at age 6. After the reform of 1997 schooling now starts at age 6 in Norway also.) Furthermore, there are also similarities concerning curricular emphases in science and mathematics. One may summarize the situation by stating that, seen from an international perspective, the focus has been more on important daily-life aspects of science and mathematics than on more advanced and abstract concepts. In mathematics, algebra and geometry have received relatively little emphasis up to age 13. On the other hand, topics such as measurement, estimation, and data presentation and interpretation have been emphasized. The similarities in the results for mathematics have been shown to be so strong that one may speak of a "Nordic profile of mathematics education" (Lie et al., 1997).

The situation for science is somewhat similar. Topics such as the concepts of cells, atoms, and molecules have received relatively little attention compared to the more daily-life aspects. Student experiments play a central role in science instruction. The focus is more on visualization for conceptual understanding and on engagement and motivation, whereas relatively little emphasis has been given to the formal aspects of scientific investigation as a logical means of creating scientific knowledge.

Science in Norway differs from that in other Nordic countries in one important respect. Up to Grade 10, science is taught as an integrated subject, whereas courses at the lower secondary level are organized in the form of separate science courses in each of the other countries. While Norway follows the English (and other English-speaking countries') tradition, their Nordic peers follow a general continental European tradition.

Norway and Iceland took part in all three TIMSS populations. In Population 3 both did the mathematics and science literacy test, and Norway also the physics test for advanced students. Denmark and Sweden participated in Population 2 and all three tests (literacy, physics, and mathematics) in Population 3.

SCIENCE ACHIEVEMENT AS A FUNCTION OF AGE

We begin our discussion of the TIMSS data by looking at overall science results for all populations, then focus on school effects and gender differences. Finally, we examine in more depth similarities and differences between countries regarding profiles of achievement across science topics and across test items.

First, we focus on trends in the science achievement scores from Population 1 to Population 3 (literacy). Figure 1 summarizes changes in science achievement across age groups for selected countries. What is shown for each country and grade level is the science score above (+) or below (−) the mean for the 12 countries that participated in all three populations. Note that all the mean values are for the same set of countries, namely those which participated in all populations, and that these averages therefore are somewhat different from the international means for *all countries* which is published in the international TIMSS reports (Beaton et al., 1996; Martin et al., 1997; Mullis et al., 1998) and used elsewhere in this chapter. The lines between the points of measurement have no meaning except as a means of simplifying the tracking of each country's changes.

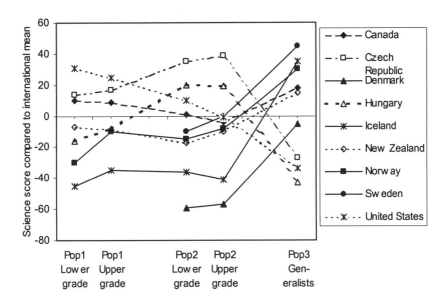

Figure 1. Science achievement in all three populations.

In Figure 1 we have, in addition to the Norwegian and Icelandic results, also displayed the results also for Denmark and Sweden even though they did not participate in all three populations. Furthermore, by including some other countries, we want to illustrate examples of some typical trends. It appears that all four Nordic countries scored relatively poorly in Populations 1 and 2. Even Sweden, as the highest-scoring country of the three groups, is below or at the mean. (It should be mentioned that Finland in the repeat of TIMSS in 1999, scored higher than its Nordic counterparts did in 1995 (see Martin et al., 2000). From the point of view of the Nordic countries, however, the main message is very clear. Nordic students tend to score relatively better as they grow older. The results for Canada and New Zealand show a similar upward trend, but not as pronounced. Also, results from the Netherlands show a weak upward trend, but substantial reference to this country has been avoided here due to its extremely low participation rate in all populations (Martin et al., 1997; Beaton et al., 1996; Mullis et al., 1998). The Eastern European countries, Hungary and the Czech Republic, both have a strong downward trend from Population 2 to Population 3, a common feature also for the Russian Federation and Slovenia (Mullis et al., 1998), but not shown here. For the United States we see that the downward trend is continuous and very pronounced.

When discussing results for the generalists of Population 3 it is important to have in mind that the samples are not easily comparable across countries. Both students' ages and the coverage of the age cohort vary from country to country (Mullis et al., 1998). However, particularly for Norway and Sweden, where more than 70 percent of the age cohort is covered by the sample, the high level of average achievement seems to be well established by the present data.

A similar trend of higher relative achievement by age for the Nordic countries may be found in the parallel results for mathematics (Mullis et al., 1998), albeit with somewhat lower Nordic results (except for Denmark). The results for the physics specialists are remarkable from a Nordic point of view. Norway, Sweden, and Denmark all rank among the four highest-scoring countries (Mullis et al., 1998). Again, since the age cohort coverage varies greatly from country to country, depending on the country's definition of a physics specialist, the rank order of countries should not be taken at face value. However, taken together, the high overall achievement level in science for the Nordic countries at the end of schooling seems to be a valid finding. It should be noted, however, that the Nordic students are older than their international peers (Mullis et al., 1998).

There are essentially two types of cultural explanations that have been put forward in relation to the results discussed here. One focuses on the Nordic emphasis on education for all, with high school-participation rates even at the upper secondary level, above 90 percent. Furthermore, theoretical subjects do play an increasing role even within the more vocational lines of study. It is worth mentioning that students within what may be regarded as vocational lines of study in all four Nordic countries score at or above the international mean for all students (Mullis et al., 1998). The national reports for each of the Scandinavian countries discuss this point thoroughly (Allerup et al., 1998; Angell et al., 1999; Skolverket, 1998) and draw attention to this positive effect of schooling for all. In particular, the Swedish (Skolverket, 1998) and Norwegian (Angell et al., 1999) national reports tend to interpret the results as an indication of an educational policy that in this respect has been fruitful.

The second type of explanation has to do with a view of childhood. There is a common tradition in Nordic countries to "let children be children," in the sense that little educational (and other) pressure is put upon them at an early age. The late start of schooling and relatively few classes per day during the first years are examples. Even more significant may be the fact that formal marks are not given in the primary school at all. Furthermore, repeating grades does not occur in the compulsory school (primary and lower secondary).

The East Asian countries Singapore, Japan, and Korea dominate the TIMSS league tables for Populations 1 and 2. (None of these countries participated in Population 3.) In these countries there is a tradition of putting strong pressure children to do their best at school. This strong pressure is embedded in the culture and seems to give a general explanation of the strong results for these countries. We believe that, insofar as views on childhood are concerned, the Nordic countries are at the opposite extreme. Some contextual TIMSS data may shed light upon this matter. The responses to two questions in the TIMSS Population 2 questionnaire are particularly relevant here. Students were asked about how important it was for them to do well in science and to have time for fun.

In Table 1 we present the mean response for each of the Nordic countries together with the international mean. Country means have been calculated from the Likert-scale responses on a scale from "Strongly disagree" (1) to "Strongly agree" (4) that the topic is important for them. The message from the table is clear. In their judgment, Nordic students seem to regard success in school science to be somewhat less important and time for fun to be clearly more important than their international peers.

Table 1. Population 2 responses to questions on importance

Country	Do well in science	Time to have fun
Denmark	3.21	3.85
Iceland	3.32	3.71
Norway	3.27	3.81
Sweden	3.07	3.77
International mean	3.30	3.58

On some other items for Population 2, students were asked how much time was used before or after school on certain activities on a normal day. Again, it is typical for Nordic students to spend a lot of time with friends (Denmark and Norway, together with Germany, at the top) and doing other non-academic activities, whereas homework takes very little of their out-of-school time (Denmark at the very bottom).

WITHIN- AND BETWEEN-SCHOOL VARIANCES

As described earlier, the Nordic countries are rather homogenous societies. This is also true for their education systems, and TIMSS data throw some light on this point. Countries vary as to how "similar" students are in the same school or classroom. Obviously, relatively low similarities between classrooms will occur in countries with a streamed system: i.e. where high and low achievers attend different classes or different schools. The opposite will happen in countries with no within or between-school streaming. If, in addition, the society is relatively homogeneous, each classroom could mirror the whole population. Generally, at Population 1 the streaming effects are very small, and at Population 3 they are dominant. In the following we discuss some data for science achievement at Population 2, upper grade. Here we expect to find the largest differences between countries. Since almost all of the countries sampled only one intact (mathematics) classroom per sampled school, one cannot partition the within-school variances into within-class and between-class components. This should be kept in mind when we discuss the within-school and between-school components of the variances in the following.

The percentages of the total variance in science achievement that occurs between schools (or rather is associated with the combined between-schools and between-classrooms-within-schools effects) are given in Martin et al. (2000). While the international average is 23 percent, all four Nordic countries appear among the eight countries with 10 percent or less of the total variance occurring between schools. (Due to shortcomings in the Danish sampling procedure, the value for Denmark is not shown in (Martin et al., 2000), but the actual value is below 10 percent). The other countries with similarly high homogeneity are Japan, Korea, Slovenia, and Cyprus. On the other extreme, more than 40 percent of the variance is associated with schools in Romania, the United States, Germany, and Singapore.

Thus, from an international perspective, science achievement in the Nordic countries appears as rather school-independent. Differences between schools seem to be relatively small, both between urban and rural areas and between high- and low-socioeconomic localities. This in itself does not mean that home background factors are not important predictors of science achievement, neither does this provide any clear explanation of the rather dramatic increase in student achievement with age for the Nordic countries discussed in the former section. But the findings can be interpreted to indicate that an important educational goal of equal opportunity for all in the Nordic countries seems to some extent to have been fulfilled.

Achievement Differences between Science and Mathematics

If we compare the TIMSS science and mathematics achievement scores for the same countries, it is striking to notice that, for all three populations, the league tables for the two subject areas look quite similar. However, a closer look reveals some characteristic differences, as we can define certain countries as "mathematics countries" and others as "science countries" according to the difference between the country's science and mathematics mean score. An interesting pattern emerges if we investigate this further and combine results for all three populations. In general, the most pronounced "mathematics countries" are the East Asian countries (Hong Kong, Japan, Korea, and Singapore) together with the French-speaking countries (France and French-speaking Belgium). On the other hand, England and most of the other English-speaking countries appear to be "science countries," and consistently so across populations.

Now, let us look at the Nordic countries from this perspective. In Table 2 we have shown the difference between the mean science and mathematics scores for the

Nordic countries together with the international extremes ("Maximum" being the most pronounced "science country," and "Minimum" as the most pronounced "mathematics country"). Norway, Sweden, and Iceland all appear consistently above the average line as "science countries." On the other hand, Denmark's results are clearly and consistently different from their Nordic counterparts, and one may speculate as to why this is so. Denmark, as a "mathematics country," seems to put relatively more emphasis on mathematics; or, formulated alternatively, relatively less focus on science. A similar picture emerges even when one compares the results of the mathematics and physics specialists in Population 3 (Mullis et al., 1998; Angell et al., 1999).

Table 2. Differences between science and mathematics mean scores

Country	Population 1	Population 2	Population 3
Denmark	n/a	−24	−38
Iceland	31	7	15
Norway	28	24	16
Sweden	n/a	17	7
Maximum	39 (England)	46 (England)	21 (Czech Rep.)
Minimum	−78 (Singapore)	−66 (Hong Kong)	−38 (Denmark)

There is no easy explanation for why countries differ along this science-mathematics dimension. Obviously, different traditions regarding structure, emphasis, and content of the curriculum may play important roles; traditions that in turn may have deep cultural antecedents. Such cultural factors may be instrumental if one tries to understand what makes English-speaking countries "science countries" and French-speaking and East Asian countries "mathematics countries." However, concerning the difference among the Nordic countries, a more "local" explanation is called for, due to the otherwise close similarities in cultural traditions (Allerup et al., 1998; Lie et al., 1997).

Geographical factors may have some influence on what is considered important and therefore emphasized in school. Wilderness in the form of forests, mountains, lakes, or coastlines is within close reach for essentially all Norwegian, Swedish, and Icelandic (as well as Finnish) people. A common extensive right for everybody to go wherever they want (except gardens and fields during the summer) exists in these countries. By using nature for all sorts of purposes, students in these countries may well have learned some science outside school that has helped them to solve TIMSS items. In this respect Denmark is somewhat different with its high population density and more cultivated landscapes. It should be noted here that from the results of TIMSS–99, Finland also appears to be a "science country," with a science-mathematics difference of 15 score points (see Martin et al., 2000 and Mullis et al., 2000).

GENDER DIFFERENCES

As an important part of the struggle for equal opportunity, gender equity has been given high priority in Nordic education and in Nordic society in general. For many years the Nordic countries have had the reputation of being the part of the world where gender equity has come the farthest. This picture may well represent one important part of the situation, in particular when it comes to strict laws against discrimination or to the number of females in the parliament or in the government. Also, a female prime minister (Norway) and female presidents (Iceland and Finland) have received international attention. In the *Human Development Report* published by the United Nations Development Program (UNDP, 2000), the five Nordic countries came out as the first five in a list of 150 countries regarding their scores on indices of gender equity.

Within science and mathematics education, it has been an important international aim to provide equal opportunities for both sexes. However, young women are generally under-represented both within higher education and in jobs that involve these subject areas, in particular mathematics and physics. The situation in the Nordic countries is not any better than in most other countries. The fact that gender equity in general has been given such high priority makes it particularly troublesome that mathematics and the "hard" sciences still have such a masculine image.

Internationally, gender related issues in science education have been the focus of much research and a number of reports have been published (e.g. Wilson, 1992; Parker et al., 1996). Since the early eighties there have also been held a number of GASAT (Gender and Science and Technology) international conferences on the topic (e.g. GASAT 8, 1996). IEA studies are very well suited for a closer inspection of gender differences, as they provide valid and reliable data on achievement as well as attitudes. The IEA First International Science Survey was the basis for an extensive discussion on gender differences (Kelly, 1981). Results from the Second International Science Study (SISS) and TIMSS have provided important data for gender-related in-depth studies (e.g., Parker et al., 1996; Mullis et al., 2000).

We start our inspection by focusing on gender differences as a function of age. In Figure 2 results in the form of differences between boys' and girls' mean science scores have been displayed for all the Nordic countries together with the international mean. These differences can be directly compared across populations due to the common metric for the standardized student scores (international mean = 500, and standard deviation = 100 for each population).

Let us first look at the results for Populations 1 and 2 (data are shown for the upper grades). For Iceland, Norway, and Sweden the gender differences are quite similar to the international situation. In all three countries boys score significantly, but not much, higher than girls. The effect size of the gender differences, also called the standardized sex difference (SSD, difference divided by the pooled standard deviation) is of the order of 0.15. For Denmark, the gender gap is already rather wide at Population 2, in fact the widest of all the participating countries after Israel.

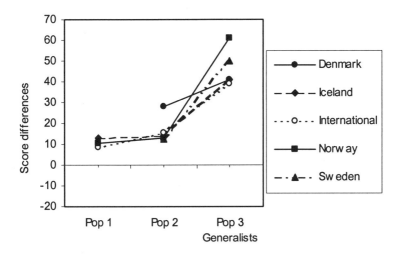

Figure 2. Science score differences in favor of boys.

Results for the TIMSS performance assessment component have not been included in the discussion so far, but it should be mentioned that no significant gender differences for the overall science part of this component were found in any of the participating countries, including Norway and Sweden (Harmon et al., 1997).

There are indications that internationally a remarkable decrease in the gender gap has occurred during the last decades. The above-mentioned SSD of 0.15 for TIMSS is much lower than that of the first IEA science study, which reported an SSD of 0.46 (Comber & Keeves, 1973) and the second science study, with an SSD of 0.36 (Postlethwaite & Wiley, 1992). Care must be taken not to draw simple conclusions from these numbers. The countries were not the same, and the Population 2 students in the earlier studies were somewhat (typically two years) older. But even among Population 1 students in the earlier studies, the SSD values were higher than 0.15 (0.23 in both studies).

When we move to science literacy for Population 3 it should be remembered that comparisons between countries must take into account which parts of the student age cohorts have been sampled. In particular, countries differ in the extent to which typical vocational lines of study have been included. Nevertheless, the general picture seems to be quite significant. Figure 2 displays a general dramatic increase in gender differences in favor of males when we go from Population 2 to science literacy in Population 3. This increase occurs in all countries (Mullis et al., 1998) and reflects, to a large extent, gender differences in curricular choices for Population 3 students. However, it is notable that, particularly for Norway, but also for Sweden, the gender gap is much wider than elsewhere. In fact, the Norwegian difference stands out as the very highest of all countries and amounts almost to an effect size of as much as 1.0 (gender difference equals the standard deviation). From Figure 2 the situation for Iceland and Denmark seems somewhat "better" than for their Nordic counterparts. However, for these two countries, the Population 3 literacy sample represented a considerably lower part of the age cohort (55 percent and 58 percent, respectively) than in many other countries. If more of the vocational lines of study had been included, one would likely have seen Danish and Icelandic gender differences more like those of Norway and Sweden. Similarly, the very high cohort coverage in Norway (84 percent) is a factor to take into account when interpreting the extreme gender difference for this country.

As a general finding, we conclude that the gender gap in science achievement increases by age in all countries, and that girls' underachievement is particularly distinct in the Nordic countries towards the end of the education system. It should also be mentioned that the international results show that the gap is particularly pronounced in physical and earth sciences (Mullis et al., 2000), this being an international and well-known trend.

In an analysis of the Norwegian data, Kjærnsli and Lie (2000) showed that there is a general tendency for boys to outperform girls, particularly in topic areas that are not commonly covered in formal school science. A comparison of marks given within the same school science courses revealed no or very small gender differences. Nevertheless, significant differences in favor of boys appeared on the TIMSS science scores. It seems that boys, to a much higher degree than girls, are interested in science and accordingly learn more from media and other out-of-school experiences. This difference in interest may be an important factor in understanding girls' underachievement despite more effort within the subject in school (Mullis et al., 2000).

The large gender gap for science achievement has caused considerable concern in the Nordic countries (e.g. Hoff, 1999; Kjærnsli & Lie, 2000). Why is it that science has an even more masculine image than in other parts of the world? From the discussion above, it seems essential to consider differences in attitudes towards the sciences.

Attitudes toward Science

In this section we draw attention to some TIMSS attitude data we consider relevant for the above discussion on gender differences. We regard attitudes at Population 2 level to be particularly crucial because these students have recently made, or (as in the Nordic countries) are within a year or two of making their first important curricular choices. In addition to achievement, students' attitudes towards the sciences obviously will have an effect on the selection, or non-selection, of science-related subjects and areas of study.

From the Population 2 questionnaire data a construct has been made based on the overall measure of students' responses on a four-point Likert scale to several items about how well they liked science and to what degree they would like a science-related job (Beaton et al., 1996). The focus here is not a comparison between countries' mean scores on this construct, but the score differences by gender. The data are somewhat complicated by the fact that these questions were included only in countries where science is taught as an integrated school subject. As explained earlier, among the Nordic countries this only applies to Norway. But the Norwegian data is very pronounced. Except for Japan, in none of the 22 countries in this category (of integrated science) was the gender difference as large as in Norway. The effect size of this difference amounts to 0.30, or about twice the magnitude of the achievement difference at the same grade level. From the point of view of gender equity we therefore regard this Norwegian "attitude gap" as more serious and having more consequences than the "achievement gap" at Population 2 (Kjærnsli & Lie, 2000).

Similarly, a comparison between countries with courses in the separate sciences reveals that, also in Denmark and Sweden, boys have a much more positive attitude towards science. However, no strong attitude gap appears in the Icelandic data. It should be noted that, for all countries, gender gaps mainly appear in the physical sciences.

It appears from the data discussed here that the large gender gap in science achievement in Population 3 in the Nordic countries has an important prerequisite in the large attitude gap already existent by Population 2. It is not surprising that large differences in attitudes toward science influence both the tendency to choose science subjects in further education and also the degree to which some science is learned outside school. A recent international study on interests and experiences within science as well as images of science and scientists (Sjøberg, 2000) has reported quite similar to those discussed here. The situation in the Nordic countries regarding gender equity is described by Sjøberg as a paradox due to the particularly large gender differences in attitudes toward science in spite of the generally high level of gender equity.

SIMILARITIES AND DIFFERENCES BETWEEN COUNTRIES

A View from Norway

Based on the similarities and common features mentioned in the introduction, one would expect the various TIMSS achievement results for Population 2, both overall and for individual items, to be rather similar for the Nordic countries. As has been described in another paper in this volume (Chapter 9), the results for Population 2 support this expectation concerning the item-by-item results that all the Nordic countries fall into the same ("Nordic") group based on patterns of responses to the science items. Furthermore, in order to investigate this from a Norwegian perspective, we calculated the correlation coefficients for *p*-value residuals (corrected for the international difficulty of each item and for overall science achievement for all countries) between Norway and all other Population 2 countries. The list of correlation coefficients can be interpreted as a list of countries sorted by the similarity of the science knowledge of the students compared to that of their Norwegian peers. Results are shown in Table 3.

Table 3. Correlation coefficients for p-value residuals between Norway and other countries

0.10 or higher	0 to 0.09	-0.11 to 0	-0.17 or lower
Sweden (0.54)	Spain (0.09)	Thailand (0.00)	Cyprus (-0.17)
Denmark (0.43)	Ireland (0.08)	Israel (-0.01)	South Africa (-0.17)
Iceland (0.31)	Belgium (Fr.) (0.08)	Slovenia (-0.05)	Kuwait (-0.17)
Switzerland (0.22)	Germany (0.06)	Latvia (-0.05)	Hungary (-0.18)
New Zealand (0.21)	Australia (0.06)	Portugal (-0.07)	Bulgaria (-0.19)
Canada (0.14)	France (0.05)	Korea (-0.08)	Lithuania (-0.19)
Belgium (Fl) (0.12)	Japan (0.04)	Greece (-0.08)	Philippines (-0.20)
Scotland (0.12)	United States (0.03)	Singapore (-0.08)	Russian Fed. (-0.20)
Netherlands (0.10)	England (0.00)	Austria (-0.10)	Czech Rep. (-0.21)
		Slovakia (-0.10)	Hong Kong (-0.22)
		Colombia (-0.11)	Iran (-0.26)
			Romania (-0.27)

As expected, Sweden, Denmark, and Iceland are clearly the top countries, and they are followed by Switzerland (0.22), New Zealand (0.21), and Canada (0.14). Seen from a Norwegian point of view, the other Nordic countries appear quite "close." Other pronounced features are that all English-speaking countries appear relatively close (at or above zero) and that all Eastern European countries appear "far away" (negative), features that may be interpreted as similar and very different emphases in science education, respectively.

Achievement across Items and Content Areas

If we go into more detail concerning subject matter topics, or rather individual items, we can also demonstrate the similarity of student achievement among the Nordic countries. In Figure 3 we have, as an example, displayed the *p*-values for each physics item for Norway (solid line) compared to the international mean (heavy dotted line) and the highest- and the lowest-scoring country (faint dotted lines). The items are sorted in descending order of international mean by item. Also shown is the (shaded) area of variation that the other three Nordic countries cover for each item. The figure (Zabulionis, 1997) shows that the Norwegian results, with very few exceptions, vary within a Nordic "river" of *p*-values. Thus, we have another strong indication that Nordic students have more or less identical strengths and weaknesses across the different science topics.

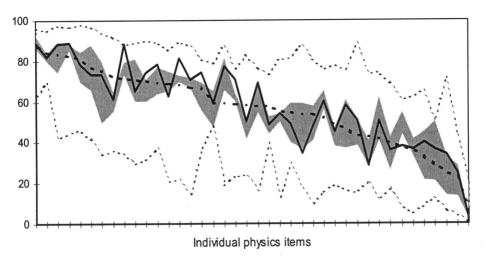

Individual physics items

Figure 3. p-values for individual physics items, Population 2 upper grade.

Achievement across Content Areas

Finally we look more closely into what sorts of similarities one can find among the Nordic countries when it comes to student achievement across content areas, and how this relates to international means. Figure 4 shows the achievement in the form of *p*-values for each of the content domains in Population 2, upper grade. The difference between each country's *p*-value and the international mean is displayed. Data are given for each domain for the Nordic countries together with the highest- ("Max") and lowest-scoring ("Min") country for each domain.

From Figure 4 one can see very strong similarities among the Nordic patterns. In particular, Norway and Sweden follow each other closely, and Denmark and Iceland even more so. The similar traditions in curricular emphases are clearly reflected, in particular the strong influence from Denmark to Iceland. There are also a few characteristic differences: Earth science seems to be more emphasized in Norway and Sweden than in the other two countries. And further, chemistry topics seem to be more emphasized in Sweden than in the Nordic counterparts, where traditionally chemistry has been given little emphasis at early age. In Norway, after the national curriculum reform of 1997, partly as a response to the TIMSS results, there is now more focus on chemistry than before.

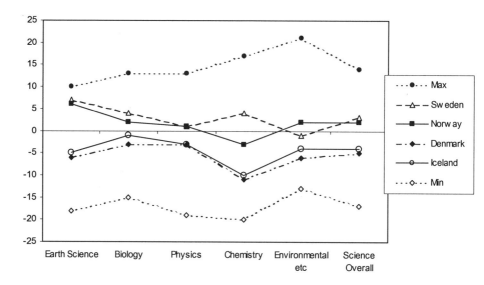

*Figure 4. Achievement score above or below international mean within content areas,
Population 2, upper grade.*

CONCLUSION

In this chapter we have analyzed the TIMSS science results from a Nordic point of view. We have focused on some characteristic features that have emerged from the data. The relatively low achievement level at Populations 1 and 2 in spite of the high level of expenditure has caused serious concerns in the Nordic countries. On the other hand, the upward trend of achievement by age has been positively received. The Population 3 data, taken as a whole and interpreted with care, indicate a relatively positive outcome for the Nordic education systems, taken as a whole. Further research may investigate what the crucial factors are for this accomplishment; and, in particular, how it is related to the Nordic tradition of putting low educational pressure on students at an early age, and strong emphasis on education for all.

Gender differences in science achievement have been shown to represent a particular problem for the Nordic countries, in spite of the high standard of gender equity more generally. Students' attitudes toward science have been discussed as a possible crucial factor in the development of gender gaps in achievement.

The last part of the chapter has focused on similarities and differences between countries regarding profiles of achievement across items and content areas. The striking similarities among the Nordic countries have been clearly demonstrated in various ways. It is also interesting to note that Nordic science education seems to be relatively "close" to the tradition in English-speaking countries, whereas Eastern European and Asian countries seem to have quite different emphases in their educational approaches.

REFERENCES

Allerup, P., Bredo, O., & Weng, P. (1998). *Matematikk og naturvitenskap i ungdomsuddanelser – en international undersøgelse.* (Danish national TIMSS Population 3 report, in Danish). Copenhagen: Danmarks Pædagogiske Institut.

Angell, C., Kjærnsli, M., & Lie, S. (1999). *Hva i all verden skjer i realfagene i videregående skole?* (Norwegian national TIMSS Population 3 report, in Norwegian). Oslo: Universitetsforlaget.

Beaton, A., Martin, M. O., Mullis I.V.S., Gonzales, E. J., Smith, T. A., & Kelly, D. A. (1996). *Science achievement in the middle school years. IEA's Third International Mathematics and Science Study (TIMSS).* Chestnut Hill, MA: International Study Center, Boston College.

Comber, L. C. & Keeves, J. P. (1973). *Science education in nineteen countries.* Stockholm: Almqvist & Wiksell

GASAT 8 (1996). Achieving the four E-s. *Proceedings and Contributions to the eighth GASAT conference, 4 volumes.* Ahmedabad, India: SATWAC Foundation.

Harmon, M., Smith, T. A., Martin, M. O., Kelly, D. L., Beaton, A. B., Mullis, I. V. S., Gonzalez, E. J., & Orpwood, G. (1997). *Performance assessment in IEA's Third International Mathematics and Science Study.* Chestnut Hill, MA: International Study Center, Boston College.

Hoff, A. (1999). *Myth or reality: What can TIMSS teach us about gender differences in the Danish science and math education? Publication 36.* Denmark: The Royal Danish School of Educational Studies.

Kelly, A. (Ed.). (1981). *The missing half. Girls and science education.* Manchester: Manchester University Press.

Kjærnsli, M. & Lie, S. (2000). Kjønnsforskjeller i realfag: Hva kan TIMSS fortelle? (Gender differences in science and math: What can be learned from TIMSS?). In G. Imsen (Ed.), *Kjønn og likestilling.* Oslo: Gyldendal Akademisk.

Lie, S., Kjærnsli, M. & Brekke, G. (1997). *Hva i all verden skjer i realfagene? (Norwegian national TIMSS Population 2 report, in Norwegian).* Oslo: Universitetsforlaget.

Martin, M. O., Mullis, I. V. S., Beaton, A., Gonzales, E. J., Smith, T. A., & Kelly, D. A. (1997). *Science achievement in the primary school years. IEA's Third International Mathematics and Science Study (TIMSS).* Chestnut Hill, MA: International Study Center, Boston College.

Martin, M. O., Mullis, I. V. S., Gregory, K. D., Hoyle, C., & Shen, C. (2000). *Effective schools in science and mathematics. IEA's Third International Mathematics and Science Study (TIMSS).* Chestnut Hill, MA: International Study Center, Boston College.

Martin, M. O., Mullis, I. V. S., Gonzales, E. J., Gregory, K. D., Smith, T. A., Chrostowski, S. J., Garden, R. A., & O'Connor, K. M. (2000). *TIMSS 1999 international science report.* Chestnut Hill, MA: International Study Center, Boston College.

Mullis I. V. S., Martin, M. O., Beaton, A., Gonzales, E. J., Kelly, D. A., & Smith, T. A. (1998). *Mathematics and science achievement in the final year of secondary school. IEA's Third International Mathematics and Science Study (TIMSS).* Chestnut Hill, MA: International Study Center, Boston College.

Mullis I. V. S., Martin, M. O., Fierros, E. G., Goldberg, A. L., & Stemler, S. E. (2000). *Gender differences in achievement. IEA's Third International Mathematics and Science Study (TIMSS).* Chestnut Hill, MA: International Study Center, Boston College.

Mullis, I. V. S., Martin, M. O., Gonzales, E. J., Gregory, K. D., Garden, R. A., O'Connor, K. M, Chrostowski, S. J., & Smith, T. A. (2000). *TIMSS 1999 international mathematics report.* Chestnut Hill, MA: International Study Center, Boston College.

Parker, L. H., Rennie, L. J., & Fraser, B. J. (Eds.). (1996). *Gender, science and mathematics. Shortening the shadow.* Dordrecht: Kluwer Academic Publishers.

Postlethwaite T. N. & Wiley, D. E. (1992). *The IEA study of science II: Science achievement in twenty-three countries.* Pergamon Press.

Sjøberg S. (2000). Interesting all children in "science for all". In Miller, R., Leach J., & Osborne J. (Eds.), *Improving science education. The contribution of research.* Buckingham: Open University Press.

Skolverket (1998). *TIMSS. Kunnskaper i matematik och naturvetenskap hos svenska elever i gymnasieskolans avgångsklasser. (Swedish national TIMSS Population 3 report, in Swedish).* Stockholm: Skolverket.

UNDP (2000). *Human development report 2000.* New York and London: Oxford University Press.

Wilson, M. (Ed.). (1992). *Options for girls. A door to the future. An anthology of science and mathematics education.* Austin, TX: Pro Ed.

Zabulionis, A. (1997). Student achievement. In Vári, P. (Ed.), *Are we similar in math and science? A study of grade 8 in nine central and eastern European countries.* Hungary: Országos Közoktatási Intézet.

4 FOCUS ON CROSS-CURRICULAR ISSUES

Chapter 14

SEPARATING SCHOOL, CLASSROOM, AND STUDENT VARIANCES AND THEIR RELATIONSHIP TO SOCIO-ECONOMIC STATUS[1]

Albert E. Beaton and Laura M. O'Dwyer

International comparative studies of educational outcomes serve many purposes, including permitting the examination of the types of classrooms in which students around the world learn. The primary purpose of this study was to explore the ways in which students were grouped into classrooms in the countries that participated in TIMSS. The fact that countries differ substantially in terms of mathematics performance has already been reported (e.g., Beaton, Mullis, Martin, Gonzalez, Kelly, & Smith, 1996). This paper addresses the homogeneity of mathematics achievement in Grade 8 classrooms. Student variance on the TIMSS middle school mathematics test was partitioned into its among-school component, between-classroom within-school component, and a within-classroom component. This research is concerned with the degree to which students in different countries are tracked or streamed into their classrooms. Since other studies have shown that school and classroom performances are related to socio-economic status (SES), we will also investigate the relationship between variance components and SES indicators.

It is important to note that we are trying to measure the degree of homogeneity in the participating countries, not to take a position for or against tracking or streaming. From the TIMSS data, we can estimate the degree of homogeneity, not its causes. That homogeneity may be a result of school actions such as an attempt to group students according to their perceived ability; it may also result from the students or their parents selecting schools or courses that are more rigorous. There are many other possible reasons and combinations of reasons for the similarity of students within classrooms. We can, to a degree, establish how differently homogeneous and heterogeneous classrooms perform.

Partitioning the variance has an important technical purpose in addition to studying international differences in performance. These partitions represent upper bounds for the amount of variance different types of variables at different levels of

D.F. Robitaille and A.E. Beaton (eds.), Secondary Analysis of the TIMSS Data, 211–231.
© 2002 *Kluwer Academic Publishers. Printed in the Netherlands.*

aggregation can "explain." A variable that has the same value for all students within a school—such as whether his or her school is in an urban or rural community or whether the school has a library—cannot help explain individual differences within that school; likewise, a teacher variable such as a pedagogical belief that is assigned to all the students within a classroom, cannot explain student differences within a classroom. As we will see below, the possibility of getting high correlations between school or teacher variables and student performance is severely limited by the size of the corresponding variance component.

A note on "explaining" variance is in order. We realize that this cross-sectional study can generate only correlations with, not causes of, student performance. The pitfalls of over-interpreting correlations are many and perilous. Some of these will be discussed below. And yet, as researchers we are looking for explanations even though we realize the inconclusiveness of our work. We will sometimes use the word "explain" here to avoid awkwardness of language and we trust that the reader will understand its limitations.

Using data from an assessment such as TIMSS, the process of partitioning of variance and interpreting it is not simple and involves a number of challenges. First, as a cross-sectional study, we are very limited in interpreting causation. The introduction of Hierarchical Linear Models (HLM) (Bryk & Raudenbush, 1992; Bryk, Raudenbush, & Congdon, 1996) has greatly improved the technology for partitioning variance but leaves problems such as negative variance estimates and the removal of missing data. Decisions about technical issues will affect statistical estimates and thus the interpretation of these data and so will be discussed here.

From our experience in developing these statistical estimates, we put forward in the final section of this paper some suggestions about how future assessments might be improved.

BACKGROUND

Partitioning of variance in educational research has been used extensively for many years. Perhaps its most famous and influential use was in the Equality of Educational Opportunity Survey (EOS) (Coleman, Campbell, Hobson, McPartland, Mood, Weinfeld, & York, 1966) in the United States. Its survey design sampled schools and tested all students within selected grades (1, 3, 6, 9, and 12). The classroom data were not separately identified and so classroom variance could not be estimated. The teachers within a school were surveyed but not associated with individual classrooms or students. As a result of this design, the EOS reported only the among-school and within-school variances. In 1966 in the United States, the percent of the total variance on the EOS mathematics test associated with schools was 15.74 percent in the sixth grade and 9.00 percent at the ninth (EOS, p. 295).

The authors of the report were not blind to the limitations of variance partitioning, and it is instructive to review their comments here:

Possible sources of school-to-school variation in achievement are—

1. Differences from one school to another in school factors affecting achievement.
2. Differences from one community to another in family backgrounds of individual[s] including abilities of students.
3. Differences from one community to another in the influences outside school, apart from the student's own family.

Possible sources of within-school variation in achievement are—

1. Differences in pupils' abilities in the same school.
2. Differences in family backgrounds of the pupils in the same school.
3. Differences in school experience among students within the same school (i.e., different teachers, different courses, etc.)
4. Different influences within the community on different students toward achievement (such as community attitudes which may discourage high achievement among lower status children, and encourage it among higher status children).

These two lists indicate that the finding of school-to-school variation in achievement is neither a necessary nor a sufficient basis for inferring the effect of school factors; other factors can cause variations among schools and school factors can cause variations within a school. (EOS, p. 295)

The original TIMSS design was expected to allow for partitioning the variance into three parts: among-schools, between-classrooms within-schools, and within-classrooms. The middle school population, Population 2, was defined as the two adjacent grades that contained the largest number of 13-year olds. In most countries, the adjacent grades were the seventh and eighth grades.

The international sampling design called for the random selection of schools, then randomly selecting two intact mathematics classrooms within those schools at the appropriate grades, and finally all of the students within those classrooms were to be assessed. The mathematics teachers of the classrooms were asked to respond to a questionnaire about their backgrounds, attitudes, beliefs, and pedagogical policies. Science teachers were also asked to fill out questionnaires; but, in most countries, the science students were in many different classrooms and so intact science classrooms were not assessed.

Due to the heavy demands on human and financial resources, not all countries were able to sample more than one classroom per target grade. Without a second classroom within a school it is not possible to examine the between-classroom within-school variance component directly. Also, some countries did not ask all of the questions needed for SES analysis and many teachers in some countries failed to return their questionnaires or to respond to all of its questions.

Performing the variance partitioning therefore required some improvisation. The technical decisions are discussed briefly here and in more detail in other TIMSS publications.

DATA ANALYSIS

The data analysis for this research involved partitioning the variance of mathematics achievement of eighth graders using three different models:

- an unconditional model that uses no information except the first mathematics plausible value and the identification of a student's school and classroom;

- a model that partitions the unconditional variance components into components associated with commonly used SES variables;

- an expanded model that partitions the unconditional variance further using SES data, information about academic press, attitudes toward mathematics learning, and classroom climate.

The models presented in this research are referred to as fixed coefficient models, that is, only the intercepts are modeled at the classroom and school levels and the random effects at the two macro levels are set to zero. Mean mathematics achievement at the individual level is allowed to vary between classrooms, and mean achievement at the classroom level is allowed to vary among schools. Therefore, mean mathematics achievement varies between classes due to classroom effects and among schools due to school effects. This means that the between-classroom within-school, and among-school effects are only modeled by the differences in the intercepts.

For the two conditional models, the student and classroom level predictors are group centered. That is, the student level predictors are centered about the classroom mean and the classroom level predictors are centered about the school mean. The school level predictors are entered into the model uncentered. Without group mean centering at the student and classroom levels, the variance within and between classrooms, and among schools would be confounded. When the variables are expressed as deviations from the group mean in the model, they are uncorrelated with all higher-level variables (Kreft & de Leeuw, 1998). With group mean centering, the coefficients are estimates of the within-classroom, between-classroom within-school, and among-school effects (Bryk & Raudenbush, 1992).

The first model measures the homogeneity of schools and classrooms in terms of mathematics achievement among the participating countries. The second model assesses how much of this variance is associated with commonly used SES indicators. The expanded model includes more information about student attitudes, pressures, and mathematics classroom climate.

Performing the data analyses involved selecting different educational systems, selecting variables and composites for analysis, and selecting computing algorithms. Forty-one education systems, typically countries, participated in TIMSS at the eighth grade level. Only four of these countries (Australia, Cyprus, Sweden, and the United States) sampled two classrooms in each school that had multiple classes and therefore had the basic data for the intended three level variance partitioning. Building on a conjecture by Hutchison (Personal Communication, June 1999),

O'Dwyer (see Chapter 22 in this volume) explored the possibility of using simulated classrooms to partition the total variance into its within-classroom, between-classroom and among-school components in the remaining 37 educational systems participating in the study.

The application of the simulated classroom method required great care in the selection of the countries to be analyzed. To be included, a country's data had to meet several tests. First, at a minimum, the country must have met the sampling requirements outlined by TIMSS in the *Technical Report, Volume 1: Design and Development* (Foy, Rust, & Schleicher, 1996). Nine of the forty-one Population 2 countries were excluded because they failed to meet the sampling participation rate guidelines, the age requirements, or failed to use the approved sampling procedures at the classroom level.

O'Dwyer (2000 & Chapter 22 in this volume) details the sample and variance structure requirements necessary for the valid application of the pseudo-classroom procedure. According to the research presented, the two-level unconditional variance structure at the adjacent grades must be sufficiently similar such that the grades can be combined to simulate a two-classroom-per-grade sample design. In addition to this, the valid application of the pseudo-classroom procedure requires that at least one intact classroom was sampled in both adjacent grades in the same school. In order to meet these requirements a further nine countries were eliminated from this research.

Finally, each country must have required students to answer background questions pertinent to the research addressed here. Three countries were excluded because they failed to ask the relevant questions. In the end, the data from the three countries that had two intact classrooms and fifteen that met the criteria for simulation were included in the analyses presented in this chapter. The complete list of countries and their dispositions is shown in Table 1. Short and concise descriptions of most of the TIMSS countries and their school systems are available in Robitaille (1997) and Robitaille, Beaton, and Plomp (2000).

The TIMSS database contains a large number of variables that were carefully screened and, in many cases, combined into composite variables. This process is documented in Raczek (2001). The variables that were finally selected for these analyses are shown in Table 2. These variables were compiled from the student data and were aggregated to the classroom and school levels for analysis. The variables are displayed in four categories:

- SES Indicators

- Academic Pressure Indicators

- Attitudes toward Mathematics Learning Indicators

- Mathematics Classroom Climate Indicators

Table 1. Suitability of participating countries for variance analyses

Country	Adherence to TIMSS sampling and age criteria at upper or lower grades (criteria not met)	Suitability according to O'Dwyer's pseudo-classroom procedure	Availability of relevant variables
Australia	No (participation rates)	—	—
Austria	No (participation rates)	—	—
Belgium (Fl)	Yes	Yes	No
Belgium (Fr)	No (participation rates)	—	—
Canada*	Yes	Yes	Yes
Colombia*	Yes	Yes	Yes
Cyprus*	Yes	Yes	Yes
Czech Republic*	Yes	Yes	Yes
Denmark	No (sampling procedures)	—	—
France*	Yes	Yes	Yes
England	Yes	No [1]	—
Germany*	Yes	Yes	Yes
Greece	No (sampling procedures)	—	—
Hong Kong*	Yes	Yes	Yes
Hungary*	Yes	Yes	Yes
Iceland	Yes	Yes	No
Iran, Islam. Rep.*	Yes	Yes	Yes
Ireland	Yes	No [2]	—
Israel	Yes	No [3]	—
Japan	Yes	Yes	No
Korea*	Yes	Yes	Yes
Kuwait	Yes	No [3]	—
Latvia*	Yes	Yes	Yes
Lithuania*	Yes	Yes	Yes
Netherlands	No (participation rates)	—	—
New Zealand	Yes	No [2]	—
Norway	Yes	No [4]	—
Philippines	Yes	No [4]	—
Portugal	Yes	No	—
Romania*	Yes	Yes	Yes
Russian Fed.	Yes	Yes	No
Scotland	No (participation rates)	—	—
Singapore	Yes	No [2]	—
Slovak Republic*	Yes	Yes	Yes
Slovenia*	Yes	Yes	Yes
South Africa	No (sampling procedures)	—	—
Spain*	Yes	Yes	Yes
Sweden*	Yes	Yes	Yes
Switzerland	Yes	No [5]	—
Thailand	No (sampling procedures)	—	—
United States*	Yes	Yes	Yes

* Included in these analyses.
[1] Sampled within classrooms
[2] Differences in 2-level variance structure
[3] Sampled only one grade
[4] Adjacent grades were in different schools
[5] Approximately half the schools missing adjacent grades

Table 2. Description of model variables

Category	Variable Name	Variable Description & Correlation Information *
SES Indicators	MFEDUC/ CLPREDUC/ SHPREDUC	**Parental Education (2 Items)** Mean: highest education level of mother (BSBGEDUM) and father (BSBGEDUF). There is a positive correlation between this composite & math achievement in all 18 countries
	BOTHPAR/ CLBTHPAR/ SHBTHPAR	**Presence of parents in the home (2 Items)** Sum: presence of mother (BSBGADU1) and of father (BSBGADU2). There is a positive correlation between this composite & mathematics achievement in all 18 countries
	TOTPOSS1/ CLTOTPOS/ SHTOTPOS	**Possessions in the home (4 Items)** Sum: Calculator (BSBGPS01), computer (BSBGPS02), study desk (BSBGPS03), and dictionary (BSBGPS04). There is a positive correlation between this composite & mathematics achievement in all 18 countries
	BSBGBOOK/ CLBGBOOK/ SHBGBOOK	**Number of books in the home.** There is a positive correlation between this variable & mathematics achievement in all 18 countries
	BSBGHOME/ CLBGHOME/ SHBGHOME	**Number of people living in the home** (>25 recoded to missing). There is a negative correlation between this variable & mathematics achievement in all 18 countries
	WORK/ CLMNWORK/ SHMNWORK	**Out-of-school work time (2 Items)** Mean: Time at paid job outside of school (BSBGPAID), time doing jobs at home (BSBGDAY4). There is a negative correlation between this variable & mathematics achievement in all 18 countries.
Academic Pressure	MATPRES/ CLMTPRES/ SHMTPRES	**Maternal Pressure (3 Items)** Mean: My mother thinks it is important for me to do well in math (BSBMMIP2), to do well in <language of test> (BSBGMIP3), to be in a high achieving class (BSBGMIP6).
	SELPRES/ CLSLPRES/ SHSLPRES	**Self Pressure (3 Items)** Mean: I think it is important for me to do well in math at school (BSBMSIP2), to do well in <language of test> (BSBGSIP3), to be in a high achieving class (BSBGSIP6).
	PEERPRES/ CLPRPRES/ SHPRPRES	**Peer Pressure (3 Items)** Mean: My friends think it is important for me to do well in math at school (BSBMFIP2), to do well in <language of test> (BSBGFIP3); to be in high achieving class (BSBGFIP6).
Student Attitudes Toward Math	LIKEMAT3/ CLLKMAT3/ SHLKMAT3	**Liking for mathematics (5 Items)** Mean: How much you like math (BSBMLIKE), enjoy learning math (BSBMENJY), think math is boring (BSBMBORE), think math is an easy subject (BSBMEASY), think would like a job that involved math (BSBMWORK).
	MATHIMP4/ CLMTIMP4/ SHMTIMP4	**Perceived importance of mathematics (5 Items)** Mean: I need to do well in math to get desired job (BSBGJOB), to get into school I prefer (BSBMSCHL), to please self (BSBMSELF), to please parents (BSBMPRNT), math is important to everyone's life (BSBMLIFE).
Class Climate	CLIMATE/ CLCLIMAT/ SHCLIMAT	**Mathematics classroom climate (3 Items)** Mean: In my math class, students often neglect their work (BSBSCLS1), students are orderly and quiet during lessons (BSBMCLS2), students do as teacher says (BSBMCLS3).

* Items reversed as needed so that higher values indicate presence of the attribute. Composites are missing if all items missing.

The SES measures are several commonly used indicators of socio-economic status. Since the indicators contain little information about wealth and none about professional standing, they might better be called socio-educational status indicators. The remaining variables were included to strengthen the SES model with other student variables.

The HLM software of Bryk, Raudenbush, and Congdon (1992) was used to analyze the data in this research. The three level model was used to separate among-school, between-classroom within-school, and within-classroom variances. Since plausible value manipulation is not available for the three level models in HLM, the first plausible value for mathematics achievement was analyzed. Since the HLM program does not allow missing data at levels two and three, the sample sizes diminished accordingly. For comparability across models, if a variable was missing for any model it was removed from the analyses for all three models. The detailed models and their parameters are shown in Table 3.

Table 3. Description of multilevel models

Model	Level 1 Model	Level 2 Model	Level 3 Model
Unconditional	Y = P0 + E	P0 = B00 + R0	B00 = G000 + U0
SES Model	Y = P0 +	P0 = B00 +	B00 = G000 +
	P1*(MFEDUC) +	B01*(CLPREDUC) +	G001(SHPREDUC) +
	P2*(BOTHPAR) +	B02*(CLBTHPAR) +	G002(SHBTHPAR) +
	P3*(TOTPOSS1) +	B03*(CLTOTPOS) +	G003(SHTOTPOS) +
	P4*(BSBGBOOK) +	B04*(CLBGBOOK) +	G004(SHBGBOOK) +
	P5*(BSBGHOME) +	B05*(CLBGHOME) +	G005(SHBGHOME) +
	P6*(WORK) +	B06*(CLMNWORK)+	G006(SHMNWORK) +
	E	R0	U0
Expanded	Y = P0 +	P0 = B00 +	B00 = G000 +
Model	P1*(MFEDUC) +	B01*(CLPREDUC) +	G001(SHPREDUC) +
	P2*(BOTHPAR) +	B02*(CLBTHPAR) +	G002(SHBTHPAR) +
	P3*(TOTPOSS1) +	B03*(CLTOTPOS) +	G003(SHTOTPOS) +
	P4*(BSBGBOOK) +	B04*(CLBGBOOK) +	G004(SHBGBOOK) +
	P5*(BSBGHOME) +	B05*(CLBGHOME) +	G005(SHBGHOME) +
	P6*(MATPRES) +	B06*(CLMTPRES) +	G006(SHMTPRES) +
	P7*(SELPRES) +	B07*(CLSLPRES) +	G007(SHSLPRES) +
	P8*(PEERPRES) +	B08*(CLPRPRES) +	G008(SHPRPRES) +
	P9*(WORK) +	B09*(CLMNWORK) +	G009(SHMNWORK) +
	P10*(LIKEMAT3) +	B010*(CLLKMAT3) +	G0010(SHLKMAT3) +
	P11*(MATHIMP4) +	B011*(CLMTIMP4) +	G0011(SHMTIMP4) +
	P12*(CLIMATE) + E	B012*(CLCLIMAT) +	G0012(SHCLIMAT) +
		R0	U0

RESULTS

National Means and Standard Deviations

Table 4 shows the mathematics means and standard deviations for students in the upper year of TIMSS Population 2. In most countries, the upper grade was Grade 8.

The statistics are reported by country. These statistics were taken directly from the TIMSS middle school mathematics report (Beaton et al., 1996).

Table 4. Means and standard deviations of mathematics achievement, Pop. 2 (upper), 1995

Country	Mean	Standard Deviation
Australia [4]	530	98
Austria [4]	539	92
Belgium (Fl) [4]	565	92
Belgium (Fr) [4]	526	86
Bulgaria [4]	540	110
Canada [2]	527	86
Colombia [2]	385	64
Cyprus [1]	474	88
Czech Republic [2]	564	94
Denmark [4]	502	84
England [4]	506	93
France [2]	538	76
Germany [2]	509	90
Greece [4]	484	88
Hong Kong [2]	588	101
Hungary [2]	537	93
Iceland [3]	487	76
Iran, Islamic Rep. [2]	428	59
Ireland [4]	527	93
Israel [4]	522	92
Japan [3]	605	102
Korea [2]	607	109
Kuwait [4]	392	58
Latvia (LSS) [2]	493	82
Lithuania [2]	477	80
Netherlands [4]	541	89
New Zealand [4]	508	90
Norway [4]	503	84
Portugal [4]	454	64
Romania [2]	482	89
Russian Federation [3]	535	92
Scotland [4]	498	87
Singapore [4]	643	88
Slovak Republic [2]	547	92
Slovenia [2]	541	88
South Africa [4]	354	65
Spain [2]	487	73
Sweden [1]	519	85
Switzerland [4]	545	88
Thailand [4]	522	86
United States [1]	500	91

[1] Country sampled two eighth grade classrooms
[2] Country passed test for simulation
[3] Country did not collect necessary data
[4] Country did not pass simulation tests or had problematic sample

A striking thing about the means and standard deviations is that they are highly correlated. The correlation between the mathematics means and standard deviations for all 41 countries is 0.791. The correlation among the 18 countries included in this study is 0.845. These correlations are unusually high for educational indicators. A scattergram showing the relationship between means and standard deviations of mathematics achievement is shown in Figure 1.

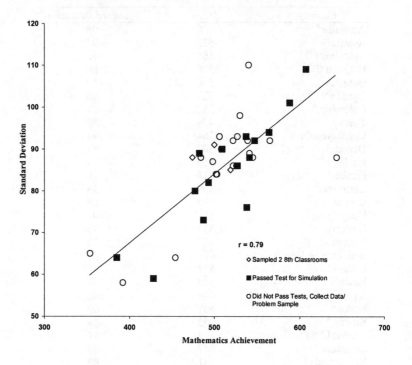

Figure 1. Standard deviations by mean mathematics achievement, Population 2, TIMSS 1995.

The large standard deviations for high-scoring countries indicate that there are quite a few students in these countries who did not do well on the TIMSS mathematics test. The small standard deviations for low-scoring countries may indicate that the test was too difficult for even their better students and was not an optimum measuring instrument for them. This position is supported by the lowest scoring countries having the lowest Cronbach Alpha reliabilities (Beaton et al., 1996). We do not have a satisfactory explanation for this correlation at this time. It is worth noting, however, that the challenge of making a test that covers the ranges of ability in all countries is very difficult and thus a test like the TIMSS test is likely to result in different variances at different levels of its scale.

Unconditional Model

Table 5 shows the unconditional partitioning of the total mathematics variance into among-school, between-classrooms within-schools, and within-classroom components. The table entries are the percentages of the total variance for each source. These percentages add up to 100 percent.

Table 5. Partitions of variance, mathematics achievement: Unconditional model

Country	Total Available Variance	Percent of Variance		
		Among Schools	Between Classrooms	Within Classrooms
Canada	6586.24	10.99	10.08	78.93
Colombia	3333.01	26.85	1.20	71.95
Cyprus	7178.83	5.43	3.04	91.53
Czech Republic	8103.24	10.96	10.81	78.23
France	5462.02	11.96	13.13	74.91
Germany	7156.72	38.15	1.85	59.99
Hong Kong	9749.80	39.69	7.42	52.89
Hungary	8244.56	14.86	4.51	80.63
Iran, Islamic Rep.	3703.73	9.98	2.18	87.84
Korea	11055.55	5.73	0.23	94.03
Latvia (LSS)	6434.27	9.70	5.94	84.36
Lithuania	6173.52	6.34	17.24	76.42
Romania	7502.46	22.02	10.31	67.66
Slovak Republic	7515.27	7.37	9.46	83.17
Slovenia	7480.83	5.75	3.83	90.42
Spain	5203.06	12.90	4.46	82.64
Sweden	5960.19	1.49	31.05	67.47
United States	7040.61	15.42	35.25	49.33

NOTE: Models weighted with HOUWGT; all countries are adjusted using pseudo-classroom procedure except for Cyprus, Sweden, and the United States. HOUWGT is the student sampling weight scaled so that the sum of the weights is equal to the sample size.

For the countries that did not sample two eighth grade mathematics classrooms within a school, the construction of this variance and its components relies on the technical innovation that has been thoroughly developed and documented by O'Dwyer (Chapter 22 in this volume). After carefully examining the data from each country, the scores of the students in the lower (usually seventh) grade were adjusted to simulate the performance of a second upper grade classroom. Knowledge about a school system, the properties of its sample, and the statistical properties of its test scores inform the judgment about whether or not the simulated data are a reasonable estimate of the data from a second eighth grade classroom. As mentioned above, fifteen countries were accepted for inclusion in this report and their simulated data were analyzed along with the data from the three countries in this report that had sampled a second eighth grade classroom.

The first column in this table is the total variance for each country as defined by the HLM algorithm. For a number of reasons, this column is not simply the square of the national standard deviations shown in the previous table. First, HLM requires that all observations have complete data, and that is not the case here. To make the samples the same for the different models, we have excluded all cases that have missing data for any of the models. Second, the samples in this research have been modified using the pseudo-classroom procedure in which mean classroom achievement was adjusted to simulate achievement at the adjacent grade, but the variance of the adjusted classroom remains unchanged. The result is that the variances in our HLM samples are smaller but the correlation between the published standard deviations and the square root of the HLM variance is very high, $r = 0.980$.

Unconditional Among-School Variance

The unconditional among-school variance is an indicator of the degree to which schools within a country differ. The percentage of among-school variance is very small in Sweden (1.49 percent) and around 5 or 6 percent in Cyprus, Korea, Slovenia, and Lithuania. These small percentages indicate that the average mathematics scores are about the same in the various schools within the countries. The percentages of among-school variance are highest in Hong Kong (39.69 percent) and Germany (38.15 percent) where students are streamed into separate schools by the eighth grade. The large percentage in Germany is anticipated by their policy of "offering differentiated teaching in accordance with student ability, talent, and inclination" (Riquarts, 1997, p.141). Beyond age 10 students participate in the *Hauptschule, Realschule* or *Gymnasium* system of education, within which there is no streaming. Although Hong Kong does not have an official policy regarding tracking (Leung & Law, 1997), Table 5 shows that tracking among schools is as common in Hong Kong as it is in Germany.

Unconditional Between-Classroom Within-School Variance

The unconditional between-classroom variance is an indicator of the degree to which average mathematics scores differ by classrooms within schools. The between-classroom variance in Korea is amazingly small (0.23 percent) indicating a conscious effort to equalize the performance of the classrooms. Eleven countries have percentages of between-classroom variances less than 10 percent. The largest between-classroom variance percentages are in the United States (36.25 percent) with Sweden close behind at 31.05 percent. No other country has more than 20 percent of its variance between classes.

Within-classroom Variance

The percentage of total variance within classrooms indicates how alike students within classrooms are: the smaller the percentage, the more homogenous the students in the classrooms. Larger percentages mean that students are taught mathematics in mixed ability classrooms. The only country that has a within-class percentage of variance less than 50 percent is the United States (49.33 percent) although Hong Kong and Germany are close behind with percentages in the 50s. In Korea, 94.03 percent of the variance is within classrooms and both Slovenia and

Cyprus have over 90 percent of their variances within classrooms. In Korea and Cyprus, no official policy exists on within school tracking, there is no tracking between schools, and students are taught in mixed ability classrooms (Kim, 1997; Papanastasiou, 1997). These policies are consistent with the large percentage of variance within classrooms in these countries. Likewise, Slovenia does not have an official policy regarding tracking students into different classrooms or schools, although some schools practice limited tracking (Setinc, 1997).

Classroom Tracking Index

To explore the effect and efficiency of the within school separation of students, we have explored a tracking index that measures the degree to which classrooms within a school differ.

Let us say that a school needs to assign students to two mathematics classrooms and has available their scores on a mathematics test. There are many ways that this could be done. To maximize the difference between the two classrooms, one would sort the students in order by their mathematics scores and then place the top half of the scorers into one classroom and the bottom half in the other. To make the two classrooms as equal as possible, one would sort the students and select them from adjacent pairs with one student placed in one classroom and the other in other classroom. Another way to form classrooms would be to assign the students willy-nilly into the classrooms using a random mechanism.

To view how students within a school are actually distributed, we have developed a tracking index. Each possible assignment of students to two classrooms would generate a mean for each classroom and thus the difference between the two averages can be calculated. The tracking index compares the mean difference of the actual assignment of students within a school to the distribution of all possible ways in which the students could have been assigned. A low value, for example 10, would indicate that only 10 percent of the possible classroom assignments would have a mean difference in mathematics scores closer zero than the actual mean for that classroom and thus the means of the two classrooms are quite close. Conversely, a value of 90 indicates that the classroom assignments are more different than 90 percent of the possible assignments and therefore relatively different. The index would, on the average, be 50 if students were assigned randomly (Beaton, 2001).

The values of the tracking index for these countries are shown in Table 6. For the fifteen countries that met the criteria for inclusion in this study but did not have a second available classroom at the same grade, the tracking index was computed using data from the adjacent grade, and adjusted using O'Dwyer's pseudo-classroom procedure. In the three countries with two classrooms per grade, the tracking index was computed using achievement information from both classes. The first column in the table is the mean mathematics score for these countries. The next column is the average of the tracking index over all schools in the country with more than one classroom. The number of schools is shown next. Finally, there is the proportion of schools with an index within certain ranges.

Table 6. TIMSS tracking index

Country	Mean Math Score	Mean Tracking Index	Total N	Number (Percent) of Classrooms at Tracking Level									
				Very Low <0.1		Low 0.1 thru 0.3		Medium 0.3 thru 0.6		High 0.6 thru 0.9		Very High >0.9	
				N	%	N	%	N	%	N	%	N	%
Canada	527	0.63	351	20	6	50	14	76	22	105	30	100	28
Columbia	385	0.56	140	4	3	25	18	49	35	37	26	25	18
Czech Rep.	564	0.65	149	9	6	20	13	29	19	41	28	50	34
Cyprus	474	0.65	55	1	2	7	13	14	25	16	29	17	31
Germany	509	0.61	130	10	8	15	12	34	26	38	29	33	25
Spain	487	0.59	152	14	9	20	13	40	26	44	29	34	22
France	538	0.67	119	11	9	8	7	24	20	34	29	42	35
Hong Kong	588	0.70	85	8	9	8	9	15	18	16	19	38	45
Hungary	537	0.60	149	7	5	24	16	38	26	48	32	32	21
Iran, Isl. Rep.	428	0.53	191	18	9	37	19	51	27	56	29	29	15
Korea	607	0.50	150	15	10	31	21	41	27	49	33	14	9
Lithuania	477	0.68	145	11	8	14	10	29	20	37	26	54	37
Latvia (LSS)	493	0.61	140	10	7	18	13	36	26	38	27	38	27
Romania	482	0.70	162	6	4	16	10	37	23	34	21	69	43
Slovak Rep.	547	0.69	145	9	6	11	8	29	20	39	27	57	39
Slovenia	541	0.62	121	8	7	12	10	32	26	38	31	31	26
Sweden	519	0.89	115	0	0	2	2	10	9	24	21	79	69
United States	500	0.89	179	0	0	5	3	17	9	21	12	136	76

The average index within these countries ranges from 0.50 in Korea to 0.89 in Sweden and the United States. That is, on the average Korea uses the equivalent of random assignment but the details suggest that 10 percent of the Korean schools attempt to equalize their classes. In the United States, on the other hand, 76 percent of the schools have a very high index, which suggests that the classrooms are separated by their mathematical skills. None of the sampled schools in the United States sample were close to having classrooms with equal means.

The tracking index is a way of measuring and understanding the amount of tracking within schools but it does not necessarily suggest that schools that track or do not track are more successful in teaching their students. For these TIMSS countries, the correlation between the mean mathematics score and the average tracking index is less than 0.10, which is not a strong relationship. However, this small correlation does not support the hypothesis that tracking actually hurts the lower performing students nor does it suggest that tracking in fact helps these students.

Socio-Economic Status Model

The question addressed here is the degree to which a few socio-educational variables are associated with mathematics achievement. The six variables presented in Table 2 were included in the SES model. These student variables are aggregated for use in the school and classroom models. The results are as follows:

Total School Variation in Mathematics Performance Associated with SES

The total variance associated with the SES indicators is shown in Table 7. This is the total over all three levels, each added in proportion to its size. SES indicators are associated with less than 10 percent of the total variance in Korea, the Slovak Republic, Cyprus, and Canada.

Table 7. Partitions of variance, mathematics achievement: SES model

		Percent of Variance		
Country	*Total Percent of Variance Modeled*	*Among Schools*	*Between Classrooms*	*Within Classrooms*
Canada	9.88	26.19	26.74	5.46
Colombia	13.95	47.16	28.44	1.31
Cyprus	9.45	58.83	61.67	4.79
Czech Republic	27.96	99.23	85.50	10.02
France	14.52	70.73	30.76	2.70
Germany	24.46	54.92	35.67	4.74
Hong Kong	25.23	51.18	50.01	2.27
Hungary	14.27	82.39	38.53	0.36
Iran, Islamic Rep.	16.96	83.19	3.16	9.77
Korea	5.00	40.12	1.72	2.86
Latvia (LSS)	17.76	87.95	-18.14	12.22
Lithuania	17.81	79.76	47.56	5.96
Romania	15.99	37.65	27.44	7.19
Slovak Republic	9.36	24.24	31.83	5.48
Slovenia	10.91	35.65	53.36	7.54
Spain	20.88	80.72	16.80	11.75
Sweden	21.46	71.41	55.98	4.47
United States	35.11	75.39	61.70	3.52

NOTE: Models weighted with HOUWGT; all countries are adjusted using pseudo-classroom procedure except for Cyprus, Sweden, and the United States.

The association is highest in the United States (35 percent) with the Czech Republic (28 percent) Hong Kong (25 percent), and Germany (24 percent) trailing somewhat behind. The percentage of variance at each level does not add to 100 percent; the percentage of variance in each column is the percentage explained at each specific level in the hierarchy. For example, in Canada, 26.19 percent of the available variance among schools is explained by the model. These results may be related to the unconditional results in Table 5. For example, again in Canada, 26.19

percent of the available 10.99 percent among schools, plus 26.74 percent of the available 10.08 percent between classrooms, plus 5.46 percent of the available 78.93 percent within classrooms make up the total percent of variance explained by the model, 9.88 percent.

Among-school Variance Associated with SES

Although SES indicators are associated to some degree with the among-school variance in all countries, the range of associated variance is startling. Those results are also shown in Table 7. At the low end, the Slovak Republic percent of mathematics variance associated with SES is about 24 percent; at the high end, the percentage for the Czech Republic is over 99 percent. This result is especially surprising since the two countries have a cultural similarity including many years in which they were a single country. In both countries, however, the amount of among-school variance was small and thus the 99 percent is a large part of a very small variance component.

Seven countries had percentages over 75 percent, which means that searching for school variables that "explain" performance in mathematics will have a difficult time in finding variables that add substantially to the predictive power. Adding any variable to the SES variables (e.g., the size of a school's library) can only explain the residual variance, and in the Czech Republic there is little variance to explain. This is not to say that other factors do not help or hurt mathematics achievement but that their effect, if any, is confounded with the SES effect.

Between-Classroom Within-School Variance Associated with SES

First, let us address a logical peculiarity of HLM. The percentage of variance for Latvia is negative, −18, which is logically impossible. In principle, adding variables to a model should never increase the residual variance; if an added variable is completely useless it should be assigned a coefficient of zero and thus automatically removed from the model. But with HLM, such negative values may result from the computing algorithm. The reasons for this are discussed in Snijders and Bosker (1994). They note that "...it sometimes happens that adding explanatory variables increases rather than decreases some of the variance components. This leads, under the stated definition of R^2, to negative values for the contribution of this explanatory variable to R^2 sometimes even to negative values for R^2" (p 343). For practical purposes, we will consider negative values as zeroes.

The percent of between-classroom variance associated with SES ranges from very small in Latvia (0 percent), Korea (2 percent), and Iran (3 percent) to very large in the Czech Republic (86 percent). Over 60 percent of the variance in the United States and Cyprus is associated with the SES variables. The high percentages indicate that even within schools, mathematics classrooms are somewhat homogeneous in terms of SES. This fact may be as a result of the selection of mathematics courses by the students or their parents, the assignment of students to classrooms by the school system, or for some other reason.

Within-classroom Variance Associated with SES

The percent of within-classroom variance associated with SES indicators is also shown in Table 7. The percentages range from very small in Hungary and Colombia, which are close to zero, to around 12 percent in Latvia and Spain. The negative percentage of between-classroom variance for Latvia makes its within-classroom variance suspect. All in all, the SES indicators do not explain much of the within-classroom variance.

Table 8. Partitions of variance, mathematics achievement: Expanded model

		Percent of Variance		
Country	Total Percent of Variance Modeled	Among Schools	Between Classrooms	Within Classrooms
Canada	20.92	62.96	33.72	13.43
Colombia	16.91	49.73	89.98	3.45
Cyprus	18.17	51.48	75.80	14.28
Czech Republic	37.59	99.68	89.62	21.71
France	21.92	76.61	39.93	10.04
Germany	32.45	64.11	41.80	12.02
Hong Kong	28.54	45.61	74.10	9.32
Hungary	23.39	88.79	63.69	9.08
Iran, Islamic Rep.	24.31	82.85	12.25	17.96
Korea	11.61	44.81	43.30	9.50
Latvia (LSS)	29.93	89.49	74.13	19.97
Lithuania	26.89	67.46	59.84	16.09
Romania	21.21	17.53	64.36	15.83
Slovak Republic	13.20	32.47	41.49	8.28
Slovenia	22.92	52.20	65.73	19.25
Spain	28.48	85.80	20.23	19.97
Sweden	29.49	74.62	65.47	11.94
United States	40.63	76.15	68.69	9.49

NOTE: Models weighted with HOUWGT; all countries are adjusted using pseudo-classroom procedure except for Cyprus, Sweden, and the United States.

The Expanded Model

The results for the expanded model are shown in Table 8. As expected, the amount of explained variance in general rises as the additional student variables are added to the model. In some countries, the rise in total explained variance is substantial; the percent explained in Latvia and Slovenia increased by 12 percent and in Canada by 11 percent. In Canada, the increased percentage of among-school variance explained was 37 percent whereas in the United States the increase was less than 1 percent. In two countries, Lithuania and Romania, the HLM estimates of the percentage of among-school variance explained decreased substantially, which is problematic. The additional variables added substantially to the percentage of between-classroom variance explained in Colombia and Latvia. The within-classroom explained variance went up somewhat for all countries. The percentages in Table 8 are interpreted in the same way as those in Table 7.

DISCUSSION

The results of these analyses suggest or support some interesting and important findings about school systems around the world and in the United States in particular.

There is substantial variation in mathematics achievement among the countries that participated in TIMSS. The countries also differ in terms of their variances. What is surprising in TIMSS is that the correlations between national mean performances and their standard deviations are very large: between 0.791 and 0.862 depending on the included countries. This analysis reflects the relationship of the total variance to the national means.

The way the variance is partitioned around the world is quite different. In some countries, the among-school variation is between 1 percent and 6 percent. This indicates substantially equal mathematics performance among the schools within these countries. The among-school variation in some other countries went above 35 percent which results from having students tracked into different types of schools at Grade 8. The between-classroom-within-school variance also varies greatly, from nearly 0 in Korea to over 35 percent in the United States. Within Korea, an effort is made to keep classes equal, but students are separated into different classes in the United States. The within-classroom variance is the residual after the among-school and between-classroom variances are removed. Relatively speaking, the most homogeneous mathematics classrooms occurred in the United States where less than 50 percent of its variance was within classrooms; in Korea, the most heterogeneous country in this analysis, almost 95 percent of its variance lay within classrooms.

The differences in the partition of variances are due to many things including the differences among communities as well as school policies. As the data show, countries vary substantially in the way that students are assigned to schools and to classrooms.

Generally speaking, SES is correlated with student performance in all countries. The degree to which the total student variances are explained ranges from about 5 percent in Korea to 35 percent in the United States. Overall, the percent of total variance explained by SES related variables is not very large. However, when examined according to the type of variance explained, the picture changes. The amount of among-school variance that is explained ranges from around 24 percent in the Slovak Republic to 99 percent in the Czech Republic. Since school differences in the Czech Republic are already associated with SES variables, there is essentially no school variable that can add to the predictive power; any additional variable that is added must be correlated with the SES variables. The amount of between-classroom variance explained by SES ranges from about 0 percent in Latvia or 2 percent in Korea to 61 percent in Cyprus and the United States. Finally, the amount of within-classroom variance explained by SES varies from nearly 0 percent to around 12 percent. The residual within-classroom variance leaves plenty of variance to be explained. The expanded model gave essentially similar results.

The relationship between SES and mathematics is important for understanding the limitations of cross-sectional studies. Variables that are constant for a school can only explain the school-to-school variance, and school variance is already explained by common SES variables. Similarly, the variables that are constant for all students in a classroom can explain only the between-classroom variance. Since the among-school and between class variance are already explained by SES variables, there is little hope for finding new school or teacher variables that improve the already very high predictive power of SES.

The United States is an interesting case in point. The among-school variance is about 15 percent, the between-classroom is 36 percent, and the within-classroom is 49 percent. The SES variables explain about 75 percent of the available among-school variance, about 62 percent of the available between-classroom variance, and only about 4 percent of the available within-school variance. For one reason or another, SES is correlated with a student's location in a school and in a classroom within a school. However, once the students are assigned to a mathematics classroom, SES indicators cease to become a substantial predictor of mathematical competence. Mathematical success as measured by the TIMSS achievement tests appear to suggest that once the students are assigned to a classroom, having highly educated parents at home or having a computer in the home has little further to do with mathematical accomplishment.

SUGGESTIONS FOR FUTURE RESEARCH

Organizing an international study such as TIMSS is very difficult and challenging. The participating countries and sponsoring agencies bring to the table, different perspectives and expectations that must be addressed. They also bring different personnel, funding, and time constraints. Study features that pursue the research questions of some countries may be unimportant and unnecessarily costly for others. A research design that answers each country's questions and purposes will never exist; the implemented design must be a compromise among competing pressures. Maintaining technical excellence throughout the assessment is necessary to assure the credibility of the results.

The design of the TIMSS study had many interesting features. We especially liked the definitions of Populations 1 and 2 as the pair of grades that had the largest number of students at a particular age level and the formal requirement of sampling two intact classes at the each level, at least at the middle school level. The selection of two adjacent grades permitted an estimate of how much students grew on the mathematics test scale between the two years and made possible an intuitively simple standard for growth. Although this design feature was largely ignored, the selection of two classrooms per grade made the partitioning of variance a straightforward matter.

Where countries had not selected two classrooms per grade, the selection of classes in adjacent grades was fortunate for this study since it permitted the lower grade scores to be adjusted to simulate the properties of a second mathematics classroom. Although this technical innovation was very valuable here, it would have been much better to collect the appropriate data in the first place and not have to use simulation.

A very real problem with using assessment to view the effects of different policies and different cultures is the upper bound of the explainable variances. The only level at which there is a large portion of variance to be explained is within classrooms. Our models implicitly assumed that what a teacher does affects all students equally and thus cannot contribute to the within-classroom variation. An answer to this problem is to try to get more information about the interaction of teachers and individual students. This would move the analysis to the within-classroom level. In this way, the upper bounds of the teacher variables would lose importance as the knowledge of students was explored.

NOTES

[1] This study is based upon work supported by the National Science Foundation under Grant #REC-9815180. Any opinions, findings, and conclusions or recommendations expressed in this paper are those of the authors and do not necessarily reflect the views of the National Science Foundation.

REFERENCES

Beaton, A. E., Mullis, I. V., Martin, M. O., Gonzalez, E. J., Kelly, D. L., & Smith, T. A. (1996). *Mathematics achievement in the middle school years: IEA's Third International Mathematics and Science Study.* Chestnut Hill, MA: Center for the Study of Testing, Evaluation, and Educational Policy, Boston College.

Beaton, A. E. (2001). An index to measure within school tracking. In progress.

Bryk, A. S., & Raudenbush, S. W. (1992). *Hierarchical linear models: Applications and data analysis methods.* Newbury Park, CA: Sage

Bryk, A. S., Raudenbush, S. W., & Congdon, R. (1996). *Hierarchical linear and nonlinear modeling with the HLM/2 and HLM/3 programs.* Chicago: Scientific Software International, Inc.

Coleman, J. S., Campbell, E., Hobson, C., McPartland, J., Mood, A. M., Weinfeld, F., & York, R. (1966). *Equality of educational opportunity.* Washington, DC: U. S. Government Printing Office.

Foy, P., Rust, K., & Schleicher, A. (1996). Sample design. In M. O. Martin and D. L. Kelly (Eds.), *Third International Mathematics and Science Study. Technical report volume 1: Design and development.* Chestnut Hill, MA: Center for the Study of Testing, Evaluation, and Educational Policy, Boston College.

Kim, J. (1997) Republic of Korea. In D. F. Robitaille (Ed.), *National Contexts for Mathematics and Science Education. An encyclopedia of the education systems participating in TIMSS.* Vancouver, Canada: Pacific Educational Press.

Kreft, I., & de Leeuw, J. (1998). *Introducing multilevel modeling.* Thousand Oaks, CA: SAGE Publications, Inc.

Leung, F. K. S., & Law, N. W. Y. (1997) Hong Kong. In D. F. Robitaille (Ed.), *National Contexts for Mathematics and Science Education. An encyclopedia of the education systems participating in TIMSS.* Vancouver, Canada: Pacific Educational Press.

O'Dwyer, L. M. (2000). Extending the application of multi-level modeling to data from the Third International Mathematics and Science Study (Doctoral Dissertation, Boston College, 2000). UMI Dissertation Abstracts Number 9995932.

Papanastasiou, C. (1997) Cyprus. In D. F. Robitaille (Ed.), *National Contexts for Mathematics and Science Education. An encyclopedia of the education systems participating in TIMSS.* Vancouver, Canada: Pacific Educational Press.

Raczek, A. E. (2001). Comparing the reliability and validity of TIMSS contextual indicators across participating countries. Paper presented at the Annual Meeting of the American Educational Research Association, Seattle, WA.

Riquarts, K. (1997) Germany. In D. F. Robitaille (Ed.), *National Contexts for Mathematics and Science Education. An encyclopedia of the education systems participating in TIMSS.* Vancouver, Canada: Pacific Educational Press.

Robitaille D. F. (Ed.). (1997). *National Contexts for Mathematics and Science Education. An encyclopedia of the education systems participating in TIMSS.* Vancouver, Canada: Pacific Educational Press.

Robitaille, D. F., Beaton, A. E., & Plomp, T. (Eds.). (2000). *The impact of TIMSS on the teaching and learning of mathematics and science.* Vancouver, Canada: Pacific Educational Press.

Setinc, M. (1997) Slovenia. In D. F. Robitaille (Ed.), *National Contexts for Mathematics and Science Education. An encyclopedia of the education systems participating in TIMSS.* Vancouver, Canada: Pacific Educational Press.

Snijders, T. A. B. & Bosker, R. J. (1994). Modeled variance in two-level models. *Sociological Methods and Research, 22* (3), 342-363.

Chapter 15

ON THE RELATIONSHIP BETWEEN MATHEMATICS AND SCIENCE ACHIEVEMENT IN THE UNITED STATES[1]

Min Li, Richard J. Shavelson, Haggai Kupermintz, and
Maria Araceli Ruiz-Primo

Curiously, we have different curricular frameworks for mathematics and science, and studies of achievement in these domains have rarely asked the obvious question, "What is their relationship?" Research with data provided by TIMSS is no exception (e.g., Schmidt, McKnight, Valverde, Houang, & Wiley, 1997; Schmidt, Raizen, Britton, & Bianchi, 1997). Rather, attention has focused on achievement in either mathematics or science—perhaps in recognition that it is difficult enough to understand achievement in one domain, let alone two domains, especially as achievement links to reform in mathematics and science education. The purpose of this chapter, in broad terms, is to provide an initial exploration of the relationship between mathematics and science achievement for the United States' seventh and eighth graders (TIMSS Population 2) overall, and for girls and boys in particular. The chapter does not attempt to explain this relationship with factors such as curricular variation across nations or teaching methods within a nation; we leave that task to those who follow. Our goal in this chapter is descriptive.

BACKGROUND

Even though the relation between mathematics and science achievement is seldom studied, there are good reasons to do so, especially because even the casual observer can see the link. To motivate our study of the relationship between middle school students' mathematics and science achievement, we briefly sketch alternative views of the relationship between mathematics and science, and the relationship between mathematics and science curricula. We do so as a means of setting the context for our analyses and motivating the kinds of analyses we carry out with no expectation that there is one right view or that any view will ultimately get reflected simply in students' achievement test scores.

D.F. Robitaille and A.E. Beaton (eds.), Secondary Analysis of the TIMSS Data, 233–249.
© 2002 Kluwer Academic Publishers. Printed in the Netherlands.

The extreme view holds that science ultimately resolves into mathematics, namely mathematical modeling. According to Harpending (1985, p. 931):

> Language does not serve science very well as an analytic device.... J. B. S. Haldane said that if someone could not use something in an equation, then he did not really know what it meant, and Haldane's principle accounts for the universal use of mathematical notation in science.

A broader view of science, however, would recognize the heterogeneity of the sciences, their commonality in attempting to model the physical or social world, and the variability of their goals as to mathematically modeling the objects of their study. Moreover, this view would recognize the interrelations between mathematics, science, and other human endeavors.

This said, our focus is on the structural and functional relations between mathematics and the natural sciences. This relationship might be viewed symmetrically—whether mathematics is a tool essential to science or science is a stimulus for new mathematical discoveries as Newton did with the calculus.

The Project Physics curriculum, for example, recognized the symmetric relation between mathematics and science. According to Halton (1973, 1978) the interconnectedness among academic subjects should be reflected in the way students are taught science and mathematics. He suggested switching the traditional presentation of topics in physics from a narrow vertical column (Figure 1a) to a connected presentation in which a specific advance in physics was linked to earlier and later achievements, not only in physics but in other fields as well (Figure 1b). For example, Figure 1b characterizes links between earlier and later achievements in Newtonian dynamics, as applied to planetary motion, a subject that is usually one of the "beads" on the physics chain (Figure 1a). Newton had studied theology and philosophy, and those ideas emerged in the *Principia* in his section about the nature of time and space (see link A). Within physics itself, Newton brought to a culmination the work of Kepler and Galileo (link B). Newton used to a large extent the mathematics of the Greeks (link C) and repaid his debt to mathematics by enriching the field with his development of the calculus (link D). Within physics all who followed Newton will use his law and approach (link E). His effects on the Deist theologians (link F), on John Dalton's atomic models in chemistry (link G), and on the artistic sensibilities of the eighteenth century in which Newton swayed the muses (link H) are quite easily documented.

From this symmetric perspective, we might ask (and we do!), "What is the relationship between mathematics and science achievement?" A more refined question might ask, "What is the relationship of performance in fractions, algebra, geometry, data analysis, and measurement to performance in earth science, life science, physics and chemistry?"—the achievement dimensions measured by TIMSS. We then offer an alternative way to conceptualize the achievement distinctions in the TIMSS test, and explore the implication of this alternative conception on the relationships between mathematics and science

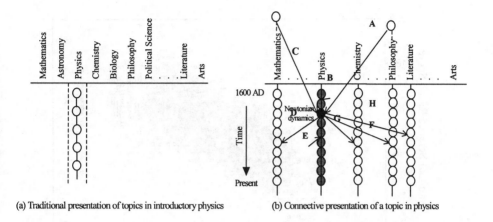

(a) Traditional presentation of topics in introductory physics   (b) Connective presentation of a topic in physics

Figure 1. Traditional (a) and connective (b) presentation of topics in introductory physics. (After Halton, 1973)

However, we might also view the relationship between mathematics and science as asymmetric, with mathematics as an essential tool of science. Biologist Paul Williams stated the view succinctly at a meeting on TIMSS: "Mathematics is part of science... a set of tools that helps scientists understand a marvelously complex world" (Paul Williams, U. S. TIMSS Policy Seminar, Miami, FL 2/12/2001).

The issue of the asymmetry between mathematics and science has been played out in science curricula, often with mathematics, much to the consternation of mathematicians, being relegated to the status of tool. For example, this asymmetric view was taken in the science curricula responding to Sputnik, the Physical Sciences Study Committee's high school physics curriculum. "PSSC used algebra and graphical substitutes for calculus (slopes, areas) as a way to analyze data and express physical laws. They used a fair amount more math than the typical U. S. physics text of that day and they used it more integrally with the text and diagrams and experiments in the lab" (Decker Walker, Personal Communication, 2/11/2001). Today we see mathematics as a tool in Advanced Placement physics and college physics for non-majors and majors in science and engineering.

RELATIONSHIP BETWEEN TIMSS MATHEMATICS AND SCIENCE ACHIEVEMENT SCORES

One version of this chapter is very short. We found a high correlation between mathematics and science achievement scores for the 10,973 middle-school students in the TIMSS U. S. sample: 0.74 (see Figure 2).[2] As can be seen from Figure 2, there is a ceiling effect on both the mathematics and science scores that may have attenuated this correlation a bit.

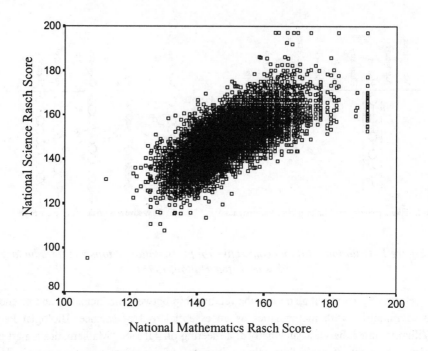

Figure 2. Relationship between students' mathematics and science achievement test scores on TIMSS (Population 2).

We also knew that the magnitude of the correlation between mathematics and science scores would be influenced by the large variability in performance across different classrooms. Differences between classrooms arise from the selectivity of the school population served, the variability in students' opportunity to learn mathematics and science, and other factors.[3] Consequently, we calculated the correlation between mathematics and science achievement scores within each classroom and then calculated the averaged ("pooled") within-classroom correlation. As we expected, the "pooled within-classroom" correlation shrunk considerably, to 0.55. The level of students' mathematics and science achievement varies considerably and systematically from one classroom to another in the United States.

The picture of the mathematics-science relationship remained the same for boys and girls.[4] But before letting the correlational cat out of the bag, we first explored whether the usual average achievement differences in mathematics and science found in the U. S. between boys and girls also appeared also in TIMSS. Boys, on average, scored higher than girls on the mathematics test with a mean difference of less than 1 point (0.64), a statistically ($p = 0.001$) but certainly not practically significant (effect size $= 0.06$)[5] difference (Table 1). The corresponding difference in science achievement was a bit larger, 1.47 points, statistically significant ($p<0.001$) but still a small effect (effect size $= 0.14$). So, the usual gender differences are reflected in TIMSS as well, but they are quite small.

Table 1. Means and standard deviations of mathematics and science scores by gender

		Mathematics		Science	
Gender	n	Mean	S.D.	Mean	S.D.
Girls	5537	149.14	9.75	148.48	9.59
Boys	5436	149.78	10.55	149.95	10.80

Now, we can let the correlational cat out of the bag. The correlation between girls' mathematics and science total scores was 0.74 and for boys it was… 0.74! We also calculated the pooled within-class correlation for girls (0.53) and boys (0.55).

RELATIONSHIP BETWEEN MATHEMATICS AND SCIENCE ACHIEVEMENT SUBSCALE SCORES

TIMSS provides a set of subscale scores derived from the test items on the mathematics and science achievement tests using a 3-parameter psychometric model (available on-line at http://timss.bc.edu/timss1995i/Database.html). In mathematics, the five subscales are: (1) fractions/number-sense ("fractions"), (2) geometry, (3) algebra, (4) data representation/analysis/probability ("data"), and (5) measurement. In science, the four subscales are: (1) earth science, (2) life science, (3) physics, and (4) chemistry. Examples of released items from each subscale are listed in Table 2.

Table 2. Examples of released items in subscales

Subscales	Item code	Description
Mathematics		
Fractions	I6	Fraction larger than 2/7.
Geometry	J15	Which two triangles are similar?
Algebra	K4	$x/2 < 7$ is equivalent to …
Data Representation	L10	Highest temperature on chart.
Measurement	M1	Weight shown on the scale.
Science		
Earth Science	J1	Description of Earth's surface.
Life Science	K11	Interdependence among aquatic organisms
Physics	L7	Communicate in space.
Chemistry	M10	Which is not a mixture?

All test scores are affected by error of measurement and TIMSS total scores are no exception. Using the TIMSS subscales, we can estimate the error-free ("disattenuated") correlation between mathematics and science achievement scores. Moreover, we use these scores to pose the question, "Even if there were not a gender difference in the relation between total mathematics and science achievement scores, might there be a difference on one or another subscale, in mathematics, science, or both?" Finally, might the prediction of each science subscale from the set of mathematics subscales shed light on the asymmetry between mathematics and science performance? We turn now to these questions.

Error-Free Correlation between Mathematics and Science Achievement Scores

Both the mathematics and science test scores contain a small amount of error, as do all test scores (and other measurements!). These measurement errors attenuate (reduce) the true magnitude of the relationship between mathematics and science achievement. There are a number of ways to estimate what the correlation would be if achievement were measured without error.

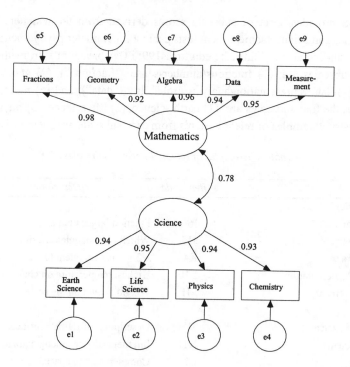

Figure 3. Error-free relationship between mathematics and science achievement.

We have chosen one particular method[6] to estimate an error-free mathematics score and an error-free science score for students, using information from the subscale scores. We then correlate the error-free mathematics score with the error-free science score to estimate the "true" correlation between mathematics and science achievement. We found that, once stripped of error, the correlation was 0.78 (in contrast to the observed correlation of 0.74).[7] Our findings are presented in Figure 3.

Gender Differences in Mathematics and Science Achievement Subscales

Although we found negligible differences between girls' and boys' average mathematics and science total scores, patterns across mathematics and science subscale scores might differ for girls and boys. To address this question, we begin by looking at differences in boys' and girls' average mathematics and science subscale scores (Table 3).

In mathematics, boys performed statistically better than girls on two of the five subscales, although the differences were quite small: fractions (effect size = 0.12) and measurement (0.12). In science, boys performed statistically better on all but the life science scale, consistent with past U. S. research (e.g., Jovanovic, Solano-Flores, & Shavelson, 1994) although the effects were, again, small: earth science (effect size = 0.22), physics (effect size = 0.26), and chemistry (effect size = 0.27).

Table 3. *Mean scores and standard deviations of mathematics and science subscale scores* by gender*

Subscales by Subject-Matter Domain	Girls		Boys		F-test	
	Mean	S.D.	Mean	S.D.	F	P value
	n= 5537		n= 5436		df = 1, 10971	
Mathematics						
Fractions/Number Sense	482.59	80.15	492.75	85.07	41.52	0.000
Geometry	457.93	76.20	460.85	81.61	3.74	0.053
Algebra	488.12	80.93	485.04	86.53	3.73	0.053
Data Representation/ Analysis/Probability	489.74	85.49	488.99	90.49	0.20	0.652
Measurement	439.45	92.21	450.40	96.08	37.13	0.000
Science						
Earth Science	481.58	84.15	500.92	89.42	136.13	0.000
Life Science	496.71	103.32	495.63	108.79	0.28	0.594
Physics	469.47	85.38	492.16	91.53	180.37	0.000
Chemistry	476.53	94.73	503.31	100.55	206.20	0.000

*Subscale scores are standardized with mean 500 and standard deviation 100.

Although mean differences might be small, it is still possible that the pattern of correlations across subscales differs for boys and girls. To this end, we correlated overall mathematics and science subscale scores (Table 4).[8] First, note that the correlations among subscale scores in mathematics and in science, and for both boys and girls, are highly correlated, and at about the same level. For example, the correlation between the fractions score and the geometry score is 0.90 for girls and 0.90 for boys.

Table 4. Overall correlations between mathematics and science subscale scores for girls (bold) and boys (italics)

Subscales	Mathematics					Science			
	F	G	A	D	M	E	L	P	C
Mathematics									
Fractions (F)	1.00	0.90	0.94	0.92	0.93	0.72	0.72	0.73	0.70
Geometry (G)	0.90	1.00	0.90	0.87	0.88	0.70	0.70	0.70	0.68
Algebra (A)	0.94	0.89	1.00	0.91	0.91	0.71	0.72	0.73	0.70
Data (D)	0.92	0.86	0.91	1.00	0.89	0.71	0.72	0.72	0.69
Measurement (M)	0.93	0.88	0.91	0.89	1.00	0.69	0.70	0.71	0.68
Science									
Earth Science (E)	0.71	0.68	0.71	0.70	0.69	1.00	0.91	0.89	0.89
Life Science (L)	0.71	0.69	0.72	0.71	0.69	0.90	1.00	0.91	0.90
Physics (P)	0.72	0.70	0.72	0.70	0.70	0.87	0.91	1.00	0.87
Chemistry (C)	0.69	0.66	0.68	0.68	0.67	0.87	0.89	0.86	1.00

Next, note that the correlation between mathematics subscale scores and science subscale scores are moderately high, around 0.70, for both girls and boys. Indeed, these correlations for boys and girls are so close as to make it plausible that they differ only due to sampling. Distinguishing between girls and boys in the correlation among mathematics and science subscale scores adds no new information.

Finally, although the correlation between any mathematics-science subscale pair is not as high as the correlation between the mathematics and science total observed scores (0.74), these correlations are only slightly lower. The difference arises because the subscale scores, containing fewer items than the total scores, contain more measurement error. Hence measurement error will tend to attenuate the subscale correlations more than the total score correlations. Moreover, total scores in mathematics and science, being aggregates across a wide content domain, may share general ability in common more than do subscale scores. This, too, would account for the magnitude difference.[9]

Differential Prediction of Science Subscale Scores from Mathematics Subscale Scores

Achievement in different domains of science might just draw upon different mathematical skills. Indeed, some science domains such as physics depend more on mathematics than other domains, such as life science, although this is rapidly changing. Moreover, all of science depends on probability, data analysis, and data representation so that scores on this subscale might predict science achievement in each of the four science domains measured by TIMSS. Finally, the pattern of relationships between mathematics subscale scores and science subscale scores might differ by gender, even though the mean differences between genders were small.

To explore these conjectures, we employed multiple regression to predict students' science subscale scores (e.g., earth science) from all five of their mathematics subscale scores and their gender. However, it just might be that for girls, data analysis plays a greater role in predicting their life science scores than for boys, but for boys, algebra might predict life science scores better than for girls. To address this hypothesis, we also included (cross-product) terms in our prediction of the science subscale scores to capture this possible "interaction" effect. Without exception, the interaction effect was not statistically significant and consequently, what we report holds for both girls and boys.

The regression results are summarized in Table 5. First, we observe that the prediction of students' science subscale scores (e.g., earth science) from the five mathematics subscales did not depend on the students' gender. That is, knowing a student's gender did not help predict her science score. This is not surprising because the gender differences in average achievement were small (see Table 3).[10] Second, the accuracy of predicting students' science subscale scores depended far more on their mathematics achievement than on their gender.

Table 5. Standardized regression coefficients for science subscale scores with the overall

Dependent Variable	Estimate of standardized regression coefficient					Gender
	Math subscale score					
	Fractions/ number sense	Geometry	Algebra	Data rep./ analysis/ probability	Measurement	
	β_1	β_2	β_3	β_4	β_5	β_6
Earth science	0.143**	0.147**	0.157**	0.231**	0.078*	0.099**
Life science	0.102**	0.152**	0.217**	0.222**	0.077**	-0.014*
Physics	0.120**	0.145**	0.242**	0.184**	0.075**	0.118**
Chemistry	0.168**	0.152**	0.176**	0.186**	0.051**	0.124**

Note. Standardized regression coefficients present the estimated effects on the science scores, which are the change of science-standardized scores due to a one-unit change in the independent standardized variables.
* Significant at 0.05 level.
** Significant at 0.01 level.

For the mathematics subscales, we expected data representation to be an important predictor of all the science subscale scores, and it is either the best or second best predictor. We also expected physics and chemistry to depend more on the algebra subscale than earth and life science but that turned out to be true only for physics. Indeed, chemistry scores were least well predicted by the mathematics subscale scores indicated by the magnitude of the adjusted R-square values. And curiously, measurement proved to be the poorest predictor of any of the science subscales.[11] Simply put, a clear pattern emerged in predicting science subscales from mathematics subscales: algebra and data representation scores consistently were the best predictors of science subscale scores. Prediction of chemistry scores was more problematic than prediction of the other subscale scores.

ANALYSIS OF THE STRUCTURE OF TIMSS MATHEMATICS AND SCIENCE TESTS

Thus far in our analyses we have used the subscale scores provided by TIMSS. These scores reflect expert judgment as to the underlying content distinctions among the pool of items comprising the mathematics and science tests. Based primarily on an analysis of item content, each item was designated to the subscale it best reflected (see Table 2). In this section we depart from this strategy to present an empirical examination of the structure of the mathematics and science tests, based on an item-level factor analysis.

Snow and his colleagues (Kupermintz et al., 1995; Hamilton et al., 1995) have demonstrated the usefulness of this approach for identifying dimensions of mathematics and science tests in a series of validity studies using the National Education Longitudinal Study of 1988 (NELS:88) achievement data. Four science and two mathematics factors were identified in the NELS data. *Everyday or Elementary Science* (ES) required knowledge of scientific concepts that could easily be learned outside of school. Although students may acquire their knowledge about science from various sources, it is likely that these particular items reflected more everyday knowledge than items loading on the other three factors. *Chemistry Knowledge* (CK) seemed to be defined primarily by subject matter. These items included concepts such as mixtures and compounds, chemical change, and solubility. They required knowledge of chemistry terms, but placed few reasoning demands on students. *Scientific Reasoning* (SR) contained items requiring manipulation of numerical equations, interpretation of graphs, or hypothesis formulation. Some of the items involved the application of knowledge of science vocabulary terms, but the primary feature that characterized these items seemed to be their reasoning requirement. *Reasoning with Knowledge* (RK) involved formal scientific concepts as well as the requirement to apply reasoning skills. In addition, in these items, the formal scientific terms appeared in the response options rather than in the stems; this might have placed less demand on science vocabulary knowledge.

Mathematical Reasoning (MR) included items that required mainly inferential reasoning, with a demand for logical argument rather than direct computations. *Mathematical Knowledge* (MK), on the other hand, was characterized by straightforward computation or the use of mathematical concepts and algorithms, usually in one-step solutions.

Would we retrieve similar dimensions on the TIMSS achievement tests? If so, would we discover similar sets of relations between mathematics and science achievement as we did with TIMSS' published scores? After presenting results from factor analyses of the TIMSS mathematics and science tests, we ask questions similar to the ones that motivated our previous analyses: What is the relationship between the obtained mathematics and science factors? Are there gender-related differences on these factors?

We analyzed TIMSS' mathematics and science test items with full-information factor analysis (Bock, Gibbons, & Muraki, 1988) as implemented in the TESTFACT computer program (Wilson, Wood, & Gibbons, 1991). The method is based on specifying both a multidimensional *item response model* for the test items and a Thurstonian factor structure for modeling the *latent ability dimensions* that affect the probability of correct responses.[12] Combining the models for latent ability structure and item response provides a strong tool for estimating item loadings on distinct abilities. Our analyses used multiple-choice items from TIMSS clusters A through H (48 "core" and "focus" items in each subject).[13] (Other TIMSS items were administered to sub-samples of too few students to include in the analysis.)

Science Factors

Factor analysis of the 48 science items indicated two factors (Table 6). The first factor, labeled *Everyday Science/Reasoning*, included items that either asked about general facts that students were likely to acquire outside their science classrooms or required students to apply reasoning to problems which placed little demand on specific or specialized scientific knowledge.

Table 6. Science factors and items with the 10 highest loadings

	Loadings		
	Everyday Science	*Scientific Knowledge*	
Item	*Factor 1*	*Factor 2*	*Item description*
H02	0.75	-0.30	What are vitamins
E08	0.74	-0.19	A son can inherit traits
F04	0.72	0.01	Areas soil washed away
F05	0.62	0.05	Why oxygen equipment necessary
B03	0.58	0.15	Which has greatest density
G09	0.58	0.01	Traits are transferred from
E09	0.54	0.00	When wind became colder
F06	0.54	0.16	Reason why paint prevents rust
G10	0.53	0.27	What left if took all atoms
A09	0.48	0.36	Fanning makes fire burn hotter
B01	-0.21	0.75	Where is hottest layer of earth

| | Loadings | | |
| | Everyday Science | Scientific Knowledge | |
Item	Factor 1	Factor 2	Item description
H05	-0.22	0.73	Where came energy in food from
E12	0.03	0.70	Most caves are formed by
D06	0.20	0.58	Seeds come from which part
C11	0.23	0.57	What possible serious effect
H06	0.12	0.55	Reaction of burning wood will
D02	0.07	0.55	Magnet & coffee
B04	0.18	0.53	Change pulse after race
C12	-0.08	0.47	Which is not a fossil fuel
F01	0.30	0.46	What shows mammal

In contrast, the second factor, labeled *Scientific Knowledge*, included items that required students to retrieve and apply specialized, formal scientific facts and terminology most likely learned in school (e.g., Suter, 2000). These factors show some resemblance to the factors found in the NELS Grade 8 sample (Hamilton et al., 1995). Everyday Science/Reasoning appears to be similar to ES, while Scientific Knowledge can be thought of as a mixture of the SR and RK factors described above. In contrast to the NELS:88 findings, no content factor (such as CK) emerged in our analyses, although TIMSS sub-scales reflect primarily content distinctions.

Mathematics Factors

The mathematics factors, more than the science factors, replicated quite closely the structure found for the NELS:88 data (Kupermintz et al., 1995). Two factors were also identified for the mathematics items (Table 7). In the first factor, labeled *mathematical reasoning*,[14] items typically required students to reason and afforded multiple solution strategies. Solutions involved performing sequential steps and translations between figural, verbal, and mathematical representations. The second factor, labeled *Mathematical Knowledge*, included items that required straightforward computations or symbolic manipulations (such as setting up a simple algebraic equation). This factor seems to involve mainly straightforward applications of classroom-based knowledge.

Mathematics—Science Correlations and Gender Differences

We begin by examining gender-related differences on the mathematics and science dimension. Table 8 presents the average score differences on these dimensions.

Table 7. Mathematics factors and items with the 10 highest loadings

| | Loadings | | |
| | Mathematical Reasoning | Mathematical Knowledge | |
Item	Factor 1	Factor 2	Item description
A04	1.09	-0.34	How many laps has Alice run
A01	0.97	-0.10	How many more squares shaded
E03	0.81	-0.12	What fraction of hour passed
E06	0.79	0.00	What is the length of rectangle
G02	0.78	-0.07	Length of pipe
H11	0.75	0.06	How many defective bulbs
D09	0.73	0.10	Which is smallest fraction
F12	0.70	-0.08	What fraction is shaded
B10	0.67	0.07	Which is smallest number
F07	0.66	0.17	What was average speed in m/sec
H09	-0.57	1.18	Which sum closest to 691+208
D10	-0.36	0.85	Which equation shows cost of
H10	0.06	0.67	Equation express relationship
C04	0.23	0.66	When 2.25 larger / smaller (b)
G01	-0.05	0.62	# In class score < 7
H08	0.09	0.60	Which shows 2/3 shade of square
C03	0.16	0.55	Which could be true
C06	0.07	0.54	Which is worst estimate
B12	0.31	0.50	Which of equations represents
E04	0.25	0.49	Difference 4-digit numbers

Table 8. Mean scores and standard deviations of mathematics and science subscale scores by gender

| | Girls | | Boys | | F-test | |
| | Mean | S.D. | Mean | S.D. | F | P value |
Dimension	n=5531		n=5432			df = 1, 10961
Mathematics						
Reasoning	0.00	0.98	0.18	1.03	87.88	<0.001
Knowledge	0.06	0.78	-0.06	0.81	57.03	<0.001
Science						
Everyday Science	-0.03	0.79	0.01	0.87	8.69	0.003
Scientific Knowledge	-0.10	0.74	0.10	0.78	192.98	<0.001

Consistent with previous results, boys outperformed girls on the mathematical reasoning dimensions (effect size = 0.18), while girls scored higher than did boys on the mathematical knowledge dimension (effect size = 0.15). These results indicate that the dimensions derived from the factor analysis are somewhat more sensitive to gender differences than the content subscales identified in the TIMSS mathematics test specification table.

A very small difference in average score was detected for the dimension of everyday science, favoring boys (effect size = 0.05). In contrast, boys achieved higher scores on the scientific knowledge dimension than girls (effect size = 0.26). Gender differences in science, then, appear to reflect differences in response to instruction as well as memory for and reasoning with specific facts and terms rather than differences in exposure to the various assemblies of facts and concepts that make up general or everyday knowledge.

Table 9. Correlations between mathematics and science dimensions for girls (bold) and boys (italics)

	Mathematics		Science	
Dimension	*MR*	*MK*	*ES*	*SK*
Mathematics				
Reasoning (MR)	1.00	*0.65*	*0.61*	*0.57*
Knowledge (MK)	**0.61**	1.00	*0.58*	*0.52*
Science				
Everyday Science (ES)	**0.59**	**0.55**	1.00	*0.62*
Scientific Knowledge (SK)	**0.55**	**0.47**	**0.58**	1.00

Examining the cross-subject correlations, we observed relatively low correlations between quantitative knowledge and scientific knowledge, and the comparable correlations of both mathematics dimensions with everyday science/ reasoning. This pattern is consistent with the fact that scientific knowledge items typically required no computation or mathematical reasoning. Rather, these items primarily asked students to retrieve various facts and definitions. In contrast, everyday science/reasoning items included tasks such as tabular data interpretation. The regression results in Table 10 exhibit this pattern more clearly.

Table 10. Standardized regression coefficients for science dimensions

	Standardized Regression Coefficient		
	Mathematics		Gender
	Reasoning	*Knowledge*	
Dependent Variable	β_1	β_2	β_3
Everyday science	0.40**	0.31**	0.02*
Scientific knowledge	0.40**	0.24**	0.11**

* Significant at .05 level.
** Significant at .01 level.

Overall, the correlations among the dimensions extracted in our factor analyses are considerably smaller than the correlations we reported for the TIMSS subscales. This is not surprising because one of the purposes of the factor analysis is to maximize the distinction among the extracted factors. Note that the correlations for boys seem to be somewhat stronger than the correlations for girls. This finding may reflect the larger variability of boys' scores. A more substantive hypothesis about the correlation differences cannot be examined with the current data, but we suspect that differences in cognitive processes, academic interests and perceptions, and course taking patterns should be on the short list of potential explanatory variables.

CONCLUDING COMMENTS

We found the usual United States mean differences between genders on both the TIMSS-provided achievement test scores, and scores we derived from factor analyses of the TIMSS mathematics and science test items. The magnitudes of the differences on the TIMSS overall and subscale scores were quite small. Because our factor analyses maximized the difference between the dimensions represented on the TIMSS tests, where gender differences were found with our factor scores, they were somewhat larger than the differences found with the TIMSS scores. We also found consistently high correlations between girls' and boys' mathematics and science scores, regardless of whether we used scores provided by TIMSS or scores derived from our factor analyses. Finally, we found only negligible differences between predictions of science achievement scores from mathematics scores, using both TIMSS-provided and our factor scores (with the exception of the fact that the mathematical reasoning factor predicted science achievement somewhat better than the mathematical knowledge factor).

Nevertheless, this is only the beginning of a related story, one of international comparison of the relationship between mathematics and science. We encourage others to pursue the tale just begun: Would this relationship hold up in countries where the mathematics and science curricula were more or less connected than the U. S. curricula? Would the gender differences hold up in other countries where women have more or less access to mathematics and science education and employment?

NOTES

[1] This research was supported by a grant from the National Science Foundation (REC-9628293). We wish to thank Larry Suter for comments on an early version of the chapter, and two anonymous reviewers for comments on the penultimate version. The findings and interpretations expressed herein do not reflect the position or policies of the National Science Foundation, or of the reviewers.

[2] We used TIMSS national Rasch scores because they have been used in government reports. Both Rasch scores, computed by standardizing mathematics and science logit scores, have a weighted mean of 150 and a standard deviation of 10 (Gonzalez & Smith, 1997).

[3] Indeed, the correlation between mathematics and science classroom mean total achievement scores is 0.93.

[4] We speak of the student's "gender" throughout this article to reflect the student's "sex"—female (girl) and male (boy)—preferring this term to that of "sex" although we recognize the cultural baggage that "gender" carries that "sex" does not.

[5] Effect size is defined as the difference between boys' and girls' mean scores divided by the pooled standard deviation for the two groups. An effect size of 0.20 is considered small, 0.50 is considered medium, and 0.80 large.

[6] Structural equation modeling with latent variables.

[7] We examined the same model for girls and boys with the same result. Indeed, the boys' and girls' models were statistically equivalent so that the model presented in Figure 3 fits for all students and for girls and boys.

[8] Pooled within-classroom correlations between mathematics and science subscale scores were about 0.25 points lower than overall correlations. Pooled within correlations among mathematics subscales and among science subscales were about 0.10 points lower. Pooled within correlations are available from the first author.

[9] We are indebted to Larry Suter for pointing out this possible interpretation.

[10] The mean gender difference was not statistically significant for life science achievement; the negative standardized regression coefficient for gender in predicting life science achievement reflects this negligible difference, *net the effect of all the mathematics subscale scores*.

[11] One possible explanation is that the measurement test items all dealt with the metric system and U. S. education clings to the British system, except in some science classes.

[12] The statistical analysis of the test data is based on maximum likelihood estimates and iterative computation, where a principal factor analysis of the tetrachoric correlation matrix provides reasonable starting points for the iterative process. Once a factor structure is decided upon, a Bayes estimate (average of the posterior ability distribution given a response pattern) generates scores on each ability dimension. For an excellent review of some alternative approaches to binary item factor analysis, see Mislevy (1986).

[13] TIMSS used a sampling scheme where a randomly selected student was assigned a randomly selected booklet that contained a subset of the mathematics, science or both items. In this way, content could be sampled broadly but no single student would be overburdened by the testing.

[14] Actually, the mathematical reasoning factor found in the TIMSS analysis was a bit more general than the corresponding factor in the NELS:88 data. The TIMSS items required application of more general reasoning and problem-solving heuristics than did the NELS:88 items. We were tempted to call the TIMSS factor, "Quantitative Reasoning," but decided that this label would be more confusing than helpful.

REFERENCES

Bock, R. D., Gibbons, R., & Muraki, E. (1988). Full-information item factor analysis. *Applied Psychological Measurement, 12*, 261-280.

Gonzalez, E. J., & Smith, T. A. (Eds.). (1997). *User guide for the TIMSS international database*. Chestnut Hill, MA: TIMSS International Study Center, Boston College.

Halton, G. (1973). *Thematic origins of scientific thought. Kepler to Einstein*. Cambridge, MA: Harvard University Press.

Halton, G. (1978). *The scientific imagination: Case studies* Cambridge: Cambridge University Press.

Hamilton, L. S., Nussbaum, E. M., Kupermintz, H., Kerkhoven, J. I. M., & Snow, R. E. (1995). Enhancing the validity of large-scale educational assessment: II. NELS:88 science achievement. *American Educational Research Journal, 32,* 555-581.

Harpending, H. (1985). Review of R. Boyd and P. J. Richardson's Culture and Evolutionary Process. *Science, 230,* 931.

Jovanovic, J., Solano-Flores, G., & Shavelson, R. J. (1994). Performance-based assessments: Will gender differences in science achievement be eliminated? *Education and Urban Society, 6(4),* 352-366.

Kupermintz, H., Ennis, M. M., Hamilton, L. S., Talbert, J. E., & Snow, R. E. (1995). Enhancing the validity of large-scale educational assessment: I. NELS:88 mathematics achievement. *American Educational Research Journal, 32,* 525-554.

Mislevy, R. J. (1986). Recent developments in the factor analysis of categorical variables. *Journal of Educational Statistics, 11(1),* 3-31.

Schmidt, W. H., McKnight, C. C., Valverde, G. A., Houang, R. T., & Wiley, D. E. (1997). *Many visions, many aims: A cross-national investigation of curricular intentions in school mathematics. TIMSS Volume 1.* Dordrecht: Kluwer Academic Publishers.

Schmidt, W. H., Raizen, S. A., Britton, E. D., & Bianchi, L. J. (1997). *Many visions, many aims: A cross-national investigation of curricular intentions in school science. TIMSS Volume 1.* Kluwer Academic Publishers, Dordrecht.

Suter, L. E. (2000). Is science achievement immutable? Evidence from international studies on schooling and achievement. *Review of Educational Research, 70(4),* 529-545.

Whiteside, D. T. (1962). Kepler the mathematician. *History of science. An annual review of literature, research and teaching, 1,* 86-90.

Wilson, D., Wood, R., & Gibbons, R. (1991). *TESTFACT: Testing scoring, item statistics, and item factor analysis.* Chicago: Scientific Software.

Chapter 16

STUDENTS' ATTITUDES AND PERCEPTIONS

Edward W. Kifer

This chapter reports on students' views of various aspects of mathematics and science, mathematics and science learning, and the place of mathematics and science in the context of schools. Investigating such views is both consistent with the history of IEA and with educational research that seeks to understand facets of the impact of schools and schooling in arenas other than academic achievement.

The first major IEA study, a mathematics study (Husén, 1967), contained five attitude scales among its measurements in the affective domain with questions about mathematics as a process, difficulties of learning mathematics, the place of mathematics in society, schools and school learning, and man and his environment. Subsequent IEA studies of mathematics (Burstein, 1994) and science (Comber & Keeves, 1973) also contained a plethora of questions seeking students' views about subject matter or content issues as well as an array of questions seeking opinions about broader issues of school and schooling.

It should be no surprise that international studies contain such questions. Parents and teachers, for example, want children to have positive attitudes toward school and what goes on there. From their points of view, children or students are doing well in school when they are engaged in positive ways in the school community.

There are attitudes that are to be imparted to students within classrooms and schools. Bronfenbrenner (1970), for example, studied Soviet pre-schools where toys in the back of the room were so large that children had to cooperate to use them. Schools and schooling are embedded in a society with norms to which persons are expected to conform; hence the importance of intentions to behave, one definition of an attitude. Since these perceptions, opinions, and attitudes are hypothesized to emerge over time through experiences in the schools embedded in social and cultural contexts, as opposed to being directly taught (Dreeben, 1968), they are ripe candidates for exploration in the diverse contexts offered in an international study. And what could be more diverse than this one with more than 40 educational systems providing the results?

D.F. Robitaille and A.E. Beaton (eds.), Secondary Analysis of the TIMSS Data, 251–275.
© 2002 *Kluwer Academic Publishers. Printed in the Netherlands.*

There are several ways to think about studying attitudes and perceptions within international comparative studies. One is to view these affective characteristics as predictors of academic achievement. Do those students, classrooms, or educational systems with the most positive attitudes also tend to have the highest test scores? A second approach is to ask whether success in academic work leads to more positive views. Do students, classrooms, or educational systems with high test scores also tend to have the most positive attitudes? A third approach is to document the responses of students and then make comparisons among them and the systems from which they come.

For reasons given in a comparable analysis (Kifer & Robitaille, 1989), I will use the third approach in reporting these results. Rather than enter the precarious arena of cause and effect or complex modeling of the relationships between and among academic achievement and affective outcomes, I will consider responses to these questions as a snapshot of what students believe about mathematics or science and issues related to mathematics or science. The intent will be to find patterns among those responses and focus on what appear to be important differences in those patterns.

This approach to the data assumes that these responses were not generated haphazardly. It presumes that perceptions about or attitudes toward, for example, the importance of mathematics or the extent to which being successful in science is a matter of natural ability and/or hard work are a reflection of experiences in schools translating societal norms and values to its citizens. What one has, then, are the collective voices of over forty educational systems telling us what they believe and value about mathematics.

METHODS

These results were gathered using questionnaires and are self-reports of the participants. Other ways of gathering evidence about attitudes, preferences and opinions (e.g., observations) might produce other patterns of results. The huge number of languages to which the questions were translated may further exacerbate such limitations of paper and pencil questionnaires.

The majority of the questions were in Likert format. For example a student would be asked:

My mother thinks it is important for me to do well in science.

The student would be asked to respond on the Likert scale:

Strongly Agree   Agree   Undecided   Disagree   Strongly Disagree

Students who either Strongly Agree or Agree to these questions are said to endorse the idea expressed in the questions. Those who Strongly Disagree or Disagree do not endorse them. The middle category is ambiguous: it can be interpreted as either being undecided about a response to a question or having no opinion at all.

There were several steps involved in the analyses of these data. First, two summary data sets were tabulated. One set contained the percent of students in an educational system that endorsed each question; the second table was the percent that did not endorse it. Although I have analyzed both sets, the results discussed in this chapter are based on the percent of students who endorsed a question.

Median Polishing

The fictitious table below shows the structure of a data set and provides a basis for discussing median polishing, an Exploratory Data Analysis technique (Tukey, 1977; Velleman & Hoaglin, 1981).

Table 1. Fictitious summary table

	Question			
System	1	2	3	System Median
1	85	65	90	85
2	60	55	70	60
3	70	80	50	70
Question Median	70	65	70	

There are nine data points in all (the shaded part of the table). For example, the first system has an 85 for the first question. The 85 is the percent of respondents who endorsed Question 1. Each of the nine data points is the percent endorsement for each system and question. Taking the median of each column summarizes the responses. That summary shows that Question 3 (70) was highly endorsed while Question 2 (65) was less likely to be endorsed. A similar summary for rows provides a comparison between systems. In the fictitious table, respondents in System 1(85) viewed the questions most favorably while those in System 3 (70) were less favorably disposed to the set of questions. The model for median polishing is:

DATA = GRAND EFFECT + ROW EFFECT + COLUMN EFFECT + UNIQUE EFFECT

which implies:

UNIQUE EFFECT = DATA – GRAND EFFECT – ROW EFFECT – COLUMN EFFECT

Based on the model, median polishing provides two additional interpretable values. First, it creates Question Effects and System Effects. Those effects are defined as the differences between the row and column medians and the median for all of the data (Grand Effect = 70). For example, the effect for question 1 is 0 (70-70). The effect for System 1 is 15 (85-70). Looking across the System Effects one can see that System 2 respondents are 10 percent less likely to endorse these items. Looking across the Question Effects indicates that Question 2 is 5 percent less likely to be endorsed. Since the Unique Effects are defined as the difference between the data and the Grand, System, and Question Effects, one has, as examples:

The unique effect for System 1, Question 1: $0 = 85 - 70 - 15 - 0$

The unique effect for System 1, Question 2: $-15 = 65 - 70 - 15 + 5$

Table 2 below gives the data and a polished table.

Table 2. A polished table with the data and each of the effects

Data				Unique Effects			System Effects
Question 1	*Question 2*	*Question 3*		*Question 1*	*Question 2*	*Question 3*	
85	65	90	System 1	0	-15	0	15
60	55	70	System 2	0	0	5	-10
70	80	50	System 3	0	15	-25	0
Question Effects				0	-5	5	GE = 70

Having completed polishing the table, I would interpret the results as follows. Since the overall endorsement rate is 70 percent (Grand Effect = 70), the respondents are favorably disposed to these questions. Respondents in System 1 are the most positive of all since they are 15 percent more likely to endorse these questions than is typical (System Effect = 15). The question that receives the highest endorsement is Question 3 (Question Effect = 5). Finally, there is at least one unusual finding in the table. Given the pattern of responses, the Unique Effect (25 percent less than would be expected given the data) for System 3 and Question 3 appears to be large. What is there about that system or that question that may have produced that response?

Interpreting the Results

In the remainder of the chapter I will use the polished results as a basis for discussing and interpreting differences between the educational systems and the sets of questions to which their students responded. I will discuss the results for Grade 4, Grade 8, and Population 3. Where it seems appropriate I will be make comparisons between the Grade 4, Grade 8, and Population 3 mathematics and physics students and between students' responses when there are comparable questions.

STUDENTS' ATTITUDES TOWARD MATHEMATICS AND SCIENCE

The target populations for this section are Grades 4 and 8 and Population 3 specialists. For most educational systems virtually all students are enrolled in school in the earlier grade levels and have had experiences in mathematics and science. The

Population 3 students included in these analyses are those classified as specialists in either mathematics or physics.

General Attitudes

There are four questions that address general attitudes toward mathematics and science:

> 1. How much do you like mathematics?
> 2. Do you think that you enjoy learning mathematics?
> 3. Do you think that mathematics is boring?
> 4. Do you think that mathematics is an easy subject?

> 1. How much do you like science?
> 2. Do you think that you enjoy learning science?
> 3. Do you think that science is boring?
> 4. Do you think that science is an easy subject?

Grade 4 Results

Table 3 below gives the percent of Grade 4 students who endorsed (Strongly Agreed or Agreed) these questions by system. In addition, the table contains the median polished effects.

Question Differences

The overall effect of 72, indicating a modestly high rate of endorsement, masks important differences among the questions. Note how these students overwhelmingly say that they like mathematics and science and enjoy learning it. They do not think either mathematics or science is boring. And, more than half of the respondents believe learning mathematics and science is not easy.

Note the patterns of Question Effects are similar across the two content areas. By that I mean the signs and the magnitudes of the Question Effects are similar for students' responses to mathematics and science. That is something that appears frequently in these results. I assume similar patterns of results means students see these subjects as being comparably important, enjoyable, not boring, rather than they are not discriminating.

These differences are rather small for most of the systems. Seventeen of the systems are within 6 percent in terms of endorsement by students. There are exceptions, however. Students from the Islamic Republic of Iran (IRN) and Greece (GRC) are most likely to endorse the questions, while students from Japan (JPN) and the Netherlands (NLD) are least likely to endorse them.

Table 3. Percent endorsement of general attitude questions – Grade 4 students

Country	Math 1: How much like math	Math 2: Think enjoy learn math	Math 3: Think math is boring	Math 4: Think math is easy	Sci. 1: How much like science	Sci. 2: Think enjoy learn science	Sci. 3: Think science is boring	Sci. 4: Think science is easy	System Effect
Australia (AUS)	83	82	27	56	82	80	21	54	-1
Austria (AUT)	76	71	26	55	79	72	20	66	-5
Canada (CAN)	89	88	22	68	80	79	24	60	0
Cyprus (CYP)	96	95	20	90	90	88	22	85	9
Czech Rep. (CSK)	83	78	20	64	83	77	17	70	-2
England (GBR)	84	84	24	51	81	79	24	46	-1
Greece (GRC)	94	96	13	79	95	95	13	87	13
Hong Kong (HKG)	81	83	26	56	90	87	17	68	1
Hungary (HUN)	84	77	29	57	84	76	24	64	-1
Iceland (ISL)	93	88	15	72	86	78	20	65	0
Iran, Isl. Rep. (IRN)	97	95	45	86	97	93	44	92	17
Ireland (IRL)	83	82	26	59	78	76	22	58	-1
Israel (ISR)	82	84	25	58	82	81	22	58	-1
Japan (JPN)	72	72	21	40	85	88	12	52	-10
Korea (KOR)	73	74	35	56	88	86	16	71	0
Kuwait (KWT)	94	94	31	84	92	89	31	83	9
Latvia, LSS* (LVA)	82	84	24	46	79	76	23	59	-3
Netherlands (NLD)	65	68	29	49	67	71	25	56	-11
New Zealand (NZL)	82	82	27	54	83	83	20	50	-1
Norway (NOR)	77	86	29	76	78	80	24	76	3
Portugal (PRT)	92	96	23	74	95	96	19	84	12
Scotland (SCO)	84	—[1]	—	—	82	—	—	—	—
Singapore (SGP)	93	92	11	58	90	88	11	54	2
Slovenia (SVN)	89	85	23	62	89	83	19	65	2
Thailand (THA)	92	78	19	68	86	73	26	65	1
United States (USA)	84	85	27	63	85	84	20	62	1
Question Effects	12	11	-47	-12	11	8	-51	-8	GE=72

* Latvia is annotated LSS for Latvian Speaking Schools only.
[1] — indicates the item was not asked in this country.

Unique Effects

There are, however, some unique effects that point to additional differences. The boxplot, Figure 1, shows the unique effects and identifies the unusual responses. The unique effects are patterned. There is more variation for the questions about mathematics and science being boring or hard than there is for liking or enjoying. That is, students in these systems are more equivocal on the former questions than the latter. The unusual effects belong to Greece and Japan. Japanese students say they like and enjoy science more than would be expected by their other rather low responses. Greek students are less likely to endorse mathematics and science being boring, given their generally high responses.

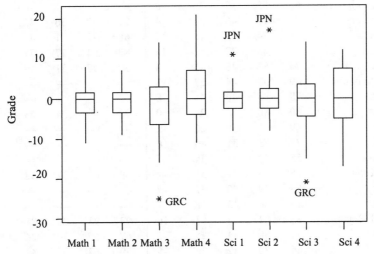

Figure 1. Boxplot of the unique effects – general attitudes: Grade 4.

Grade 8 Results

I present the Grade 8 results separately for mathematics and science because a number of the educational systems did not administer the general science items. These are shown in Table 4.

Again, despite the differences in the number of systems that administered the questions, there is remarkable consistency between the results for science and mathematics. The Grand Effects are within two percent and the pattern of the Question Effects is comparable. Grade 8 students like and enjoy mathematics science as do their Grade 4 counterparts, and they do not believe that either is boring or easy.

The System Effects are comparable, too. Figure 2 shows a plot of the System Effects for systems that participated in both Grade 4 and Grade 8. The relationship between these effects for the two grades is high: the Pearson Product-Moment correlation is about 0.83. Although Korea appears to be an outlier, the pattern of results suggests that students at these grade levels are responding in comparable ways within an educational system.

Table 4. Summary results for Mathematics and Science – Grade 8 general attitudes

	Grade 8 Mathematics					Grade 8 Science				
Country	Q1	Q2	Q3	Q4	System Effects	Q1	Q2	Q3	Q4	System Effects
Australia	64	62	52	29	-4	60	62	46	34	-9
Austria	57	42	41	30	-8	—	—	—	—	—
Belgium-Fl.	68	55	35	25	-8	—	—	—	—	—
Belgium-Fr.	70	76	46	35	4	71	77	36	42	2
Canada	74	74	45	48	7	68	71	41	43	0
Colombia	78	88	27	57	16	87	94	17	77	21
Cyprus	73	83	28	45	8	70	77	31	49	3
Czech Republic	49	38	44	27	-13	—	—	—	—	—
Denmark	78	85	30	41	8	—	—	—	—	—
England	80	80	28	24	0	78	82	24	23	0
France	69	65	38	26	-2	—	—	—	—	—
Greece	74	73	31	40	6	—	—	—	—	—
Germany	55	40	41	34	-7	—	—	—	—	—
Hong Kong	65	65	41	31	-2	69	68	31	38	-3
Hungary	58	39	40	31	-7	—	—	—	—	—
Iceland	79	66	33	38	1	—	—	—	—	—
Iran, Isl. Rep.	85	82	34	63	16	93	93	26	81	23
Ireland	73	68	34	25	-2	67	69	30	26	-3
Israel	66	70	32	26	-5	59	67	33	43	-3
Japan	53	46	35	13	-17	56	53	33	15	-16
Korea	58	41	40	20	-12	59	40	33	20	-17
Kuwait	84	84	29	61	17	89	87	24	74	18
Latvia (LSS)	67	51	35	25	-8	—	—	—	—	—
Lithuania	53	45	48	17	-16	—	—	—	—	—
Netherlands	58	56	46	32	-6	—	—	—	—	—
New Zealand	72	74	43	35	3	68	71	38	35	0
Norway	63	76	49	43	8	67	75	37	48	4
Philippines	80	91	35	53	15	83	94	35	64	17
Portugal	72	82	47	28	5	—	—	—	—	—
Romania	71	73	32	41	5	—	—	—	—	—
Russian Fed.	73	54	30	23	-12	—	—	—	—	—
Scotland	74	81	0	29	0	78	25	0	80	-12
Singapore	82	78	28	34	5	92	90	15	42	9
Slovakia	60	58	32	32	-8	—	—	—	—	—
Slovenia	66	48	37	23	-7	—	—	—	—	—
South Africa	72	84	38	61	11	74	87	40	71	11
Spain	63	56	42	35	-3	73	69	31	41	-2
Sweden	60	75	50	44	9	—	—	—	—	—
Switzerland	68	57	34	35	-4	67	59	34	47	-1
Thailand	82	74	24	33	3	90	87	16	44	8
United States	71	70	47	45	5	71	73	39	53	4
Question Effects	15	13	-13	-19	GE=54	13	15	-23	-13	GE=56

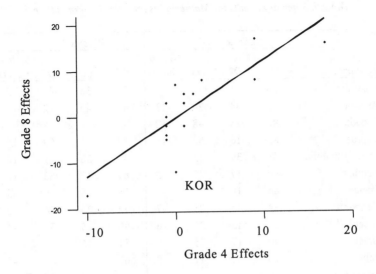

Figure 2. Plot of Grade 4 and Grade 8 system effects with a least squares regression line.

Population 3 – Mathematics and Physics students

There are different samples for the Population 3 part of TIMSS. For this chapter I am using data from students who are either in the mathematics or physics specialist samples. Although these questions are worded slightly differently and have different responses options for the Population 3 students as opposed to their counterparts in Grades 4 and 8, they are aimed at similar constructs. The questions are:

1. How much do you like mathematics (physics)?
2. Mathematics (Physics) is boring.
3. Mathematics (Physics) is an easy subject.

The data describing the student responses is contained in Table 5. Notice that the Grand Effects are low, a result of two questions that students do not endorse. Yet, the pattern of endorsements is similar to the patterns of the Grade 4 and Grade 8 students. Table 6 below shows the percentages.

In both science and mathematics each of the populations endorses liking the subject areas at a higher rate. With the exception of Grade 4 students, where over half of the responses suggest that the subjects are easy, the general results suggest that neither mathematics nor science is boring and that, at least for two populations, neither is easy.

The System Effects are comparable across mathematics and science. Those systems with positive effects in one subject area tend also to have positive effects in the other. That is true, too, of the negative effects.

Table 5. Summary results for Mathematics and Physics–Population 3

Country	Mathematics				Physics			
	Q1	Q2	Q3	System Effects	Q1	Q2	Q3	System Effects
Australia	80	31	38	7	82	36	41	9
Austria	43	44	20	-13	43	46	22	-16
Canada	82	25	48	12	—	—	—	—
Cyprus	85	16	52	15	89	12	59	16
Czech Republic	43	39	19	-14	47	38	21	-16
Denmark	90	17	36	3	89	17	53	16
France	84	18	29	-4	87	18	30	-5
Germany	54	35	25	-7	72	23	44	0
Greece	87	13	26	-6	90	15	33	-4
Israel	78	17	35	2	81	16	43	6
Italy	70	33	17	0	67	38	7	-6
Latvia (LSS)	—	—	—	—	73	27	32	0
Lithuania	77	28	37	4	—	—	—	—
Norway	—	—	—	—	86	23	37	0
Russian Fed.	85	17	25	-7	80	23	26	0
Slovenia	63	29	20	-7	78	19	36	-1
Sweden	80	25	42	10	82	21	47	9
Switzerland	63	31	22	-7	62	34	23	-11
United States	83	29	46	13	78	30	46	8
Question Effects	37	-8	0	GE=33	36	-13	0	GE=37

Table 6. Endorsement percent for mathematics and science across populations and questions

	Mathematics			Science		
	Like	Boring	Easy	Like	Boring	Easy
Grade 4	84	25	60	83	21	64
Grade 8	69	33	43	69	41	35
Population 3	70	25	33	73	25	37

DOING WELL IN MATHEMATICS AND SCIENCE

Doing well in mathematics and science is a general theme of a number of the questions in TIMSS. Students in Grades 4 and 8 were asked whether they usually do well in mathematics and science. Students in Population 3 were asked whether they usually have done well in mathematics and science. In each of the populations, students were asked how important doing well was to the students themselves, their mothers, their friends, and, depending on the population, to their fathers.

Comparable questions were asked about the importance of being good at sports, having fun, and, again depending on the population, being placed in a high achieving class. Finally, there were questions about why students think doing well is important and what it takes to do well.

Doing well or having done well

The first set of results deals with the questions of either doing well in science and mathematics in Grades 4 and 8 or, for Population 3, having done well in mathematics and science.

The overall response to these questions is positive for each of the student populations. The medians for Grade 4 are 89 in mathematics and 85 in science. Comparable values for Grade 8 are 73 and 77; for Population 3 mathematics students, 76 and 81; and for Population 3 physics students, 74 and 82. Although students' perceptions of doing well in mathematics versus science are barely different (the effects are one percent), there are large differences between systems on these questions. Table 7 shows the effects for Grades 4 and 8 as well as for the mathematics students in Population 3.

There are several generalizations that can be drawn from the display. First, the range of effects is large in each of the populations but it is much smaller in Grade 4 than Grade 8. Notice that Grade 4 differences go from −20 to 10 while the Grade 8 mathematics differences, for example, range from less than −30 to 20. There is consistency among the effects as well. Note the Hong Kong (HKG) and Japan (JPN) large negative effects in Grade 4 become more negative in Grade 8. Greece (GRC) and the Islamic Republic of Iran (ISR) have positive effects at Grade 4 that get more positive at Grade 8.

There may be a geographic effect in these results. Asian and Eastern European systems tend to have negative effects while Middle Eastern, Western European, and American systems have positive effects.

Confounded with the geographic differences, perhaps, are ones that indicate that some of the highest scoring systems in TIMSS in terms of cognitive achievement have the most negative effects in terms of students perceiving themselves as doing well. The Czech Republic (CSK), Hong Kong (HGK), Japan (JPN), Korea (KOR), and Singapore (SGP) are both high scoring and have relatively large negative effects.

Finally, having done well may or may not be a different perception than doing well. The range of Population 3 effects is narrower than those of Grade 8 and appears to be similar to Grade 4. But there are many fewer systems in the Population 3 sample and it excludes some of the systems that have the most pronounced effects. Population 3 is, as everyone knows, an extremely difficult one to categorize and, therefore, difficult to compare to the younger samples.

Table 7. System Effects across populations of students' perceptions of doing well

System Effects	Educational Systems
Grade 4	
-21 to -30	
-11 to -20	SGP THA
-6 to -10	CSK HGK JPN LVA
-0 to -5	AUT HUN NLD NZL PRT
0 to 5	AUS CAN GBR ISL IRL ISR KWT NOR SCO SVN USA
6 to 10	CYP GRC IRN
11 to 20	
21 to 30	
Grade 8 Math[1]	
LT - 30	HGK JPN KOR
-21 to -30	LTU LVA
-11 to -20	BFL CSK PRT RUS SGP THA
-6 to -10	DEU FRA ROM
-0 to -5	AUT CHE CYP HUN NLD SLV SVN
0 to 5	AUS BFR CAN ESP IRL NOR ZAF
6 to 10	GRC ISL ISR NZL SWE USA
11 to 20	COL DNK GBR IRN KWT PHL SCO
21 to 30	
Gr. 8 Science[1]	
LT - 30	HKG JPN KOR
-21 to -30	
-11 to -20	THA
-6 to -10	IRL SGP
-0 to -5	AUS CHE CYP
0 to 5	BFR CAN ESP ISR NOR NZL SCO
6 to 10	GBR KWT PHL USA ZAF
11 to 20	COL IRN
21 to 30	
Pop. 3 Math	
-21 to -30	
-11 to -20	AUT CHE CSK DEU SVN
-6 to -10	CYP SWE
-0 to -5	AUS FRA
0 to 5	DNK GRC ISR ITA LTU RUS
6 to 10	CAN
11 to 20	USA
21 to 30	

[1] A number of systems did not administer the science questions so these effects are the differences between the systems' responses and the overall median.

Who and what influences doing well?

There is a set of questions seeking to ascertain students' perceptions of who influences them in what facets of their lives. These questions ask whether it is the student (1), mother (2), father (3), or friend (4) who thinks it is important to do well in science (1), mathematics (2) and the language of the test (3), be good at sports (4), have time to have fun (5), or be placed in a high achieving class (6). Although there are potentially 20 questions, not every question was asked in all samples. The table below shows the common questions:

Table 8. Patterns of doing well influence items across populations

	Mathematics	Science	Sports	Have Fun
Mother	All	All	All	All
Friends	All	All	All	All
Me	All	All	All	All

These questions were highly endorsed by students in each of the samples. The Grand Effects were for Grade 4, 90; for Grade 8, 95; and for Population 3 mathematics, 90. Students at all levels apparently believe that everyone wants them to do well regardless of what activity they are pursuing. There are differences, however, in terms of the Question Effects. Those are displayed in Table 9 below.

Table 9. Item effects about influence

	Mathematics	Science	Sports	Having Fun
Mother				
Grade 4	2	6	-12	-2
Grade 8	-1	2	2	-12
Pop. 3	0	3	3	-37
Friends				
Grade 4	-12	-5	-7	2
Grade 8	-23	-9	-9	2
Pop. 3	0	5	0	-27
Me				
Grade 4	3	6	-2	3
Grade 8	-3	2	1	2
Pop. 3	0	2	-2	-13

Most of the effects for Grades 4 and 8 are modest. The possible exceptions are the negative effects attributed to friends especially in regard to mathematics. The most stunning effects, however, are for Population 3 mathematics students who apparently believe that no one thinks they should have time to have fun. The System Effects are displayed below in Figure 3.

As depicted in the boxplot, the majority of the System Effects are small for Grades 4 and 8. Most effects for those populations are 5 percent or less. There are, however, outliers in both samples. Grade 4 students in the Asian countries of Thailand (THA), Hong Kong (HKG), Japan (JPN), and Korea (KOR) are less likely to endorse these questions as are Grade 8 students in Germany (DEU). The System

Effects are wider in Population 3 mathematics but there are no unusual or outlying values.

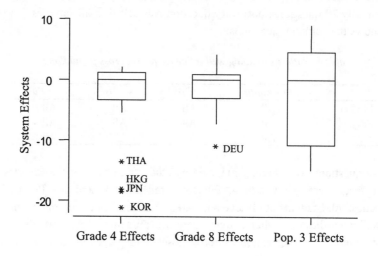

Figure 3. System Effects for influence on doing well.

Why do well in mathematics and science?

Grade 8 students were asked why it was important to do well in mathematics and science. They could endorse the following:

      1. I need to do well in mathematics to get desired job.

      2. I need to do well in mathematics to please my parents.

      3. I need to do well in mathematics to get into the school I prefer.

      4. I need to do well in mathematics to please myself.

      1. I need to do well in science to get desired job.

      2. I need to do well in science to please my parents.

      3. I need to do well in science to get into the school I prefer.

      4. I need to do well in science to please myself.

Table 10 gives a summary of the results. The Grand Effects suggest that students are more willing to endorse reasons for doing well in mathematics than for doing well in science. There is very strong overall endorsement for mathematics (81) and a more modest number (67) for science. It must be remembered, however, that there are different systems in these two sets of results.

Table 10. Summary results for Why one needs to do well in mathematics and science

Country	8th Math					8th Science				
	Q1	Q2	Q3	Q4	System Effects	Q1	Q2	Q3	Q4	System Effects
Australia	79	72	78	85	0	52	66	59	75	-6
Austria	64	54	63	82	-15	38	48	48	76	-18
Belgium-Fl.	57	68	74	87	-5	—	—	—	—	—
Belgium-Fr.	71	77	77	86	-3	53	73	59	83	-2
Canada	85	68	92	85	3	63	63	81	77	2
Colombia	85	77	94	93	9	74	75	87	92	14
Cyprus	87	70	82	87	2	57	65	68	80	-1
Czech Republic	83	84	85	78	2	—	—	—	—	—
Denmark	71	41	86	71	-11	—	—	—	—	—
England	80	64	86	93	1	62	63	75	87	4
France	71	59	83	75	-9	—	—	—	—	—
Germany	70	57	65	82	-10	—	—	—	—	—
Greece	83	75	85	92	5	—	—	—	—	—
Hong Kong	76	59	83	72	-6	55	56	74	65	-5
Hungary	77	64	75	56	-7	—	—	—	—	—
Iceland	79	43	93	80	-2	—	—	—	—	—
Iran, Isl. Rep.	90	95	95	96	11	90	95	93	94	27
Ireland	80	62	82	90	-1	50	56	66	79	-5
Israel	79	56	96	88	2	51	47	83	79	-5
Japan	55	34	91	78	-15	40	33	86	75	-12
Korea	47	56	86	68	-15	44	53	80	67	-11
Kuwait	85	92	88	95	8	85	93	86	93	21
Latvia (LSS)	85	80	89	85	5	—	—	—	—	—
Lithuania	87	53	84	50	-8	—	—	—	—	—
Netherlands	53	43	67	81	-21	—	—	—	—	—
New Zealand	83	66	80	89	1	55	61	60	80	-3
Norway	72	52	89	80	-6	47	48	64	75	-9
Philippines	90	83	90	86	8	90	83	90	86	20
Portugal	77	66	83	84	-2	—	—	—	—	—
Romania	78	76	82	69	-1	—	—	—	—	—
Russian Fed.	82	71	83	68	1	—	—	—	—	—
Scotland	88	66	84	90	3	65	60	71	81	1
Singapore	85	66	95	84	2	71	68	93	84	8
Slovakia	80	71	93	93	6	—	—	—	—	—
Slovenia	78	44	88	86	0	—	—	—	—	—
South Africa	85	74	84	83	2	82	76	80	82	12
Spain	71	82	88	94	6	65	83	78	93	10
Sweden	71	46	82	88	-5	—	—	—	—	—
Switzerland	66	57	72	89	-12	33	42	43	77	-24
Thailand	96	98	98	99	15	94	98	97	98	31
United States	85	80	96	88	9	66	79	89	84	11
Question Effect:	-2	-13	2	4	GE=81	-9	-3	3	13	GE=67

The Question Effects are consistent, at least by the signs of the effects, across both content areas, and they indicate that students are more willing to do well for themselves and to get into a preferred school than to please their parents or to get a good job. The effects are, however, relatively small.

There are larger effects reflecting system differences. When one compares Norway to Thailand in mathematics, there are startling differences between the two systems. In science, a comparable comparison is between the Islamic Republic of Iran and Switzerland.

More interesting still are the Unique Effects of this analysis. The boxplot in Figure 5 depicts those effects for the different content areas. Table 11 contains all of the Unique Effects with the largest ones shaded.

The majority of the effects are responses to questions about whether students believe they need to do well in order to get into a preferred school versus to please their parents. Students in Denmark, Iceland, Israel, Japan, and Norway in mathematics and Israel and Japan in science have similar response patterns: two Unique Effects of the same sign. In each case, they indicate they need to do well in order to get into a preferred school rather than to please their parents.

Students in other systems, for example Korea, Hong Kong, and Singapore, also more strongly endorse getting into a preferred school. The largest effect of all is in Switzerland with unusually strong endorsements of doing well for oneself in both mathematics and science.

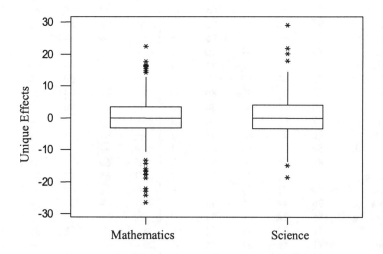

Figure 5. Boxplot of unique effects for why one needs to do well, Grade 8.

While patterns across the questions are discernible, it is not clear how one ought to interpret the system differences. One tendency might be that some of the higher scoring systems in terms of cognitive achievement have comparable patterns of results where students emphasize the need to get into a preferred school.

Table 11. Unique effects of why do well in mathematics and science, grade 8

Country	Mathematics				Science			
	Q1	Q2	Q3	Q4	Q1	Q2	Q3	Q4
Australia	0	4	-5	0	0	7	-5	1
Austria	0	1	-6	12	-2	2	-3	13
Belgium-Fl.	-17	5	-4	7	—	—	—	—
Belgium-Fr.	-5	12	-4	4	-3	10	-9	4
Canada	3	-3	6	-3	3	-3	10	-5
Colombia	-3	0	1	-1	3	-3	4	-2
Cyprus	5	0	-4	0	0	1	-1	0
Czech Republic	2	14	-1	-9	—	—	—	—
Denmark	3	-16	13	-3	—	—	—	—
England	0	-5	1	7	0	-5	1	3
France	1	-1	8	-1	—	—	—	—
Germany	2	-1	-9	8	—	—	—	—
Greece	-2	2	-4	2	—	—	—	—
Hong Kong	3	-3	5	-7	2	-3	10	-10
Hungary	5	2	-2	-22	—	—	—	—
Iceland	2	-23	11	-2	—	—	—	—
Iran	0	15	0	0	5	4	-4	-14
Ireland	2	-5	-1	5	-2	-3	2	4
Israel	-1	-13	10	1	-2	-12	18	3
Japan	-8	-19	22	8	-6	-19	29	6
Korea	-18	2	17	-2	-2	0	22	-2
Kuwait	-2	17	-3	2	6	8	-4	-8
Latvia (LSS)	1	7	0	-5	—	—	—	—
Lithuania	16	-7	8	-26	—	—	—	—
Netherlands	-5	-3	4	18	—	—	—	—
New Zealand	3	-3	-4	3	0	0	-6	2
Norway	-1	-11	11	1	-2	-7	3	4
Philippines	3	7	-2	-8	12	-1	0	-15
Portugal	-1	0	1	1	—	—	—	—
Romania	1	10	0	-14	—	—	—	—
Russian Fed.	2	2	-1	-18	—	—	—	—
Scotland	6	-5	-2	2	6	-5	0	0
Singapore	3	-4	9	-3	5	-5	14	-5
Slovakia	-5	-3	3	2	—	—	—	—
Slovenia	-1	-24	5	1	—	—	—	—
South Africa	3	4	-2	-5	13	0	-2	-10
Spain	-14	8	-2	3	-3	8	-2	2
Sweden	-3	-17	4	8	—	—	—	—
Switzerland	-1	1	0	16	-2	2	-3	20
Thailand	2	15	-1	-2	5	3	-4	-13
United States	-2	4	3	-5	-3	3	8	-8

Is it natural ability, hard work, or luck?

Students in each of the populations were asked whether doing well in mathematics and science was a matter of hard work, natural ability, or good luck. The questions were:
      1. To do well in mathematics you need lots of natural ability
      2. To do well in mathematics you need good luck
      3. To do well in mathematics you need lots of hard work studying at home

      1. To do well in science you need lots of natural ability
      2. To do well in science you need good luck
      3. To do well in science you need lots of hard work studying at home

The results for each population are in Tables 12, 13, and 14. I present the results for each of the populations together because the patterns are so similar. That is especially true for the Question Effects where the patterns suggest that the majority of students within these systems do not believe that doing well in mathematics and science is a matter of luck. That view seems to strengthen over time since the Question Effects move from -19 to -29 to -45.

The System Effects are reasonably constant over the populations. Figure 6 shows that relationship among the systems that participated in both the Grade 4 and Grade 8 studies.

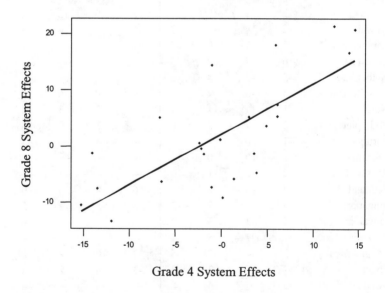

Figure 6. Plot of Grade 4 and Grade 8 system effects with a least squares regression line.

Table 12. Summary results for how to do well in mathematics and science – Population 3

Country	Mathematics			Science			System Effects
	Q1	Q2	Q3	Q1	Q2	Q3	
Australia	72	16	95	71	16	96	4
Austria	77	25	80	50	27	66	-1
Canada	74	16	92	72	16	96	5
Cyprus	58	20	83	69	22	95	-2
Czech Republic	68	42	75	46	38	86	0
Denmark	94	17	87	91	19	90	5
France	46	10	89	40	11	93	-13
Germany	73	18	71	67	16	81	-3
Greece	57	26	94	53	28	97	4
Israel	85	18	89	0	0	0	-14
Italy	86	20	93	63	19	91	4
Lithuania	75	41	71	60	35	76	0
Russian Fed.	80	27	87	76	31	81	6
Slovenia	86	24	78	73	25	83	2
Sweden	57	9	73	47	9	87	-14
Switzerland	73	15	67	66	14	75	-7
United States	57	16	89	52	17	91	-7
Question Effects	7	-45	17	-7	-44	19	GE=67

Table 13. Summary results for how to do well in mathematics and science– Grade 4

Country	Mathematics			Science			System Effects
	Q1	Q2	Q3	Q1	Q2	Q3	
Australia	85	58	84	84	59	82	-2
Austria	72	54	83	72	54	84	-7
Canada	81	49	90	82	50	89	1
Cyprus	68	58	96	69	58	96	-3
Czech Republic	61	68	87	57	65	87	0
England	—	—	—	—	—	—	—
Greece	62	48	87	63	46	87	-14
Hong Kong	74	20	95	72	19	96	-7
Hungary	96	75	88	95	75	89	14
Iceland	79	63	90	78	61	88	0
Iran, Isl. Rep.	96	74	94	96	75	93	13
Ireland	87	65	91	87	65	89	4
Israel	66	43	96	69	44	96	-12
Japan	79	53	91	79	53	88	-1
Korea	90	62	95	90	61	94	7
Kuwait	92	76	87	92	78	88	12
Latvia (LSS)	77	86	92	75	84	91	4
Netherlands	64	28	77	65	28	73	-15
Norway	93	58	89	93	59	88	1
New Zealand	84	64	87	84	65	86	3
Portugal	86	66	95	85	67	96	6
Scotland	—	—	—	—	—	—	—
Singapore	88	36	95	88	37	94	7
Slovenia	84	63	92	81	62	93	3
Thailand	78	65	68	77	63	70	-3

United States	62	46	93	62	46	93	-15
Question Effects	0	-19	8	0	-19	7	GE=80

Table 14. Summary results for how to do well in mathematics and science – Grade 8

	Mathematics			Science			System Effects
Country	Q1	Q2	Q3	Q1	Q2	Q3	
Australia	66	30	92	66	33	91	0
Austria	70	27	78	61	31	78	-8
Belgium-Fl.	58	22	85	53	24	85	-10
Belgium-Fr.	69	23	93	67	25	94	2
Canada	61	26	87	61	30	89	-5
Colombia	91	62	97	91	64	97	25
Cyprus	51	34	92	51	34	93	-4
Czech Republic	61	57	81	45	55	82	-5
Denmark	90	28	87	89	35	82	-2
England	45	23	93	47	25	93	-14
France	40	21	90	38	23	88	-16
Germany	59	25	76	57	28	82	-9
Greece	54	26	95	58	27	96	-9
Hong Kong	77	38	95	74	38	96	7
Hungary	95	56	79	88	56	79	18
Iceland	37	24	92	36	26	90	-13
Iran, Isl. Rep.	95	51	96	95	51	97	13
Ireland	72	31	95	70	32	95	5
Israel	55	17	96	53	19	95	-12
Japan	82	59	96	82	60	97	16
Korea	86	63	98	85	62	98	20
Kuwait	87	76	83	90	78	83	23
Latvia (LSS)	61	63	91	50	61	87	1
Lithuania	85	69	83	76	68	76	15
Netherlands	44	23	89	46	25	93	-14
New Zealand	62	27	92	63	29	92	-3
Norway	86	19	92	84	22	92	4
Philippines	86	75	87	86	76	88	20
Portugal	72	39	97	72	39	98	7
Romania	66	59	88	64	59	86	0
Russian Fed.	79	51	89	77	53	87	12
Scotland	—	—	—	—	—	—	—
Singapore	84	41	92	86	40	98	7
Slovakia	69	52	90	61	52	92	3
Slovenia	81	38	82	75	41	90	2
South Africa	77	72	84	78	72	84	12
Spain	66	35	89	66	35	96	0
Sweden	48	24	83	45	26	87	-13
Switzerland	60	22	71	56	25	75	-14
Thailand	69	34	77	69	35	80	-3
United States	50	32	90	51	34	90	-5
Question Effects	0	-29	22	0	-27	23	GE=66

Despite the fact that there are comparable Question Effects and System Effects for these results, the most interesting aspect of the analyses is the Unique Effects. Table 15 contains the Unique Effects for each system, mathematics and science in Grades 4 and 8, and mathematics for Population 3.

Table 15. Unique effects for ability, luck, and hard work, Grade 4, Grade 8, and Population 3

	Grade 4						Grade 8						Pop. 3					
	Math			Science			Math			Science			Math			Science		
Country	M1	M2	M3	S1	S2	S3	M1	M2	M3	S1	S2	S3	M1	M2	M3	S1	S2	S3
Australia	8	0	-1	6	0	-3	-1	-7	4	0	-6	2	-13	1	-17	-16	0	0
Austria	-1	0	2	-2	0	3	12	-1	-1	3	1	-3	7	-5	8	10	0	-14
Belgium-Fl.	—	—	—	—	—	—	2	-4	8	-3	-4	6	—	—	—	—	—	—
Belgium-Fr.	—	—	—	—	—	—	1	-16	3	-1	-16	3	—	—	—	—	—	—
Canada	0	-13	1	0	-12	0	0	-6	4	0	-5	5	-8	5	-13	-13	0	4
Switzerland	—	—	—	—	—	—	8	-1	-3	4	0	0	5	13	0	-1	-10	-3
Colombia	—	—	—	—	—	—	0	0	-15	0	1	-16	—	—	—	—	—	—
Cyprus	-9	0	11	-9	0	11	-11	1	9	-11	-1	8	-15	12	0	2	0	12
Czech Rep.	-19	7	0	-22	5	0	1	26	-1	-16	22	-1	-8	-15	19	14	-11	0
Germany	—	—	—	—	—	—	3	-3	-2	1	-1	2	2	11	0	-2	-10	0
Denmark	—	—	—	—	—	—	26	-6	2	26	-1	-5	13	25	-11	-9	-4	-1
Spain	—	—	—	—	—	—	-1	-2	1	0	-3	7	—	—	—	—	—
France	—	—	—	—	—	—	-10	0	18	-12	0	15	-17	-8	0	0	16	20
England	0	19	-8	0	19	-7	-7	0	19	-5	0	19	—	—	—	—	—	—
Greece	-4	1	13	-3	-1	13	-3	-2	17	1	-2	17	-23	-11	0	1	5	7
Hong Kong	1	-34	14	-1	-35	15	4	-6	0	1	-8	0	—	—	—	—	—	—
Hungary	2	0	-14	1	0	-12	10	1	-26	5	-1	-28	—	—	—	—	—	—
Ireland	2	0	-2	2	0	-3	0	-11	2	-1	-12	1	—	—	—	—	—	—
Iran, Isl. Rep.	3	-1	-8	3	1	-7	16	1	-5	16	0	-5	—	—	—	—	—	—
Iceland	-1	2	1	-2	0	0	-16	0	18	-17	0	14	—	—	—	—	—	—
Israel	-1	-6	20	1	-5	20	0	-8	20	-1	-8	18	22	-48	8	-10	16	-73
Italy	—	—	—	—	—	—	—	—	—	—	—	—	8	0	-5	-6	5	3
Japan	0	-6	4	0	-7	2	-1	6	-8	0	5	-9	—	—	—	—	—	—
Korea	3	-5	0	3	-7	0	0	6	-9	-1	3	-11	—	—	—	—	—	—
Kuwait	0	4	-13	0	5	-12	-2	16	-27	1	16	-29	—	—	—	—	—	—
Lithuania	—	—	—	—	—	—	4	17	-20	-4	15	-27	0	0	19	12	-14	-9
Latvia (LSS)	-7	21	0	-9	19	0	-6	25	3	-17	21	-3	—	—	—	—	—	—
Netherlands	-1	-17	5	1	-18	1	-8	0	16	-5	0	18	—	—	—	—	—	—
Norway	13	-3	0	12	-2	0	16	-21	1	14	-20	-1	—	—	—	—	—	—
N. Zealand	0	-1	-5	0	1	-4	-1	-7	8	0	-7	7	—	—	—	—	—	—
Philippines	—	—	—	—	—	—	-1	18	-21	0	17	-22	—	—	—	—	—	—
Portugal	0	-1	2	-1	0	3	0	-4	3	0	-6	2	—	—	—	—	—	—
Romania	—	—	—	—	—	—	0	22	0	-1	20	-3	—	—	—	—	—	—
Russian Fed.	—	—	—	—	—	—	0	1	-11	-1	1	-14	0	11	0	4	-3	-9
Scotland	0	19	-8	0	19	-7	0	29	-22	0	27	-23	—	—	—	—	—	—
Singapore	1	-32	0	1	-31	0	11	-3	-2	13	-6	2	—	—	—	—	—	—
Slovakia	—	—	—	—	—	—	0	12	-1	-8	10	0	—	—	—	—	—	—
Slovenia	1	-1	1	-2	-2	3	13	0	-8	7	0	-1	8	11	0	0	-10	-4
Sweden	—	—	—	—	—	—	-6	0	8	-8	0	11	-6	-1	-1	-1	0	14
Thailand	0	6	-17	0	4	-15	6	1	-7	7	0	-6	—	—	—	—	—	—

South Africa	—	—	—	—	—	—	-1	24	-15	0	22	-17	—	—	—	—	—	—
Un. States	-4	0	20	-3	0	20	-11	0	7	-11	0	5	-12	-2	-1	0	10	11

The Unique Effects are extraordinarily varied. To get some idea of what the results suggest, I created two more displays. One contains correlations across the effects; the other is a boxplot to identify unusual values. These tables are based on data from systems that participated in both the Grade 4 and Grade 8 studies. I did this to look for patterns across these Unique Effects. Table 16 contains the correlations.

Table 16. Correlations of unique effects for ability, luck, and hard work, across populations

	MA4	ML4	MW4	SA4	SL4	SW4	MA8	ML8	MW8	SA8	SL8
Math Ability, Gr 4											
Math Luck, Gr 4	-29[1]										
Math Work, Gr 4	-24	-30									
Science Ability, Gr 4	98	-32	-16								
Science Luck, Gr 4	-25	99	-32	-27							
Science Work, Gr 4	-26	-29	99	-19	-31						
Math Ability, Gr 8	47	-18	-35	39	-16	-30					
Math Luck, Gr 8	-64	43	-16	-68	39	-16	-16				
Math Work, Gr 8	-15	-09	69	-07	-11	64	-41	-26			
Science Ability, Gr 8	71	-38	-31	69	-33	-29	86	-44	-33		
Science Luck, Gr 8	-62	46	-16	-67	43	-17	-17	99	-25	-45	
Science Work, Gr 8	-14	-20	69	-07	-22	64	-36	-32	98	-27	-31

[1]Decimal points have been omitted.

I highlighted those cells of the correlation matrix that are correlations between comparable measures even though they may be from different subject areas or populations. That is, the highlighted correlations are for all those that are relationships among ability effects, among luck effects, and among the work effects.

The patterns are clear. Regardless of the whether one is looking at students' perceptions of ability, luck, or work, those perceptions are consistent across populations and grade levels. Each correlation that is highlighted is both positive and reasonably high. The correlations among the measures and subject area within a population are stronger than the correlations between the populations, although those are still substantial.

All of the remaining correlations are negative. That suggests that if students' perceptions are high on one of the measures they will be low on another. I believe these findings suggest that there is consistency within systems on what students consider most influential. Systems that uniquely endorse ability tend not to endorse either luck or work.

The boxplots in Figure 7, one for Grade 4 and one for Grade 8, identify systems that have extreme values on the various dimensions. I put these two boxplots together because there are interesting differences between the two grade levels. The spread of Unique Effects for Grade 8 students is much greater than Grade 4 students in terms of perceptions of ability and hard work. There is less agreement in those areas. On the other hand, the spread of effects for luck is much narrower for Grade 8 students than for Grade 4 students. It is as though by Grade 8 most students in most systems have decided about the luck dimension but are undecided about whether doing well in mathematics is a matter of natural ability or hard work.

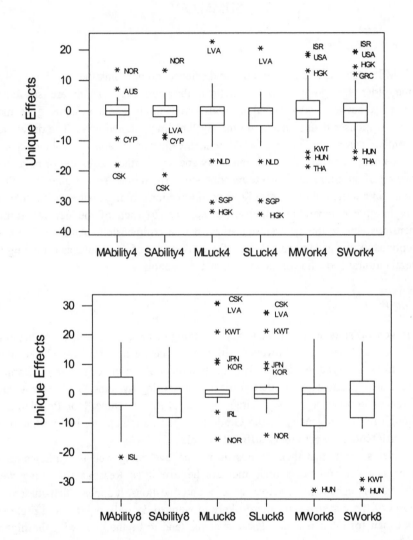

Figure 7. Boxplots of unique effects for ability, luck, and hard work, Grades 4 and 8.

The unusual values tend to come in pairs. If, at a particular grade level, a system is an outlier on a question about, say, ability in mathematics, it tends to be an outlier on the same question in science. Latvia (LVA) and Kuwait (KWT) are the only two systems that have both high and low unusual values. Grade 4 students in Latvia tend not to endorse ability in science but highly endorse luck in both mathematics and science. Grade 8 students in Kuwait do not endorse hard work in science but produce large positive Unique Effects for luck in both science and mathematics.

SUMMARY

General Attitudes

TIMSS students like mathematics and science, do not think they are boring, and, at the older age groups, do not believe they are easy. There are substantial differences between educational systems; students in some systems give more positive responses than students in other systems in patterned ways. The differences are similar across the TIMSS population, the questions, and the subject matter. Students in systems that respond, relatively speaking, either positively or negatively to one set of questions tend also to respond similarly to other sets of questions. So, if students in a system either strongly or weakly endorse liking mathematics, they are likely to give comparable responses to science, to each of the general attitude dimensions, and to the different questions within a dimension. Although there are exceptions to this general rule, I view the consistency of the results as reflecting the general cultural and ethos of the system and its schools.

Doing Well

In general students report that they are doing or have done well in mathematics and science. There are, however, differences between systems and the three populations surveyed in TIMSS. Systems that did well on the cognitive tests appear to have less positive views that some lower scoring systems. There are greater differences between systems at Grade 8 than at either Grade 4 or Population 3. Although the heterogeneity of the Grade 8 responses is notable, interpretations of those differences are extremely difficult to make.

Students report that their friends do not believe doing well in mathematics or science is as important as their mothers believe it is. Reasons for doing well, however, vary substantially across systems and questions. It appears that students in a subset of these systems are more likely to want to do well to get into a preferred school than for other reasons. These systems tend to be the ones with the highest cognitive scores.

Finally, most students believe doing well in mathematics and science is a matter either of natural ability or hard work but not luck. There are, however, a large number of substantial Unique Effects for these questions. In systems where students give unusually high endorsements to natural ability they tend to give low endorsements to hard work. The hard work systems likewise tend to place less emphasis on natural ability. There are, of course, students in each system who believe luck is a strong factor and affirm, one hopes, that luck favors the prepared.

REFERENCES

Bronfenbrenner, U. (1970). *Two worlds of childhood: U.S. and U.S.S.R.* New York: Russell Sage Foundation.

Burstein, L. (Ed.). (1994). *The IEA study of mathematics III: Student growth and classroom process in early secondary school.* London: Pergamon Press.

Comber, L. C. & Keeves, J. R. (1973). *Science education in nineteen countries.* Stockholm: Almqvist and Wiksell.

Dreeben, R. (1968). *On what is learned in school.* Reading, MA: Addison-Wesley Publishing Company.

Husén, T. (Ed.). (1967). *International study of achievement in mathematics (Vols. 1 and 2).* Stockholm: Almqvist and Wiksell.

Kifer, E. & Robitaille, D. F. (1989). Atttitudes, preferences and opinions. In D. F. Robitaille & R. A. Garden (Eds.),. *The IEA study of mathematics II: Contexts and outcomes of school mathematics.* London: Pergamon Press.

Tukey, J. W. (1977). *Exploratory data analysis.* Reading, MA: Addison-Wesley Publishing Company.

Velleman, P. F. & Hoaglin, D. C. (1981). *Applications, basics, and computing of exploratory data analysis.* Boston: Duxbury Press.

Chapter 17

ANALYZING GENDER DIFFERENCES FOR HIGH-ACHIEVING STUDENTS ON TIMSS

Ina V. S. Mullis and Steven E. Stemler

The data from TIMSS provide a rich source for examining gender equity both within and across countries. Although a number of options for exploring gender differences are available, time and resources typically allow only for reporting the results obtained from one or two methods. While much of the information obtained by various analytic methods overlaps, each research question and its corresponding analytic approach contribute something unique to a policy discussion related to gender equity.

Considerable research, including the findings from TIMSS, has shown that, as students get older, gender differences favoring males increase both in mathematics and science achievement (Campbell, 1995; Gray, 1996; Mullis, Martin, Fierros, Goldberg, & Stemler, 2000). In addition, many studies have found that gender differences in achievement are not distributed uniformly across the ability range (Manger & Gjestad, 1997; Willingham & Cole, 1997). With these issues in mind, the present chapter discusses the policy implications of three different analyses of gender differences using data from high-performing students in the final year of secondary school on TIMSS.

Overview of TIMSS 1995 Results by Gender

The TIMSS data from students in the final year of secondary school provide the best opportunity to describe results by gender for both mathematics and science, since gender differences in mathematics achievement were relatively minor for students at the primary and middle school years (Beaton et al., 1996; Mullis et al., 1997). In *Mathematics and Science Achievement in the Final Year of Secondary School* (Mullis et al., 1998), TIMSS reported that achievement differences in mathematics and science literacy favoring males were found in every country except South Africa. In advanced mathematics, gender differences favoring males were found in 11 of the 16 participating countries. Furthermore, in some countries,

D.F. Robitaille and A.E. Beaton (eds.), Secondary Analysis of the TIMSS Data, 277–290.
© 2002 *Kluwer Academic Publishers. Printed in the Netherlands.*

substantially more males than females had taken advanced mathematics courses. Similarly, males had significantly higher physics achievement than females in all countries except Latvia, and in several countries the proportion of males having taken physics outnumbered females by two or three to one.

A subsequent report, *Gender Differences in Achievement* (Mullis et al., 2000), provided an in-depth look at the TIMSS results by gender through several lenses, including one approach to exploring gender differences for high-performing students discussed later in this chapter. That report also presented the results of an item-by-item analysis showing that males tended to have higher mathematics achievement than females on items involving spatial representation, proportionality, measurement, and problems with no immediate formula. For Population 3, there were no items on which females outperformed males, on average, internationally. In science, males had higher achievement on items involving earth science and physical science, particularly if the item stimulus involved a diagram. Females, however, tended to have higher achievement on items involving health and nutrition. Selected background questionnaire data from the final-year students related to motivating factors for high achievement revealed that more males than females felt it was important to well in mathematics and science, and that significantly more males reported that they would like to have a job involving mathematics.

Analyses and Research Questions Described

When discussing gender differences in achievement, the most common analytic approach reported in the literature is an independent means *t*-test for differences in achievement between the two groups. This technique is useful for answering general questions related to the existence of gender differences in achievement; however, there are other specific research questions that require alternative analytic strategies. Three specific research questions about high-achieving students will be addressed in this chapter:

1. Did an equal percentage of male test-takers and female test-takers reach a particular benchmark for their country?

2. Are there equal percentages of males and females among high-achieving students?

3. Are gender differences in achievement favoring males more prevalent for high-performing students than for low-performing students?

Each of these questions deals with a slightly different issue and consequently requires a unique analytic strategy to answer. The first question will be addressed by comparing the percentage of male final-year students who reached an upper benchmark for their country, defined as the 75th percentile of performance for all students in the country, to the percentage of female final-year students who reached that same benchmark. This question explores the nature of the distribution of students within a given gender group meeting a particular standard. The expectation is that 25 percent of the males taking the test would meet or exceed the set standard of achievement in their country, and that 25 percent of the females taking the test would meet or exceed that same threshold. The most important policy issue here is the achievement of specific target groups, as defined by their gender.

The second research question explores the extent to which gender diversity is present among high-performing students in each country. This question will be explored by examining the final-year students who score in the top quarter of the distribution in each country and calculating the percentage of those who are male and female. This is not unrelated to the first question; however, each analysis provides unique information. If the primary group of interest is defined not by gender, but rather in terms of achievement (e.g., high-performing students), then the second research question is more germane. The expectation is that, in a world characterized by gender equity, we would find that, of the high-achieving students, 50 percent would be male and 50 percent would be females.

The third research question pertains to the nature of gender differences in achievement across the performance distribution. The third question about achievement differences for different segments of the distribution typically would be addressed using a two-way analysis of variance. Because we already know that gender differences do exist and that they uniformly favor males, however, the analysis presented herein uses a series of independent means *t*-tests that have been corrected for multiple-comparisons using the Bonferroni technique.

The third analysis relies upon a procedure for identifying high- and low-performing males and females previously used in IEA's Reading Literacy Study. In *Are Girls Better Readers? Gender Differences in Reading Literacy in 32 Countries* (Wagemaker, 1996), the approach used was different from either the benchmarking or high-performing analyses outlined above. The analysis of high- and low-performing students in that report involved first identifying the high-performing males and the high-performing females in each country and then comparing their performance on the literacy test (Martin, 1996). The male students who scored above the 75th percentile for male students on the overall literacy score in each country (that is, the top 25 percent of male students) were assigned to the male high-scoring group. The female students scoring above the 75th percentile for female students in each country (that is, the top 25 percent of female students) were assigned to the female high-scoring group. In other words, the 75th percentile cut point was not the same for males and females, but varied based upon the gender group. The mean scores of male and female students in their respective high-performing groups were then computed for each country. Similarly, for purposes of comparison, the corresponding analyses were conducted for the bottom 25 percent of male students and the bottom 25 percent of female students.

METHODS

Given the great diversity found among the educational systems of the countries participating in TIMSS, assessing students at the final year of secondary school required special consideration. Many countries had a system in which students were tracked into particular educational programs after the eighth grade. These educational programs usually took the form of an academic track or a vocational track, each with its own requirements and each requiring students to study for different amounts of time (Dumais, 1998). For example, Grade 10 could be the final year of a vocational program, and Grade 12 the final year of an academic program. Both of these grade/track combinations were considered to be part of the target population for the final year of secondary school, but Grade 10 in the academic track was not. Nevertheless, even across these various programs, "One of the goals of TIMSS was to assess the mathematics and science literacy of all students while also assessing the advanced mathematics and physics knowledge of students with preparation in these subjects" (Dumais, 1998). In attempting to accomplish this goal, TIMSS stratified the population of students at the final year of secondary school into four groups:

1. students who had studied neither advanced mathematics or physics,
2. students who had studied physics but not advanced mathematics,
3. students who had studied advanced mathematics but not physics, and
4. students who had studied both advanced mathematics and physics.

The sample of students taking the mathematics and science literacy assessment was designed to be representative of the entire population of final-year students. Furthermore, the design of the assessment was such that mathematics and science literacy could be reported separately (see Mullis et al., 1998; Mullis et al., 2000) or as a combined mathematics and science literacy score, which is the focus of the present study. The complete sample design for Population 3 is documented by Dumais (1998).

Criteria for Inclusion in the Analyses

Certain criteria were established regarding the countries that would be included in the present analysis. In particular, for ease of comparison, the countries retained for the discussion here included only those that participated in the mathematics and science literacy portion of the TIMSS assessment of final-year students. This meant that Greece, which participated only in the advanced mathematics and physics assessments, was not included. In addition, Denmark, Israel, Italy, and South Africa were excluded from this investigation because of a variety of issues related to their not having followed the guidelines for sampling or having low participation rates.

Table 1 lists the countries included in the analyses presented in this chapter and identifies their participation according to the various components of the final-year assessment, including the mathematics and science literacy assessment, the advanced mathematics assessment, and the physics assessment. The table also provides a breakdown of the total sample size of test-takers along with the percentages by gender for each of the final-year tests.

Table 1. Sample sizes for countries included in this chapter

	Mathematics and Science Literacy			Advanced Mathematics			Physics		
	Number Students	Percent Male	Percent Female	Number Students	Percent Male	Percent Female	Number Students	Percent Male	Percent Female
Australia	1844	42 (2.9)	58 (2.9)	548	55 (5.5)	45 (5.5)	564	66 (3.8)	34 (3.8)
Austria	1779	39 (3.2)	61 (3.2)	599	38 (4.1)	62 (4.1)	594	38 (3.5)	62 (3.5)
Canada	4832	47 (1.4)	53 (1.4)	2381	53 (1.6)	47 (1.6)	1967	57 (3.2)	43 (3.2)
Cyprus	473	45 (2.1)	55 (2.1)	330	61 (1.6)	39 (1.6)	307	63 (2.5)	37 (2.5)
Czech Republic	1899	51 (5.1)	49 (5.1)	833	41 (2.5)	59 (2.5)	819	38 (2.4)	62 (2.4)
France	1590	47 (3.1)	53 (3.1)	796	63 (2.0)	37 (2.0)	835	61 (2.0)	39 (2.0)
Germany	2182	56 (5.2)	44 (5.2)	2189	43 (2.4)	57 (2.4)	616	69 (3.0)	31 (3.0)
Hungary	5091	52 (2.5)	48 (2.5)	—	—	—	—	—	—
Iceland	1703	48 (0.8)	52 (0.8)	—	—	—	—	—	—
Lithuania	2887	35 (3.0)	65 (3.0)	734	51 (1.9)	49 (1.9)	—	—	—
Netherlands	1470	52 (2.3)	48 (2.3)	—	—	—	—	—	—
New Zealand	1763	49 (1.7)	51 (1.7)	—	—	—	—	—	—
Norway	2518	51 (2.0)	49 (2.0)	—	—	—	1048	74 (1.8)	26 (1.8)
Russian Federation	2289	38 (1.0)	62 (1.0)	1364	52 (2.4)	48 (2.4)	985	54 (2.0)	46 (2.0)
Slovenia	1387	51 (3.3)	49 (3.3)	1301	50 (4.2)	50 (4.2)	512	72 (3.7)	28 (3.7)
Sweden	2816	49 (2.5)	51 (2.5)	749	69 (3.4)	31 (3.4)	760	67 (3.4)	33 (3.4)
Switzerland	2976	56 (2.5)	44 (2.5)	1072	54 (2.4)	46 (2.4)	1039	51 (1.8)	49 (1.8)
United States	5371	50 (1.3)	50 (1.3)	2349	51 (2.6)	49 (2.6)	2678	52 (2.4)	48 (2.4)

( ) Standard errors appear in parentheses. Because results are rounded to the nearest whole number, some totals may appear inconsistent.

RESULTS

Research Question 1 - Gender Benchmarking Analysis

The aim of the first analysis was to look at gender differences in achievement for final-year students in relation to an established criterion, or benchmark, within each particular country. Table 2 presents the results of the gender benchmarking analyses for mathematics and science literacy, advanced mathematics, and physics. It presents the percentage of females achieving at or above the benchmark, the percentage of males achieving at or above the benchmark, and any statistically significant differences. The benchmark chosen was the upper quartile of achievement that was specific to each country in the particular component of the final-year assessment. For example, in Table 2, the upper quartile of achievement in mathematics and science literacy in the United States was 490, while the upper quartile of achievement in Australia was 555. The percentage of females reaching the upper-quarter benchmark was calculated based upon Equation 1 below:

$$P_F = \frac{N_{HPF}}{N_F} \times 100 \qquad (1)$$

where:

P_F = Percent of female test-takers reaching the upper-quarter benchmark
N_{HPF} = Number of high-performing students who are female, and
N_F = Total number of females taking the test

Similarly, the percentage of males reaching the benchmark was derived from Equation 2 presented below:

$$P_M = \frac{N_{HPM}}{N_M} \times 100 \qquad (2)$$

where:

P_M = Percent of male test-takers reaching the upper-quarter benchmark
N_{HPM} = Number of high-performing students who are male, and
N_M = Total number of males taking the test

The strength of this analysis lies in its straightforward approach and interpretation. In a hypothetical world with no gender differences in achievement, 25 percent of males in each country and 25 percent of females in each country would score at or above the upper-quarter benchmark (i.e., the point reached by 75 percent of all test takers score). In the analysis of males and females reaching the top 25 percent benchmark, we examined the extent to which the results differed from the equity value of 25 percent. In addition, testing was conducted to see if the percentage of males reaching the benchmark was significantly different from the percentage of females reaching the benchmark.

The results in Table 2 show that, in most countries (14 out of 18), significantly greater percentages of males than of females reached the upper-quarter benchmark for mathematics and science literacy in their country. In advanced mathematics, statistically significant differences between the percentages of males and females reaching the upper quartile were found in about half the countries (7 out of 13). In physics, significant differences were found in 10 of the 13 countries. The international average for physics revealed that only 13 percent of females compared to 34 percent of males reached the upper-quarter benchmark.

Table 2. Percentages by gender of high-performing students in mathematics and science literacy, advanced mathematics, and physics – Final year of secondary school

	Mathematics and Science Literacy		Advanced Mathematics		Physics	
	Males	Females	Males	Females	Males	Females
	Percentage At or Above Country's Upper Benchmark	Percentage At or Above Country's Upper Benchmark	Percentage At or Above Country's Upper Benchmark	Percentage At or Above Country's Upper Benchmark	Percentage At or Above Country's Upper Benchmark	Percentage At or Above Country's Upper Benchmark
Australia	*34 (4.4)	19 (3.2)	27 (3.8)	23 (6.2)	30 (4.3)	15 (3.4)
Austria	*39 (3.7)	17 (2.1)	*43 (4.3)	14 (3.1)	*46 (4.4)	13 (3.6)
Canada	*32 (2.2)	19 (1.6)	*32 (3.1)	17 (2.6)	*34 (2.3)	13 (2.3)
Cyprus	30 (2.7)	21 (2.1)	28 (2.9)	20 (3.2)	32 (3.9)	15 (3.3)
Czech Republic	33 (5.0)	17 (4.1)	*44 (4.5)	12 (2.4)	*50 (4.4)	10 (1.6)
France	*35 (4.3)	16 (2.6)	30 (3.2)	18 (3.2)	*31 (2.3)	17 (3.2)
Germany	30 (3.9)	20 (3.0)	*33 (3.2)	20 (3.2)	*32 (4.6)	10 (2.6)
Hungary	*30 (2.0)	20 (1.6)	—	—	—	—
Iceland	*36 (1.6)	15 (1.0)	—	—	—	—
Lithuania	30 (3.7)	23 (3.0)	*35 (2.0)	15 (2.6)	—	—
Netherlands	*33 (3.0)	17 (2.7)	—	—	—	—
New Zealand	*31 (2.3)	19 (1.8)	—	—	—	—
Norway	*36 (2.1)	14 (1.6)	—	—	*29 (2.7)	13 (2.4)
Russian Federation	*34 (3.4)	20 (2.8)	—	—	*33 (4.2)	15 (4.0)
Slovenia	*36 (5.5)	14 (3.4)	29 (5.6)	21 (3.7)	31 (7.8)	9 (5.0)
Sweden	*36 (2.5)	15 (1.2)	28 (2.7)	18 (3.2)	*33 (2.7)	9 (2.6)
Switzerland	*30 (2.4)	18 (2.2)	*36 (2.1)	13 (2.0)	*42 (2.8)	8 (1.1)
United States	*29 (2.0)	21 (1.6)	28 (3.1)	22 (2.7)	*34 (3.0)	16 (2.1)
International	*36 (0.7)	22 (0.5)	*32 (0.9)	18 (0.8)	*34 (1.0)	13 (0.9)

* represents a statistically significant gender difference at the 0.05 level. ( ) Standard errors appear in parentheses. Because results are rounded to the nearest whole number, some totals may appear inconsistent.

Research Question 2 - High-Performing Students by Gender

The second analysis focused on those students scoring in the top quarter of the achievement distribution within each country, and explored how closely the representation of males and females within the high-achieving group was to its expected value. For example, if the population of secondary students in Austria is composed of 55 percent males and 45 percent females, then the expectation would be that the representation of males and females among the high-performing students would mirror their prevalence in the population. One complication arises from the fact that a sample of these students was drawn, and the sampling error itself may have introduced an imbalance into the results. In most countries, the expectation was that 50 percent of the students would be male and 50 percent would be female. Consequently, in order to facilitate comparisons, an adjustment factor was incorporated to account for male and female imbalances that may have come about due to sampling.

Table 3 presents the percentage of males and females found among the high-performing students within each country on the mathematics and science literacy, advanced mathematics, and physics assessments. Table 3 shows that males are significantly over-represented among high-performing final-year students in mathematics and science literacy in 13 of 18 countries that participated in TIMSS. The Scandinavian countries—Norway, Iceland, and Sweden—showed some of the largest differences.

Table 3. Percentages by gender of high-performing students in mathematics and science literacy – Final year of secondary school

	Mathematics and Science Literacy		Advanced Mathematics		Physics	
	Percent Males	*Percent Females*	*Percent Males*	*Percent Females*	*Percent Males*	*Percent Females*
Australia	*65 (4.1)	35 (4.1)	54 (8.3)	46 (8.3)	67 (8.0)	33 (8.7)
Austria	*70 (3.2)	30 (3.2)	*75 (4.6)	25 (4.6)	*78 (5.6)	22 (5.6)
Canada	*63 (3.2)	37 (3.2)	*65 (4.6)	35 (4.6)	*72 (4.5)	28 (4.5)
Cyprus	59 (4.5)	41 (4.5)	59 (6.2)	41 (6.2)	68 (8.4)	32 (8.4)
Czech Republic	66 (6.1)	34 (6.1)	*79 (2.5)	21 (2.5)	*84 (2.3)	16 (2.3)
France	*68 (3.5)	32 (3.5)	63 (7.1)	37 (7.1)	*65 (4.6)	35 (4.6)
Germany	60 (5.3)	40 (5.3)	*63 (2.9)	37 (2.9)	*77 (7.1)	23 (7.1)
Hungary	*60 (3.1)	40 (3.1)	—	—	—	—
Iceland	*71 (1.9)	29 (1.9)	—	—	—	—
Lithuania	57 (4.0)	43 (4.0)	*69 (4.2)	31 (4.2)	—	—
Netherlands	*67 (3.7)	33 (3.7)	—	—	—	—
New Zealand	*62 (3.0)	38 (3.0)	—	—	—	—
Norway	*73 (2.7)	27 (2.7)	—	—	*69 (5.1)	31 (5.1)
Russian Federation	*64 (3.3)	36 (3.3)	*67 (3.2)	33 (3.2)	*69 (6.4)	31 (6.4)
Slovenia	*71 (5.3)	29 (5.3)	58 (5.8)	42 (5.8)	*78 (9.3)	22 (9.3)
Sweden	*71 (2.4)	29 (2.4)	62 (9.5)	38 (9.5)	*80 (5.6)	20 (5.6)
Switzerland	62 (4.2)	38 (4.2)	*74 (3.2)	26 (3.2)	*84 (2.4)	16 (2.4)
United States	*58 (2.7)	42 (2.7)	57 (4.5)	43 (4.5)	*68 (3.7)	32 (3.7)
International	*65 (0.9)	35 (0.9)	*65 (1.5)	35 (1.5)	*74 (1.7)	26 (1.7)

* represents a statistically significant gender difference at the 0.05 level. ( ) Standard errors appear in parentheses. Because results are rounded to the nearest whole number, some totals may appear inconsistent. Percentages have been adjusted to account for male-female imbalances in the total sample.

In advanced mathematics, the disparity in representation was often more extreme than in mathematics and science literacy. For example, in Austria, 70 percent or more of the high-performing final-year students were male across all three subject areas. Statistically significant differences between the percentages of male and female high-performing students in advanced mathematics were observed in 7 out of the 13 participating countries. In addition, Table 3 shows that, in physics, males were disproportionately represented in nearly every participating country (11 out of 13). On average, internationally, three out of four high-performing students in physics were males.

Research Question 3 - Achievement Gaps across the Performance Distribution

The data presented in this section provide differences in the mean achievement between males and females classified as high-achievers for their gender and then compares the magnitude of this difference to the magnitude of difference in mean achievement between male and females classified as low-achievers for their gender. Although the actual differences in achievement are not specified in the tables, the magnitudes of the differences are plotted in Table 4.

Table 4. Average mathematics and science literacy achievement of high-performing (upper quarter) and low-performing (lower quarter) females and males

	Average Achievement for Low-Performing Students		Average Achievement for High-Performing Students		Difference in Average Achievement Between Females and Males Among High-Performing an Low-Performing Students
	Females	Males	Females	Males	Males Scored Higher
Norway	416 (3.2)	• 454 (4.2)	608 (3.6)	• 681 (4.3)	
Sweden	434 (2.6)	• 458 (2.9)	637 (3.5)	• 705 (3.9)	
Czech Rep.	353 (5.4)	• 396 (2.6)	564 (4.8)	• 630 (3.9)	
Austria	415 (3.4)	• 442 (3.9)	596 (4.0)	• 657 (7.5)	
Slovenia	398 (3.9)	430 (12.8)	583 (5.0)	• 644 (6.8)	
Iceland	434 (1.9)	• 467 (3.1)	615 (3.1)	• 663 (3.6)	
International Avg.	383 (1.0)	• 412 (1.2)	588 (1.3)	• 634 (1.1)	
France	402 (3.7)	• 429 (4.2)	575 (2.6)	• 620 (3.5)	
Australia	400 (5.3)	414 (7.5)	621 (6.7)	• 667 (6.8)	
New Zealand	403 (5.2)	414 (7.2)	618 (3.5)	• 661 (3.8)	
Canada	412 (4.0)	• 440 (2.6)	611 (3.3)	• 653 (3.3)	
Switzerland	403 (5.4)	• 440 (4.8)	619 (3.2)	• 661 (3.6)	
Hungary	375 (2.1)	377 (2.7)	567 (3.2)	• 608 (4.1)	
Netherlands	431 (2.8)	• 483 (4.7)	641 (5.9)	• 681 (4.9)	
Russian Fed.	364 (2.4)	• 400 (2.3)	570 (6.1)	• 606 (5.1)	
Germany	363 (5.7)	• 407 (5.1)	590 (5.4)	• 624 (6.7)	
Cyprus	358 (4.7)	362 (7.1)	527 (4.4)	• 559 (6.4)	
United States	358 (3.2)	367 (2.4)	572 (3.3)	• 602 (3.5)	
Lithuania	352 (3.9)	• 389 (4.7)	560 (4.2)	• 581 (3.3)	

0 2 50 75 100 12£

• = Gender difference statistically significant at .05 level

◨ Low-Performing Students

■ High-Performing Students

( ) Standard errors appear in parentheses. Because results are rounded to the nearest whole number, some totals may appear inconsistent.

The differences in achievement for students in the upper quarter are represented by solid bars and the gender differences in mean achievement for students scoring in the lower quarter are represented by cross-hatched bars.

Table 4 presents the gender differences in achievement for high- and low-performing final-year students on the mathematics and science literacy assessment. The results reveal that gender differences favoring males were more prevalent among high-performing students than low-performing students. Specifically, significant gender differences in mathematics and science literacy favoring males were found for high-performing students in all 18 countries analyzed. By contrast, only about two-thirds of the countries had significant differences favoring males for low-performing students.

Table 5, presenting the results for gender differences in achievement for high- and low-performing final-year students on the advanced mathematics and the physics assessments, shows a pattern similar to that in Table 4. For high-performing students, significant gender differences in advanced mathematics achievement favoring males were found in 11 out of 13 countries. For low-performing students, however, statistically significant gender differences in achievement were found in 9 out of 13 countries.

Similarly, statistically significant gender differences among high-performing final-year students in physics were found across all participating countries, while slightly fewer countries (10 out of 13) had statistically significant gender differences among low-performing students. One of the largest differences for high-performing students was found in the Czech Republic, where male high-performers scored a full standard deviation (100 points) higher than females who were also classified as high-performers for their gender.

Table 5. Average advanced mathematics achievement of high-performing (upper quarter) and low-performing (lower quarter) females and males

	Advanced Mathematics				Physics			
	Average Achievement for Low-Performing Students		Average Achievement for High-Performing Students		Average Achievement for Low-Performing Students		Average Achievement for Low-Performing Students	
	Males	Females	Males	Females	Males	Females	Males	Females
Australia	394 (12.0)	374 (15.1)	662 (10.2)	651 (10.5)	430 (4.9)	395 (7.9)	638 (8.6)	584 (9.4)
Austria	390 (9.3)	293 (10.8)	582 (5.6)	511 (4.2)	378 (6.0)	321 (4.6)	584 (8.7)	500 (9.1)
Canada	400 (4.3)	374 (3.4)	661 (4.1)	600 (4.5)	398 (7.6)	363 (5.2)	621 (6.6)	552 (5.5)
Cyprus	409 (6.9)	408 (8.3)	632 (6.7)	601 (6.6)	384 (6.3)	353 (10.8)	647 (12.6)	586 (9.9)
Czech Republic	404 (5.2)	330 (5.5)	667 (18.4)	545 (9.3)	405 (4.3)	346 (2.5)	614 (7.5)	498 (6.1)
France	478 (3.0)	458 (4.4)	654 (4.9)	627 (3.9)	393 (7.2)	374 (4.1)	562 (5.1)	528 (5.6)
Germany	375 (5.7)	351 (5.6)	592 (4.6)	555 (2.4)	428 (10.6)	375 (8.4)	666 (7.3)	577 (6.0)
Lithuania	442 (6.0)	396 (4.8)	651 (7.6)	589 (12.3)	—	—	—	—
Norway	—	—	—	—	483 (4.4)	431 (7.7)	707 (4.7)	657 (6.2)
Russian Federation	428 (5.7)	383 (7.6)	710 (5.3)	650 (10.3)	444 (6.5)	376 (9.8)	705 (5.9)	647 (12.7)
Slovenia	366 (6.6)	349 (4.8)	611 (5.2)	576 (3.5)	420 (8.6)	321 (14.3)	667 (14.8)	587 (11.7)
Sweden	408 (5.7)	396 (7.7)	631 (5.5)	590 (4.7)	466 (5.8)	439 (5.6)	705 (4.5)	632 (8.0)
Switzerland	448 (4.2)	409 (6.9)	681 (7.3)	599 (4.6)	427 (5.9)	360 (4.3)	644 (4.6)	532 (3.7)
United States	344 (5.2)	308 (4.5)	582 (6.4)	554 (6.9)	367 (2.8)	339 (2.2)	516 (7.4)	472 (2.4)
International	410 (1.8)	382 (2.0)	641 (1.8)	592 (1.6)	414 (1.4)	371 (1.7)	635 (1.9)	569 (2.3)

* represents a statistically significant gender difference at the 0.05 level. ( ) Standard errors appear in parentheses.

DISCUSSION

The first research question in this study explored the nature of the distribution of students within a given gender group meeting a particular standard. The findings revealed that a significantly greater proportion of males met the criterion in mathematics and science literacy. Specifically, the difference was statistically significant in 14 out of 18 participating countries. In advanced mathematics, this pattern of disproportionate rates of achievement by males was also seen. For example, in Austria, 43 percent of males taking the advanced mathematics assessment met the standard while only 14 percent of females did. Finally, the results from the physics assessment showed the greatest range in representation. In the Czech Republic, for example, fully 50 percent of the males met the criterion, while only 10 percent of females did so. In concrete terms, this means that one out of every two boys studying physics achieved above the benchmark while only one out of ten of their female classmates did. Internationally, significant differences in the proportion of males and females reaching the benchmark appeared in 10 out of 13 countries.

The findings from the first analysis raise several interesting policy issues. Because classrooms are not usually tracked by gender, the findings point to the need to investigate the extent to which particular instructional styles may have a differential impact for males and females. A second area warranting exploration is the extent to which curricular topics may or may not be presented in a fashion that is relevant for females. Prior research (Mullis et al., 2000) has shown that females tend to spend significantly more time on homework than males; however, the nature of this relationship is not fully understood. It is perhaps the case that females spend more time on homework; however, the quality of time may not be the same. In addition, the influence of differential expectations of parents or teachers about female students' capacity for achievement has also emerged as a tenable hypothesis for achievement differences (see Steele, 1997).

The second research question explored the extent to which gender diversity was present among the high-performing students. In pursuing the second research question, the primary focus shifted away from a concern with achievement within a particular gender group and toward a concern with the gender diversity that is found among students within a certain achievement group (e.g., high-achieving students). The expectation was that in a world with gender equity, we would find that of the high-achieving students, 50 percent would be male and 50 percent would be female.

The findings from the second research question revealed that a greater proportion of high-achieving students in mathematics and science literacy were male. The difference was statistically significant in 13 out of 18 participating countries. In advanced mathematics, this pattern of disproportionate representation of males was also seen. In Austria, Czech Republic, and Switzerland, for example, more than 70 percent of the high-achieving students were males while less than 30 percent were females. As with the first analysis, the results from the physics assessment showed the greatest range in representation. In physics, France had the greatest representation of females among high-achieving students, yet females still represented only 35 percent of the high-achieving students. In concrete terms, this means that if we were to walk into a physics classroom in any of the participating countries that was filled only with high-achieving students, we would find that, at best, 3 out of 10 students in that classroom would be females.

While classrooms are rarely tracked by gender, they are often tracked by ability, particularly in the areas of advanced mathematics and physics. As a result, the relevant policy issues addressed by the second analysis are slightly different than those in the first analysis. The results of this investigation reveal that males tend to be disproportionately represented among high-achieving students, particularly in physics. Females are represented among this group, however, not at the rates one would expect. This finding might be related to the extent to which students are tracked into particular academic programs in a particular country. Perhaps it is the case that the evidence and procedures used for tracking students tend to favor males. Other possible areas for exploration relate to expectancy effects, school climate, instructional approaches, and student attitudes toward the subject matter (Mullis et al., 2000).

The focus of the first two analyses was the composition of specific high-performing target groups by gender. What we could not determine from these approaches, however, was the level of achievement of the students. The third analysis explored the nature of gender differences in achievement across the performance distribution. A procedure for identifying high- and low-performing males and females previously used in analyzing gender differences in IEA's Reading Literacy Study was described. The type of information to be gained from comparing gender differences in achievement for high- and low-performing students was exemplified using the TIMSS procedure of comparing results for males and females in the top and bottom 25 percent of the distribution within each country. The findings from the third analysis showed that gender differences in achievement favoring males tended to be larger and more prevalent among high-performing students than low-performing students. Strikingly, high-performing males in the Czech Republic outscored high-performing females by more than one full standard deviation.

The results of the third analysis are related to the previous research questions, but offer a slightly different perspective. From a policy standpoint, the results of the third analysis serve to highlight the fact that differential achievement occurs at all points along the achievement distribution. In other words, the gender gap in achievement is not something that is just a problem for low-performing students. Indeed, the results of this analysis suggest the opposite is true. Achievement differences by gender are often more pronounced among those students who are the highest-performers for their gender. These results point to the need to explore particular instructional strategies that are being implemented in classrooms with high-performing and low-performing students. If classrooms are, in fact, tracked by ability, then the results of this analysis suggest an unfortunate irony. In particular, the gender gap in achievement is smallest (although still prevalent) among low-performing students. It is the high-track classrooms that may have some lessons to learn from the lower-track classrooms with regard to practices that foster greater gender equity.

Gender differences in achievement can be examined using a variety of analytic approaches. It is sometimes the case that one analytic approach paints a more optimistic portrait than another. The results from this chapter reveal that the prevalence of gender differences in mathematics and science achievement is problematic from a number of different perspectives. The purpose of this chapter was to extend the exploration of gender differences on TIMSS using a variety of analytic procedures and discussing their connection to relevant policy questions. TIMSS data for the final-year students were used as the basis for the discussion, since these results showed pervasive gender differences across all three components of the assessment: mathematics and science literacy, advanced mathematics, and physics. Illustrative results were presented for the final-year students for three different analytic approaches targeted at answering three different research questions.

The findings from the analyses discussed in this chapter show that the gender differences in mathematics and science achievement found for final-year students in TIMSS were particularly prevalent among high-performing students. Males tended to be disproportionately represented among high-performing students, and males tended to outscore females at all points on the performance distribution. Overall, the findings from this investigation have revealed that males are not only over-represented among high-performing students, but also that male high-performing students tend to have mastered a more complex set of skills and understandings than the average high-performing female student.

Taken together, the results of each analysis presented in this chapter paint a bleak picture of gender equity in mathematics and science achievement internationally at the final year of secondary school. Yet the results also indicate that females are capable of achieving at high-levels in mathematics and science literacy, advanced mathematics, and physics. The reasons for their under-representation are likely to be related to a number of factors associated with instructional strategies, study skills, school climate, and expectancy effects. One potential avenue for future research would be to explore the extent to which variability in the gender gap exists between schools and classrooms within countries. If some schools exhibit few gender differences while others exhibit large gender differences, explanatory models may then be constructed to investigate the extent to which variables related to instructional strategies, study skills, school climate, and expectancy effects are associated with gender differences in achievement.

REFERENCES

Beaton, A. E., Mullis, I. V. S., Martin, M. O., Gonzalez, E. J., Kelly, D. L., & Smith, T. A. (1996). *Mathematics achievement in the middle school years: IEA's Third International Mathematics and Science Study (TIMSS)*. Chestnut Hill, MA: Boston College.

Campbell, P. B. (1995). Redefining the "girl problem" in mathematics. In W. G. Secada, E. Fennema, & L. B. Adjian (Eds.), *New directions for equity in mathematics education*. Cambridge: Cambridge University Press.

Dumais, J. (1998). Implementation of the TIMSS sampling design. In M. O. Martin & D. L. Kelly (Eds.), *Third International Mathematics and Science Study technical report volume III: Implementation and analysis at the final year of secondary school*. Chestnut Hill, MA: Boston College.

Gray, M. (1996). Gender and mathematics: Mythology and misogyny. In G. Hanna (Ed.), *Towards gender equity in mathematics education: An ICMI study*. Boston, MA: Kluwer Academic Publishers.

Manger, T., & Gjestad, R. (1997). Gender differences in mathematical achievement related to the ratio of girls to boys in school classes. *International Review of Education, 43*(2-3), 193–201.

Martin, M. O. (1996). Gender differences among high and low performers. In H. Wagemaker (Ed.), *Are girls better readers?: Gender differences in reading literacy in 32 countries*. Amsterdam: International Association for the Evaluation of Educational Achievement.

Mullis, I. V. S., Martin, M. O., Beaton, A. E., Gonzalez, E. J., Kelly, D. L., & Smith, T. A. (1997). *Mathematics achievement in the primary school years: IEA's Third International Mathematics and Science Study (TIMSS)*. Chestnut Hill, MA: Boston College.

Mullis, I. V. S., Martin, M. O., Beaton, A. E., Gonzalez, E. J., Kelly, D. L., & Smith, T. A. (1998). *Mathematics and science achievement in the final year of secondary school: IEA's Third International Mathematics and Science Study (TIMSS)*. Chestnut Hill, MA: Boston College.

Mullis, I. V. S., Martin, M. O., Fierros, E. G., Goldberg, A. L., & Stemler, S. E. (2000). *Gender differences in achievement: IEA's Third International Mathematics and Science Study (TIMSS)*. Chestnut Hill, MA: Boston College.

Steele, C. (1997). A threat in the air: How stereotypes shape intellectual identity and performance. *American Psychologist, 52*, 613-629.

Wagemaker, H. (Ed.). (1996). *Are girls better readers?: Gender differences in reading literacy in 32 countries*. Amsterdam: International Association for the Evaluation of Educational Achievement.

Willingham, W., & Cole, N. (1997). *Gender and fair assessment*. Princeton, NJ: Educational Testing Service.

Chapter 18

INVESTIGATING CORRELATES OF MATHEMATICS AND SCIENCE LITERACY IN THE FINAL YEAR OF SECONDARY SCHOOL

Jesse L. M. Wilkins, Michalinos Zembylas, and Kenneth J. Travers

The concepts of mathematics and science literacy have been ongoing subjects of debate in both the mathematics and science education communities, at large, and within TIMSS, in particular. While many reform initiatives in mathematics and science education around the world point to mathematics and science literacy as an important goal of education (e.g., Australian Education Council, 1991; National Council of Teachers of Mathematics [NCTM], 2000; National Research Council, 1989, 1996; New Zealand Ministry of Education, 1992), there is no clear consensus as to what constitutes mathematics and science literacy, because the notion of "literacy" has a complex and dynamic nature, and it is not easily defined (Bybee, 1997; Koballa, Kemp, & Evans, 1997; Shamos, 1995; Steen, 1997). Many scholars have discussed the significance of literacy as an overarching goal of education (Apple, 1993; Ferdman, 1990) and the problems associated with defining and developing mathematics and science literacy (Lee, 1997; Pollack, 1994; Pool, 1990; Steen, 1997). Further, there is a need to understand the factors that influence the development of mathematics and science literacy. With this in mind this chapter outlines one conceptualization of mathematics and science literacy employed in TIMSS—facility with mathematics and science as it may occur in real life (Orpwood & Garden, 1998)—and attempts to identify factors related to the development of this literacy in 16 countries that participated in the literacy component of TIMSS.

An important step towards a richer understanding of literacy is exploring its sociocultural aspects. This suggests that the notions of mathematics and science literacy do not only refer to educational achievements but also to social ones; consequently, such notions evolve over time. Moreover, what constitutes and influences literacy may differ from one region of the world to another. Therefore, an international exploration of the sociohistorical circumstances under which mathematics and science literacy evolve—such as the student or school level factors

D.F. Robitaille and A.E. Beaton (eds.), Secondary Analysis of the TIMSS Data, 291–316.
© 2002 *Kluwer Academic Publishers. Printed in the Netherlands.*

that we analyze in this chapter—will provide relevant indications about the potential of literacy throughout the world.

We use the notion of the potential of literacy to reiterate a useful Aristotelian distinction between actuality (what really is the case) and potentiality (what merely has the power to change or come to be the case). The notion of the potential of literacy directs our attention to the power of literacy to transform the quality of people's lives. If mathematics and science literacy are life-long processes that evolve constantly inside and outside the school, then, in addition to characteristics of students, measures related to educational experiences, peer and family relationships, opportunities related to the schooling process, and other factors should have an impact on empowering or disempowering students to achieve higher levels of literacy. The attitudes and the values established toward mathematics and science by students—especially in their final year of secondary school—may potentially be carried out into their daily lives and influence their future interactions as citizens and workers.

With the impetus for increasing the level of mathematics and science literacy comes a need to understand the factors that influence such development. Our intent in this chapter is to explore some of these factors and to point to some of the potential implications for thinking about the issues that are paramount to increased mathematical and scientific literacy. It is also important to identify the source of differences in student achievement, for example, are these differences in mathematics and science literacy related more to the opportunities and experiences associated with the schools that students attend or are these differences more related to the individual characteristics and experiences of the student? In addition to considering this general question, this chapter investigates the specific relationships between students' experiences and characteristics, both background and educational, and their status or level of mathematics and science literacy. In addition, this chapter considers opportunities and contextual effects associated with the school community and their influence on literacy. Specifically, this chapter considers the following four questions:

(1) What are the sources of variance in mathematics and science literacy in students' final year of secondary school?

(2) What factors related to student aptitude, curriculum and instruction, and environment affect mathematics and science literacy?

(3) How do opportunities associated with the school community relate to mathematics and science literacy?

(4) How do these relationships differ or remain consistent across the 16 different countries in the study?

It is our hope that this work will provide a starting point for educators around the world to consider understandings of mathematics and science literacy from a sociocultural point of view, and that these understandings may provide guidance for curricular decision making.

THE LITERACY COMPONENT OF TIMSS

The goal of the literacy component of TIMSS was to assess the mathematics and science that "school leavers" (students in their final year of secondary school) around the world have retained and are prepared to apply to the everyday experiences of being a citizen and worker (Orpwood and Garden, 1998). Beyond content knowledge associated with mathematics and science, the framework employed for TIMSS Population 3 allowed for the investigation of reasoning skills, knowledge of the social impact of mathematics and science, and the ability to apply this knowledge in society. (For a full account of the development of the literacy component of TIMSS see Orpwood and Garden, 1998).

The mathematics and science literacy (MSL) test was divided into three parts: mathematics literacy, science literacy, and reasoning and social utility (RSU). These three parts were based on the importance of addressing content (what was taught to the students while in school), context (whether content knowledge was contextualized in real-world situations), and process (whether students had the ability to think in logical and reasonable ways when confronted with real-world problems) (see Orpwood and Garden, 1998). The items developed for the MSL component were motivated by the following question: To what degree can students understand and apply mathematics and science concepts taught to them earlier in their school careers? (Orpwood and Garden, 1998). The MSL test was designed to test students' general knowledge and understanding of mathematical and scientific principles. The items for this test were developed to represent situations likely to arise in real life and were not connected to any particular curriculum (Orpwood and Garden, 1998). The emphasis was on measuring how well students can use their knowledge in addressing real-world problems (Martin, Mullis, Gonzalez, Smith, and Kelly, 1999). However, the items were designed to map into several general content areas associated with mathematics and science.

The mathematics literacy portion contained 38 items organized into three areas of mathematics: number sense, algebra sense/data, and measurement and estimation. The science literacy portion contained 26 items organized into three areas of science: earth science, life science and physical science. The reasoning and social utility portion contained 12 items and focused less on specific areas of content and more on certain types of performance expectations. The MSL test involved three item types: multiple choice, short answer, and extended response. Four items (TIMSS, 1998; also see Orpwood and Garden, 1998) are presented in Figures 1 to 4 and serve as illustrations of the conceptualization of mathematics and science literacy outlined in TIMSS.

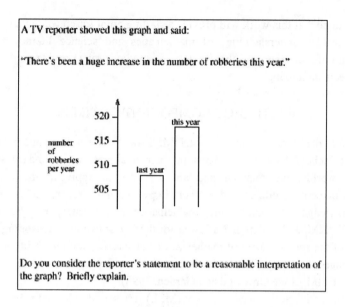

A TV reporter showed this graph and said:

"There's been a huge increase in the number of robberies this year."

Do you consider the reporter's statement to be a reasonable interpretation of the graph? Briefly explain.

Figure 1. Item from the Mathematics Literacy portion of the MSL test.

A school club is planning a bus trip to the wildlife park. A bus which will hold up to 45 people will cost 600 centros (units of money) and admission tickets cost 30 centros each.

If the cost of the trip, including bus and admission ticket, is set at 50 centros per person, what is the minimum number of people who must participate to ensure that these costs are covered?

    A. 12
    B. 20
    C. 30
    D. 45

Figure 2. Mathematics item from the Reasoning and Social Utility portion of the MSL test.

José caught influenza. Write down one way he could have caught it.

Figure 3. Science item from the Reasoning and Social Utility portion of the MSL test.

CFCs (chrorofluorocarbons) revolutionized personal and industrial life for 30 years. They were the coolant in refrigerators and the propellants in aerosols, pressure packs, and fire extinguishers. There are now very strong international moves to stop the use of these substances because

A. they are chemically inert.
B. they contribute to the greenhouse effect.
C. they are poisonous to humans.
D. **they destroy the ozone layer.**

Figure 4. Item from the Science Literacy portion of the MSL test.

The fundamental issue of concern for the MSL component of TIMSS was to clarify the extent to which selected concepts in mathematics and science had been understood by students and could be applied in real-world contexts, and to identify whether students had the ability to reason carefully and understand the social, political, and economic impacts of mathematics and science.

MODELING CORRELATES OF ACHIEVEMENT

An important component in understanding how to increase the level of mathematics and science literacy is to identify those factors that influence its development. In the next sections, we discuss two frameworks that help structure possible factors into comparable models for investigation.

Predictors of Student-level Achievement

Walberg's model of educational productivity (Walberg, 1981, 1984, 1992) identifies nine key factors that relate to students' affective, behavioral, and cognitive development. These factors can be classified into three general groups: (1) personal variables such as prior achievement, age, and motivation or self-concept; (2) instructional variables such as amount or quality of instruction; and (3) environmental variables related to the home, teacher/classroom, peers, and media exposure. Results of about 8000 studies provide evidence of the consistent influence of these factors on student achievement (Walberg, 1992). Subsequent studies testing these factors provide further evidence of their predictive power for achievement in mathematics and science (Walberg, 1984, 1992; Young, Reynolds, and Walberg, 1996; Reynolds and Walberg, 1991; 1992; Tsai and Walberg, 1983). This model of educational productivity serves as the framework for investigating factors related to mathematics and science literacy in the present study.

Predictors of School-Level Achievement

Although the educational productivity theory provides a testable set of factors related to student achievement this is clearly only one set of factors that might relate to student achievement. Factors at many levels (school, district, community, nation) can have an influence on student achievement. Wilkins (1999) outlined, in his study of demographic opportunity structure, four forms of capital—financial, human, cultural, and geographic—that together characterize the source of many opportunities that directly or indirectly affect educational achievement.

The community that surrounds a school is an example of a demographic opportunity structure. Within a demographic opportunity structure students are afforded different opportunities based on the levels of existing capital. For example, financial capital (Coleman, 1988) relates to fiscal resources that allow parents (or communities) to provide educational resources for students. Human capital (Coleman, 1988) is often associated with the educational level of the community or parents' level of education and thus may provide evidence of opportunities related to home environment. Parents' involvement in their child's homework or parents being able to provide more educationally stimulating environments would represent evidence of human capital. Cultural capital (Bourdieu, 1986) is related to opportunities afforded students based on their cultural background (this component is not considered in the present study). Geographic capital (Ghelfi & Parker, 1997) is related to opportunities students might have as a result of geographic proximity. For example, in some countries students that live in rural areas may not have access to museums, libraries, or other educational resources. Wilkins (1999) found that all four forms of capital were related to school-level achievement.

Other groups of factors related to the schooling process have also been posited as predictors of school outcomes (e.g., Willms, 1992). One group of factors relates to the characteristics of the school, for example, school size. Another group of factors relates to the inner workings of the school or school climate, for example, the disciplinary climate of the school. These groups of factors have been shown to be related to school-level outcome variables (see Willms, 1992).

Opportunities related to school-level factors may not be directly related to the status of individual students' mathematics and science literacy. However, by studying these opportunities educators are better able to understand the overall educational experience and ultimately better able to identify needs and provide students with the necessary opportunities to learn.

DATA

The TIMSS literacy study surveyed students in their final year of secondary school in 22 countries. The intent of the study was to survey those students that were about to leave secondary school and enter postsecondary education or the workforce (Mullis et al., 1998). Six of the 22 countries (Denmark, Germany, Israel,

Netherlands, Slovenia, and South Africa) were not included in the present study due to sampling problems (Gonzalez, Smith, and Sibberns, 1998). The main problem for most of these countries was low school participation resulting in inadequate school information. Countries included in the study are listed in Table 1.

Table 1. Sample sizes for TIMSS literacy study

	Students	Schools
Australia	1941	87
Austria	1962	169
Canada	5232	337
Cyprus	534	28
Czech Republic	2167	150
France	1590	56
Hungary	5091	204
Iceland	1703	30
Italy	1616	101
Lithuania	2887	142
New Zealand	1763	79
Norway	2518	131
Russian Federation	2289	163
Sweden	3068	145
Switzerland	3308	383
United States	5807	211

MEASURES

Student and school measures were created using information from the TIMSS student-level and school-level surveys and questionnaires. Descriptions for all variables are provided in Table 2.

Outcome Measures

TIMSS used plausible-values methodology to create estimates of student achievement for the mathematics literacy, science literacy, and RSU scales (for further details see Gonzalez et al., 1998). In order to assess a wider range of content areas, and because of time constraints, individual students were administered different subsets of items from a total pool of items. Plausible values are estimates of individual student scores that would have been obtained if they had been administered a test that included all of the items. Because of the measurement error associated with this technique, five plausible values for each scale were estimated for each student. Based on the recommendations of Gonzalez et al. (1998) all five plausible values for each scale were used in subsequent analyses.

Student-Level Measures

This section lists the student-level variables used in the predictive models. In the case that variables were content specific, individual measures related to mathematics and science were created. In addition, composite measures combining information from mathematics and science were created for predicting reasoning and social utility (RSU).

Student Aptitude and Background

Student aptitude and background variables included student gender, age, educational aspirations, and mathematics and science self-concept.

Curriculum and Instruction

Although the TIMSS literacy tests were not linked to a specific curriculum (Orpwood and Garden, 1998), student course taking patterns and areas of specialization provided indicators of experiences related to curriculum and instruction. These measures include whether students were taking a mathematics or science course in their final year of secondary school and whether they were classified as specialists (having taken advanced mathematics or physics) or non-specialists.

Environmental Variables

Information about students' home environment included parent education, parent college push, parent subject push, the number of books in the home, and the availability in the home of a calculator, computer, place to study, and a dictionary. Peer influence was measured by peer college push and peer subject push. Exposure to the media was based on student reports of time watching television and reading a book for enjoyment.

School-Level Variables

An indicator of opportunities associated with financial capital was created using a school-level aggregate of the number of books and other educational resources in students' homes. Similarly, an indicator of opportunities associated with human capital was created using a school-level aggregate of parents' education. An indicator of opportunities associated with geographic capital was created based on principals' reports of the type of community in which the school was located. Other school-level measures included school size and school climate (in terms of behavior problems).

Table 2. Description of student-level and school-level variables

Student-Level Variables	Description
Mathematics Literacy	Plausible values for mathematics literacy
Science Literacy	Plausible values for science literacy
Reasoning and Social Utility	Plausible values for RSU
Student Aptitude and Background	
Gender	Male = 1, Female = 0
Age	Student age in years.
Self-Concept	
I have usually done well in mathematics.	4 =Strongly Agree; 1 =Strongly Disagree
I have usually done well in the sciences.	4 =Strongly Agree; 1 =Strongly Disagree
Math-Science self-concept	Average of math and science self-concept
Educational Aspirations	
University	1 = yes; 0 = no
Vocational or Other	1 = yes; 0 = no
None	1 = yes; 0 = no
Curriculum and Instruction	
Taking a math course in final year	1 = yes; 0 = no
Taking a science course in final year	1 = yes; 0 = no
Taking math or science course in final year	1 = yes; 0 = no
Specialists vs. Non-Specialists [a]	1 = Specialists (students who have taken advanced mathematics or physics); 0 = Non-Specialists
Home Environment	
Parents' Education	Highest education of mother or father; 1 = Finished University; 0 = Otherwise; Students who reported "don't know" were treated as missing.
Parent College Push	
What do parents think you ought to do immediately after you finish secondary school?	1 = Full-time University; 0 = Other. Average of parents. [b]
Parent Subject Push	
Important to do well in math at school.	1 = yes; 0 = no; Average of parents. [b]
Important to do well in science at school.	1 = yes; 0 = no; Average of parents. [b]
Combination of the two questions.	Average of math and science.
Books in the Home	1 = More than 100 books in the home; 0 = Less than 100 books in the home.
Other Items in the Home (Calculator, Computer, Study Desk, Dictionary)	Total number of items (0-4).
Peer Environment	
Peer College Push	
What do your friends think you ought to do immediately after you finish secondary school?	1 = Full-time University; 0 = Other.

Table 2. Description of student-level and school-level variables (cont'd)

Student-Level Variables	Description
Peer Subject Push	
Important to do well in mathematics at school.	1 = yes; 0 = no.
Important to do well in science at school.	1 = yes; 0 = no.
Combination of the two questions.	Average of math and science.
Media	
Time Watching TV	Time watching TV on normal school day.
Time Reading Books	Time reading a book for enjoyment on a normal school day.

School-Level Variables	Description
Demographic Opportunity	
Financial Capital	School-level aggregate of the number of books in students' home (1 = more than 100; 0 = less than 100), plus number of other items in the home (calculator, computer, study desk, dictionary).
Human Capital	School-level aggregate of parent education.
Geographic Capital	
City	Close to the center of a town/city: 1 = yes, 0 = no.
Rural	Combined village or rural (farm) area and a geographically isolated area: 1 = yes, 0 = no.
Outskirts	On the outskirts of a town/city: 1 = yes, 0 = no.
Other	Not reported or missing: 1 = yes, 0 = no.
Climate [c]	Composite variable created by multiplying principal-reported frequency (1="rarely" to 4 ="daily") and severity (1="not a problem" to 3="serious problem") of 11 behaviors in a school. Scores for behaviors higher than 6 were coded 1 (problem behavior) and 0 (otherwise). Summing the values of the 11 behaviors provided a measure of school climate.
School Size [c]	Total number of students in school.

[a]In some cases countries did not differentiate between "specialists" and "non-specialists." Although Russia and Lithuania did administer the advanced math test these students were either not identified or not included as part of the literacy sample. Other countries did not take part in the advanced mathematics and physics testing (see Mullis et al. 1998).

[b]When information was available for only one parent this was used for the measure without averaging.

[c] Not reported or missing school-level data was imputed using country grand mean plus a normal deviate multiplied by the standard deviation.

DATA ANALYSIS

The TIMSS data are stratified, with students nested within schools, therefore hierarchical linear modeling techniques (HLM), Bryk and Raudenbush, 1992) were used to analyze the proposed educational models investigating the correlates of achievement as well as to partition the variance in achievement into school and student components. The stratified design of TIMSS with students selected from within schools creates a situation where selection of individual students is not truly independent which may result in cluster effects and possible underestimation of standard errors. However, HLM incorporates a unique random effect for each school into the model that corrects for possible design effects. In addition, HLM (Version 5) provides parameter estimates based on robust standard errors. Following recommendations of the TIMSS *User Guide* (Gonzalez et al., 1998), student and school sampling weights were used in the analyses. Weights were normalized to maintain the effective sample size of the countries (Raudenbush et al., 2000).

Variance attributable to schools and students for each country was determined using an unconditional hierarchical linear model. These models were estimated using the full sample of students involved in the TIMSS literacy study before listwise deletions (see Table 3).

A two-level hierarchical linear model was used to investigate the correlates of literacy. Students with missing data were deleted listwise to create working samples for each country. An unconditional model similar to that used for the analysis of variance components was estimated first using the working sample to ascertain the amount of variance attributable to schools (see Table 3). A within-school equation regressed each literacy scale (mathematics, RSU, and science) on variables related to student background and aptitude, instructional experiences, and environment. These predictor variables were all grand-mean-centered and effects were fixed across schools. Provided that the unconditional model found evidence of significant between-school variance, a between-school model was estimated regressing school means on the characteristics of the demographic opportunity structure as well as school size and school climate.

RESULTS

Variance Components for Literacy Scales

Table 3 presents variance components for the mathematics, RSU, and science literacy scales. Discussion of the results in this section refers to components estimated using the total sample. The differences among countries in total variance suggest the existence of differences in the make-up of the students and schools that may depend on differing experiences, resources, and opportunities. For example, Australia, Canada, the Czech Republic, and Switzerland have a large amount of total variance compared to other countries. These large variance components may be due

Table 3. Sources of variance for Mathematics and Science Literacy

		Total Sample			Working Sample		
		Total Variance	% School	% Student	Total Variance	% School	% Student
Australia	Math	11,689.9	19	81	10,882.0	15	85
	RSU	10,546.6	22	78	9,473.1	18	82
	Science	12,591.2	19	81	11,705.8	15	85
Austria	Math	8,068.9	31	69	8,035.5	28	72
	RSU	9,533.1	28	72	8,806.7	28	72
	Science	11,548.0	20	80	13,294.9	14	86
Canada	Math	20,958.9	3	97	14,732.4	7	93
	RSU	15,687.9	4	96	11,724.7	6	94
	Science	15,897.0	4	96	15,627.6	4	96
Cyprus	Math	5,451.6	9	91	5,237.0	1	99
	RSU	7,019.4	13	87	6,664.7	3	97
	Science	7,092.2	12	88	6,751.7	2	98
Czech Rep.	Math	14,691.8	33	67	13,140.6	32	68
	RSU	14,205.6	25	75	11,781.5	25	75
	Science	14,917.5	22	78	13,146.7	22	78
France	Math	7,051.3	33	67	6,635.5	31	69
	RSU	5,488.7	39	61	4,950.6	36	64
	Science	7,051.9	34	66	6,672.3	32	68
Hungary	Math	8,057.2	52	48	7,797.9	49	51
	RSU	6,801.2	48	52	6,440.2	47	53
	Science	7,122.7	44	56	7,032.5	43	57
Iceland	Math	7,735.2	12	88	7,195.0	12	88
	RSU	5,412.7	13	87	5,021.7	10	90
	Science	5,760.4	13	87	5,430.3	12	88
Italy	Math	7,466.7	57	43	6,920.8	52	48
	RSU	7,863.3	60	40	7,274.8	59	41
	Science	7,300.1	51	49	6,770.0	48	52
Lithuania	Math	6,844.7	52	48	6,546.7	49	51
	RSU	8,582.9	49	51	8,125.5	46	54
	Science	6,867.6	49	51	6,712.0	47	53
New Zealand	Math	9,442.5	10	90	8,614.9	9	91
	RSU	7,786.6	11	89	6,913.9	12	88
	Science	8,624.6	13	87	7,747.7	13	87
Norway	Math	8,582.6	25	75	8,053.9	26	74
	RSU	7,352.1	31	69	6,476.5	32	68
	Science	8,037.0	29	71	7,274.8	31	69
Russian Fed.	Math	7,612.3	42	58	7,369.6	43	57
	RSU	8,024.7	39	61	7,667.6	42	58
	Science	8,974.8	37	63	8,997.6	38	62
Sweden	Math	9,198.8	23	77	8,016.7	21	79
	RSU	6,444.3	26	74	5,585.3	26	74
	Science	7,216.7	24	76	6,692.6	23	77
Switzerland	Math	15,011.7	21	79	12,048.6	26	74
	RSU	13,008.7	22	78	11,207.2	26	74
	Science	17,018.0	23	77	15,402.1	24	76
United States	Math	8,927.9	17	83	8,549.1	17	83
	RSU	8,027.1	18	82	7,430.2	17	83
	Science	9,291.6	18	82	8,628.5	18	82

to curricular experiences within the country. For example, in Canada and Switzerland, the two countries with the greatest amount of within-country variance, a relatively large proportion of students reported not taking mathematics and science in their final year (Mullis et al. 1998). Similarly, in the Czech Republic, a relatively large number of students reported not taking science in their final year (Mullis et al., 1998). Such large variance components could also be attributable to program differences. Three of the four countries in TIMSS Population 3 (Australia, Canada, and Switzerland) that track students into academic, vocational, and general tracks (those not fitting into academic, technical, or vocational) had the highest within-country variance. In comparison, small amounts of total variance may be due to more homogeneous programs or a result of more homogeneous sampling. For example, in Cyprus, France, Hungary, Iceland, and Lithuania, five countries with lower total variance, all students were either taking mathematics or science or all were considered non-specialists. Therefore it is not surprising that these countries, in which sampled students have had more similar experiences and opportunities, would have smaller variance in achievement (i.e., students with similar achievement levels).

When considering differences in variance due to schools or students there was relatively little within-country difference across the three literacy scales. That is, within a given country the amount of variance attributable to schools was relatively equal for mathematics, science, and RSU. However, there were noticeable between-country differences in the amount of variance due to schools (and students). For example, Hungary, Italy, Lithuania, and the Russian Federation had large variance components, roughly 40 percent or more, due to schools, with as much as 60 percent of the variance in RSU due to schools in Italy. At the same time, six countries, Australia, Canada, Cyprus, Iceland, New Zealand, and the United States, had roughly 20 percent or less of the total variance in achievement due to schools, with as little as 3 percent of the variance in mathematics literacy due to school differences in Canada. Several factors may help explain the small amount of between-school variance in Cyprus. First, the small number of schools sampled in Cyprus as well as the homogeneous nature of the sample of schools (private and vocational schools excluded) could contribute to the small amount of variance. In addition, Cyprus has both a nationally centralized curriculum and textbooks. However, the other five countries do not have consistent national policies, although Canada does have regional polices related to curriculum and textbooks. Australia, Canada, and the United States have educational programs that tend to track a large proportion of students into classes with very little preparation for postsecondary education or other vocational programs. If this policy is consistent across schools it may contribute to large within-school differences in student opportunities and thus greater within-school variance in achievement.

Modeling Correlates of Literacy

For each country, three separate hierarchical linear models were estimated. Except for the cases noted in Table 4, 16 student-level variables and seven school-level variables were included in the models. However, it is important to note that while theoretically each of the three models was the same, depending on the type of literacy outcome the independent predictors were changed to be consistent with the outcome measure. For example, mathematics, science and mathematics or science self-concept were included respectively in the mathematics, science, and RSU predictive models. Because of the inordinate number of regression coefficients that were estimated it would be unreasonable to report specifically the effects for each of them. Therefore, results are discussed in terms of general trends found across the 16 systems with additional important predictors discussed for each country. It is important to understand that relationships reported on any particular variable are controlled for all other variables in the model. Table 4 presents a summary of the results from the three separate models for each country (mathematics literacy [M], science literacy [S], and reasoning and social utility [R]). The following two sections highlight the student- and school-level variables that were the most salient both within and between the different school systems.

Student-Level Effects

It is apparent from Table 4 that the model of educational productivity does hold up across the 16 systems. All of the variables were significant predictors in at least one country. However, it is also apparent that the consistency with which this model of productivity predicts achievement differs greatly across the 16 systems. For example, in Italy, Lithuania, and Switzerland very few of the student-level variables consistently predicted achievement, while in Australia, New Zealand, and the United States a majority of the variables were found to significantly predict achievement. Although, the level of predictive power of the model tended to differ across countries, within-country findings tended to be more consistent across the three literacy scales.

Student-level variables related to student background and aptitude seem to be more consistent predictors of achievement both between countries and within countries (i.e., across the different scales). For example, with the exception of France, students' gender consistently predicted achievement in all countries. In addition, self-concept consistently predicted student achievement in all 16 countries and for the most part across the three literacy scales. The level of students' mathematics and science curriculum and instruction (i.e., specialist vs. non-specialists) as well as continued exposure to mathematics and science (taking courses in final year) had consistent positive relationships with achievement in all three literacy scales. While environmental variables, as a whole, were found to predict achievement, the particular important variable changed depending on the

Table 4. *Final models for predicting Mathematics and Science Literacy by country*

Student-Level Effects	Australia M	Australia R	Australia S	Austria M	Austria R	Austria S	Canada M	Canada R	Canada S	Cyprus M	Cyprus R	Cyprus S	Czech Rep. M	Czech Rep. R	Czech Rep. S	France M	France R	France S	Hungary M	Hungary R	Hungary S	Iceland M	Iceland R	Iceland S
STUDENT																								
Gender	/	/	Ɵ	Ɵ	/	Ɵ	+	/	/				Ɵ	Ɵ	Ɵ				/	+	+	+	+	/
Age	/	+																				+	+	
Educational Aspirations																								
University vs. None	Ɵ	Ɵ	Ɵ	Ɵ	Ɵ	Ɵ	+			Ɵ	Ɵ	Ɵ	+	+	/	+	/	Ɵ		Ɵ		Ɵ	Ɵ	Ɵ
Vocat./Other vs. None																+	/	+				/	/	/
Self-concept	Ɵ	Ɵ	Ɵ	/	Ɵ	Ɵ	+	Ɵ	Ɵ	Ɵ	/	Ɵ	Ɵ	Ɵ	+	Ɵ	Ɵ	Ɵ	Ɵ	Ɵ	Ɵ	Ɵ	Ɵ	Ɵ
INSTRUCTION																								
Taking Course Final Year	/	/	/	/	/		+	+	Ɵ	all	all	all				all	all	all	all	all		non	non	non
Specialist	Ɵ	Ɵ	Ɵ	Ɵ	Ɵ	Ɵ	/	/	Ɵ	Ɵ	Ɵ	Ɵ	Ɵ	Ɵ	/	Ɵ	Ɵ	Ɵ	non	non	Ɵ	non	non	non
ENVIRONMENT																								
Parent's Education	Ɵ				/									+		Ɵ	/			/	/		/	/
Books in the Home		+	+	+		+	+	+			+			+	+	+	/	+		/	/	Ɵ	/	/
Other Items in Home						/		+	Ɵ							+	nc	nc			+			Ɵ
Parent College Push				+			nc	nc	nc	+	+		/	+		nc	nc	nc			+			Ɵ
Parent Subject Push	+						nc	nc	nc				+	+		+	nc	nc			+			
Peer College Push	+						nc	nc	nc					ς		nc	nc	nc						
Peer Subject Push																			H	ς	ς			
Time Watching TV										ς				ς	−				−	ς	−			
Time Reading Books			+				+		+												Ɵ			
School-Level Effects	**M**	**R**	**S**	**M**	**R**	**S**	**M**	**R**	**S**	**M**	**R**	**S**	**M**	**R**	**S**	**M**	**R**	**S**	**M**	**R**	**S**	**M**	**R**	**S**
DEMOGRAPHIC OPPORTUNITY																								
Financial Capital	+	+	/			−					nv	nv	Ɵ	Ɵ	/		+	+	Ɵ	Ɵ	Ɵ	+		
Human Capital								/	/		nv	nv									+			
Geographic Capital																								
Rural vs. City								+	/	nv	nv	nv												
Outskirts vs. City								+	+	nv	nv	nv												
Other vs. City										nv	nv	nv					−							
Climate						−				nv	nv	nv	−	−										
School Size			+	−		−				nv	nv	nv	−	−	−	+	+			−	−			−

Table 4. Final models for predicting Mathematics and Science Literacy by country (continued)

Student-Level Effects	Italy			Lithuania			New Zealand			Norway			Russian Fed.			Sweden			Switzerland			United States		
	M	R	S	M	R	S	M	R	S	M	R	S	M	R	S	M	R	S	M	R	S	M	R	S
STUDENT																								
Gender	⊖	⊖	⊖	⊖	⊖	⊖	⊖	+	⊖	⊖	⊖	⊖	⊖	⊖	⊖		/	⊖			+	⊖	+	⊖
Age							+												H	H	ç			
Educational Aspirations																								
University vs. None						+	⊖	⊖	/	+						+		+	/	/	/	/	⊖	/
Vocat./Other vs. None							⊖	⊖	/										/	/	/	/	+	+
Self-concept	⊖	+		⊖		⊖	⊖	⊖	⊖	⊖		⊖	⊖		⊖	⊖	⊖	⊖				⊖	⊖	⊖
INSTRUCTION																								
Taking Course Final Year	⊖	⊖	⊖		+		⊖	⊖	⊖	⊖		⊖	all	all	all	⊖	+	⊖				⊖		+
Specialist	⊖	⊖	⊖	non	non	non	non	non	non	⊖		⊖	non	non	non	⊖	⊖	⊖			+	⊖		/
ENVIRONMENT																								
Parent's Education				+			⊖																	
Books in the Home							⊖	/	/															
Other Items in Home							/	+	/															
Parent College Push																			+	+	+			
Parent Subject Push																			+	+	+			
Peer College Push																								
Peer Subject Push											−	−			ç								ç	
Time Watching TV					−	ç				ç	−	ç					−							ç
Time Reading Books	−				+	+			+															
School-Level Effects	M	R	S	M	R	S	M	R	S	M	R	S	M	R	S	M	R	S	M	R	S	M	R	S
DEMOGRAPHIC OPPORTUNITY																								
Financial Capital	+	+	+	+	+	+	+		+			+		+		+	/	+	⊖	⊖	Φ	+	+	/
Human Capital				/					+					+					/	/			/	
Geographic Capital																								
Rural vs. City	−				ç	−							na	na	na	H						/		+
Outskirts vs. City		ç	ç	ç	−	ç							na	na	na									
Other vs. City	na	na	na		−								na	na	na		−	−					/	+
Climate													na	na	na	−	−	−						
School Size																								

Note: **M** = Mathematics Literacy; **R** = Reasoning and Social Utility; **S** = Science Literacy. + = positive effect: < .05; / = $p < .05$; Φ = $p < .01$; ⊖ = $p < .001$. Minuses represent a negative effect: − $p < .05$; ç $p < .01$; H $p < .001$. non=Sample included only non-specialist; all=All students taking math or science; na=Not applicable; nc=Information not collected; nv=School-level variance not sig. $p < .01$.

country and literacy scale. What follows is a discussion of some of the particular variables and trends highlighted above.

Gender. The results for gender show that overall boys tend to have higher scores on all three literacy scales. However, in several countries these effects were inconsistent (Australia, Cyprus, Sweden, Switzerland) or non-existent (France).

Self-concept. Probably the most salient predictor in the models was students' self-concept or belief that they had usually done well in mathematics or science. The positive relationship between self-concept and achievement, while not surprising, suggests that it is an important variable for predicting student success in mathematics and science literacy irrespective of country differences.[1] In comparison, a student's educational aspirations, or desire to pursue education beyond secondary school, another measure of student aptitude, did not consistently predict student literacy across countries. However, within several countries it consistently predicted success on all three literacy scales. This between-country difference in the effect of aspirations may be due to between-country differences in postsecondary opportunities. The relationship between literacy and aspirations may reflect motivation to succeed in order to be able to attend university, whereas a student's self-concept, while it may be affected by opportunities, is based on beliefs about self, independent of external factors.

Curriculum and instruction. Not all countries tested students from different levels of specialization in mathematics and science (Mullis et al., 1998; see Table 4, "non" = sampled only non-specialists). However, for the countries that did sample students from different curricular backgrounds, the opportunity to take either advanced mathematics, physics, or both had a positive effect on student literacy. With the exception of the Czech Republic and Switzerland this effect was consistent across the three literacy scales. In Switzerland and the Czech Republic the influence of curriculum was only important for science literacy.

After controlling for level of curricular specialization the continued exposure to mathematics and science in the final year did not have a consistent effect on student success. Only in Australia was there a consistent effect due to exposure across the three scales. In Canada and Sweden continued exposure to mathematics in the final year was important in predicting success in mathematics literacy. Also in Sweden, continued exposure in mathematics and science was important in predicting RSU. In Norway and the United States the exposure to science had a significant effect on success in science literacy. The consistency of the effect of taking courses in the final year was further clouded by the fact that in several countries all students take either mathematics or science, or both in their final year (Cyprus, France, Hungary, and the Russian Federation). In Iceland, New Zealand, and Lithuania where only generalists were surveyed, but where not all students take mathematics or science, the effect of continued exposure was consistently positive across the three scales in Iceland and New Zealand but only positive for RSU in Lithuania. Because of the different sampling structures of the different countries it is hard to identify any general pattern. However, it seems that continued exposure to mathematics and science is important for heightened mathematics and science literacy, with the study of advanced mathematics and science providing the greatest impact.

Environmental variables. Although many of the environmental variables in the model significantly predicted student success for all three literacy scales, there were no consistent trends across the 16 countries. This finding alone provides important evidence to suggest that between-country differences in the role of the home, peers, and parents are many and that it is hard to understand, internationally, the effects of these variables except that cultural differences do exist. In some countries it may not make sense to talk about, for example, home resources or parent push as these variables do not truly represent the nature of the home environment in these countries, or in some countries there may be little variance in these variables.

School-Level Effects

It is apparent from Table 4 that the model of demographic opportunity is predictive of school-level achievement. All of the components of demographic opportunity significantly predicted school success in at least one country, and some component of the model predicted school-level success in at least one of the literacy scales. However, as in the case of the student-level variables, it is apparent that the consistency with which this model predicts school-level achievement differs across the 16 systems. Interestingly, financial capital and human capital tended to flip-flop, that is, either one or the other predicted school-level achievement but not both simultaneously. This may be a result of a correlation between these two variables that ultimately washes out the effects of the weaker predictor. However, in Australia and the Russian Federation financial capital was predictive of school-level science literacy and human capital was related to mathematics literacy. The fact that in most countries either financial or human capital predicts school success suggests that internationally the influence of community resources and education level is important for the promotion of literacy.

Country-Specific Effects

Specific student and school-level effects particular to each country in addition to those discussed earlier (gender, self-concept, and curriculum/instruction) are highlighted in this section.

Australia. Older students in Australia were found to have higher scores in mathematics literacy and RSU. Students who reported having more than 100 books in their home were found to have higher scores in mathematics and science literacy. Parent education and reading books were also positively related to science literacy. Parent subject push was found to be positively related to RSU. At the school level, financial capital was positively related to school-level achievement in science, while human capital was positively related to mathematics literacy and RSU.

Austria. Parental influence on student success was apparent in Austria, although parental effects were not consistent across the literacy scales. Parents' education and parents' belief in the importance of science were positively related to science literacy. Parents' belief in the importance of attending university was positively related to mathematics and RSU. At the school level, parent education and school climate was found to have a negative effect on science literacy.

Canada. The number of items (e.g., calculator) in the home had a consistent positive effect on all three literacy scales in Canada. Reading books was found to have a positive relationship with students' mathematics literacy. At the school level, rural schools were found to have higher achievement than those in the city in mathematics and science literacy and schools on the outskirts of a town or city were found to do better in science literacy and RSU. Parental education consistently predicted higher school achievement for mathematics and science literacy.

Cyprus. Students who intend to go to university were found to have higher achievement for all three literacy scales. Parental influence was found to be related to mathematics literacy and RSU. Interestingly, reading books in Cyprus was found to be negatively related to success in mathematics literacy. Preliminary unconditional models suggested an insignificant amount of school-level variance in Cyprus, thus, parameters for a between-school equation were not estimated.

Czech Republic. Students who intended to go to university were found to have higher scores in science literacy and RSU. Neither self-concept nor aspirations were related to mathematics literacy. Parents' education was found to be positively related to mathematics literacy, while items in the home were positively related to science literacy and RSU. Parents' belief in the importance of mathematics (and science) was found to be related to higher mathematics literacy and RSU. Peer college push was found to have a positive relationship with mathematics literacy, although peer subject push was negatively related to RSU. The influence of television was found to have a negative relationship with RSU. Consistent school-level effects related to the opportunities associated with financial capital were found in the Czech Republic as well as a consistent negative effect related to school climate.

France. As noted earlier, there were no gender differences found across the three literacy scales in France. There was, however, a consistent positive effect related to the educational aspirations of students intending to continue their education, whether at the university level, vocational, or otherwise. There was also a consistent positive effect related to resources in the home (books and otherwise). The importance of mathematics for parents was also found to be positively related to mathematics literacy. At the school level, the opportunities associated with the demographic opportunity structure were not related to mathematics literacy. However, opportunities associated with human capital were found to be related to science literacy and RSU. Schools located on the outskirts of a town or city were found to have lower RSU scores than those in the city. However, students in larger schools were found to have higher RSU.

Hungary. Parent education and parent belief in the importance of science, as well as the number of books in the home were positively related to science literacy. Time spent reading books was also positively related to science literacy. Time watching TV was found to have a negative relationship to mathematics and science literacy. Peer belief in the importance of mathematics and science consistently had a negative relationship across the literacy scales. At the school level, opportunities associated with financial capital consistently predicted higher school-level literacy. In addition, human capital predicted higher science literacy scores. School size had a consistent negative relationship with RSU and science literacy.

Iceland. The only additional consistent predictor of literacy was parents' belief in the importance of doing well in mathematics and science, which had a positive relationship with all three literacy scales. The number of books in the home was positively related to RSU and science literacy. Financial capital was found to be positively related to school-level mathematics literacy and school size was negatively related to science literacy.

Italy. There was a negative relationship between mathematics literacy and time spent reading books in Italy. However, no other additional relationships besides those discussed in general were found. At the school level, opportunities associated with financial capital had a positive relationship with all three literacy scales. Schools on the outskirts of a town or city had lower scores than schools in the city for all three scales.

Lithuania. Students who intended to go to university were found to have higher science literacy scores. The number of books in the home was positively related to mathematics literacy and reading books was found to be positively related to science literacy. It is interesting to note that although only non-specialists were included in the literacy sample, taking mathematics or science in the final year only had a positive relationship with RSU. At the school level, opportunities associated with financial capital had a positive relationship with all three literacy scales. Schools on the outskirts of a town or city had consistently lower scores than schools in the city for all three scales. Rural schools had lower scores for science than schools in the city.

New Zealand. New Zealand had one of the highest number of significant regression coefficients at the student level. Students intending to continue their education, whether at the university level, vocational or otherwise, had significantly higher literacy scores than students who had no intention of furthering their education. Students with more than 100 books in the home and whose parents felt that it was important to go to college consistently had significantly higher literacy scores. There was also a positive relationship between the existence of other items in the home and mathematics literacy. Students who spent more time reading books tended to exhibit higher science literacy scores. At the school level, opportunities associated with financial capital and geographic capital were positively related to science literacy.

Norway. Students who intended to go to university were found to have higher mathematics literacy than those who had no intention of continuing their education. Time spent watching television was found to have a consistent negative relationship with students' literacy. Peer belief in the importance of mathematics and science was also found to have a negative relationship with RSU. At the school-level, only financial capital was found to have a significant positive relationship with science literacy.

Russian Federation. The Russian Federation had very few significant coefficients. The students sampled were all non-specialists and all took mathematics and science in their final year. Beyond gender and self-concept, only peer beliefs in the importance of science had a significant (negative) relationship with science literacy. Schools with greater opportunities associated with financial capital had higher RSU scores and schools with greater opportunities associated with human capital had higher mathematics literacy scores.

Sweden. There were no gender effects related to mathematics literacy. Taking mathematics and science in the final year was found to have a positive effect on mathematics literacy and RSU after controlling for taking advanced mathematics or physics. Time watching television was found to have a negative relationship with mathematics literacy and RSU. Opportunities associated with financial capital were found to have a positive relationship with all three literacy scales.

Switzerland. Gender effects in Switzerland were found only in science. Effects associated with level of curriculum specialization were apparent only in science literacy. Student age was found to have a negative relationship with mathematics literacy. Parental belief in the importance of mathematics and science had a significant relationship with all three scales and peer college push had a significant relationship with mathematics literacy. Opportunities associated with financial capital were found to have a positive relationship with all three literacy scales.

United States. The United States had the greatest number of significant student-level relationships. Student age had a consistent negative relationship with literacy. Students intending to continue their education, whether at the university level, vocational, or otherwise, had significantly higher literacy scores than students that had no intention of furthering their education. Students with more than 100 books in the home consistently had higher literacy scores than those with fewer books. Parent belief in the importance of mathematics and science had a positive relationship with RSU and science literacy and peer belief in the importance of mathematics was negatively related to mathematics literacy. Science literacy was found to be negatively related to time watching television and positively related to reading books for enjoyment. Opportunities associated with financial (for RSU and science) and human capital were consistently related to school literacy, while rural schools tended to do better in RSU and science literacy than city schools.

CONCLUSIONS

The goal of this chapter was to investigate correlates of mathematics and science literacy in 16 countries. Specifically, the study attempted to understand the source of variance in achievement whether school or student, identify sociocultural factors, both student level and school level, that predict student achievement, and finally compare these factors and sources of variance across the different countries. In essence, our effort was to identify the processes by means of which the potential of mathematical and scientific literacy is constructed within a particular sociocultural context. Following are some of the more salient findings and conclusions drawn from the study.

Sources of Variance

Based on the results of this study, and reflective of results discussed in other international reports (e.g., Schmidt, McKnight, Cogan, Jakwerth, and Houang, 1999; Schmidt, Wolfe, and Kifer, 1992), the majority of variance in mathematics and science literacy in the final year of secondary school was found to be within schools, that is, more related to differences in individual student characteristics and experiences within school. Within countries there was little difference in the source of variance across literacy scales. It is important to point out, however, that there was variation across countries in this attribution, suggesting that in some countries there may be greater or lesser disparity between schools. Because of the design of TIMSS, intact classrooms were not surveyed in Population 3. The additional knowledge of between-classroom differences would have added additional insights to the sources of within-school differences.

Student-level Correlates of Literacy

With regard to student-level correlates of mathematics and science literacy, although many variables were found to be related to achievement, three variables consistently stood out as important: gender, self-concept, and exposure to mathematics and science. With the awareness of everyone's need for a functional level of mathematics and science literacy to be a well-informed citizen and worker it is important to recognize the gender gap that exists across the world. This gap represents an example of the potential disparity in opportunities and success between different groups of people as they venture out into everyday life beyond secondary school.

In order to function in a technological society not only is it necessary to have an everyday understanding of mathematics and science but a person must also possess a certain level of confidence and willingness to take part in such situations. Consistent with this idea, a person's self-concept was found to be the most salient predictor of success across the different countries. This finding suggests a strong relationship between a person's feelings about their ability to do well and their eventual success. Again, this shows that an important component of a person's ability to make a valuable contribution to society and themselves is a belief in their own mathematical and scientific potential.

A student's continued exposure to mathematics and science seems to be an important predictor of mathematics and science literacy. With the different sampling frames and curricular programs across the countries it is hard to find a consistent pattern. However, it is apparent that whether or not students take more mathematics and science courses, in particular more advanced mathematics and science, predicts achievement. It is also important to point out that for those students not studying advanced mathematics or physics (see Table 4, Hungary, Iceland, and New Zealand), taking additional coursework did predict higher achievement. This is an important finding that needs further consideration. The level of mathematics and science content on the literacy test was not beyond that most often introduced in early secondary school (Orpwood and Garden, 1998) and it is therefore important to consider the types of courses that non-specialists need to take in order to maintain and further develop their mathematics and science literacy. As suggested by Wilkins (2000) an important part of literacy is an awareness of the nature of mathematics and science and gaining with this awareness a sense of inquiry and investigation characteristic of mathematics and science, as well as an understanding of the impact of mathematics and science on society. Maybe these are the types of understandings and "habits of mind" that could be developed in continued study of mathematics and science that could contribute to the idea of literacy as a lifelong pursuit.

For example, currently the nature of mathematics and science and their impact on society are issues that are not often incorporated into courses at the secondary level. However, an understanding of these issues is important to be a well-informed citizen and worker (Wilkins, 2000). Therefore, it will be important for educators to carefully link the objectives of mathematics and science literacy with specific curricula and instructional strategies that promote these ideas and goals.

The only conclusion that can be drawn from the investigation of environmental factors is that the particular important variables differ by country. However, this finding alone suggests that there are many differences in culture that can affect students' development in equally yet different ways. For this reason, it is important to note that the environmental measures used in this study may only be appropriate for some countries and meaningless when used for other countries.

School-Level Correlates of Literacy

When considering the school community, it was found in most countries that opportunities associated with financial, human, or geographic capital played an important role in the status of school-level literacy. This finding suggests that schools situated in communities with greater resources whether physical (books, computers, etc.), personal (parental education), or proximal, have higher levels of mathematics and science literacy. However, no one factor consistently predicted status across all countries, although opportunities associated with financial and human capital seem to be the most salient. Between-country differences in both cultural and geographic make-up make it difficult to draw general conclusions. For example, opportunities associated with geographic capital may not be consistent across the world as schools in rural areas in one country may have similar opportunities or at least equally important opportunities as city schools in another country. However, in general, it was found that the characteristics of the school community do have an impact on school-level literacy.

Final Thoughts

We believe that the analyses in this chapter identify a number of possibilities in examining social and school structures that may support or inhibit the development of the potential of mathematics and science literacy. What kinds of conditions are likely to support the development of this potential? Either considering individual factors or external ones, the personal or public contexts seem to be mutually constitutive and are dependent on creating opportunities for extensive and intensive experiences in which mathematics and science become meaningful ways of knowing and living in everyday life. To transform this potentiality into actuality, mathematics and science educators around the world need to work collaboratively in developing systems, programs, and practices that are authentic and contextually relevant and that promote the underlying goals of mathematics and science literacy.

NOTES

[1] Findings from Kifer (Chapter 16, this volume) suggest a negative relationship between self-concept and achievement. However, the analyses in Chapter 16 were conducted at the country level, while the analyses in this chapter were conducted within countries.

REFERENCES

Apple, M. (1993). *Official knowledge: Democratic education in a conservative age.* London: Routlege.

Australian Education Council (1991). *A national statement on mathematics for Australian schools.* Carlton, Victoria: Curriculum Corporation.

Bourdieu, P. (1986). The forms of capital. In J. G. Richardson (Ed.), *Handbook of theory and research for the sociology of education.* New York: Greenwood Press.

Bryk, A. S., & Raudenbush, S. W. (1992). *Hierarchical linear models.* Newbury Parks, CA: Sage Publications.

Bybee, R. W. (1997). *Achieving scientific literacy: From purposes to practices.* Portsmouth, N.H.: Heinemann Publishing.

Coleman, J. S. (1988). Social capital in the creation of human capital. *American Journal of Sociology, 94* (Supplement), S95-S120.

Ferdman, B. (1990). Literacy and cultural identity. *Harvard Educational Review, 60*(2), 181-204.

Ghelfi, L. M., & Parker, T. S. (1997). *A county-level measure of urban influence* (Staff Paper No. 9702). Washington, DC: U. S. Department of Agriculture.

Gonzalez, E. J., Smith, T. A., & Sibberns, H. (1998). *User guide for the TIMSS international database: Final year of secondary school.* Chestnut Hill, MA: TIMSS International Study Center.

Koballa, T., Kemp, A., & Evans, R. (1997). The spectrum of scientific literacy. *Science Teacher, 64*(7), 27-31.

Lee, O. (1997). Scientific literacy for all: What is it, and how can we achieve it? *Journal of Research in Science Teaching, 34,* 210-222.

Martin, M. O., Mullis, I. V. S., Gonzalez, E., J., Smith, T., & Kelly, D. L. (1999). *School contexts for learning and instruction: IEA's Third International Mathematics and Science Study (TIMSS).* Chestnut Hill, MA: Boston College.

Mullis, I. V. S., Martin, M. O., Beaton, A. E., Gonzalez, E. J., Kelly, D. L., & Smith, T. A. (1998). *Mathematics and science achievement in the final year of secondary school: IEA's Third International Mathematics and Science Study (TIMSS).* Chestnut Hill, MA: Boston College.

National Council of Teachers of Mathematics. (2000). *Principles and standards for school mathematics.* Reston, VA: Author.

National Research Council. (1989). *Everybody counts: A report to the nation on the future of mathematics education.* Washington, DC: National Academy Press.

National Research Council. (1996). *National Science Education Standards.* Washington, DC: National Academy Press.

New Zealand Ministry of Education. (1992). *Mathematics in the New Zealand curriculum.* Wellington: Learning Media.

Orpwood, G., & Garden, R. A. (1998). *Assessing mathematics and science literacy.* [TIMSS Monograph No. 4.] Vancouver, CA: Pacific Educational Press.

Pollack, V. L. (1994). Science education. II: Scientific literacy and the Karplus taxonomy. *Journal of Science Education and Technology, 3,* 89-97.

Pool, R. (1990). Science literacy: The enemy in us. *Science, 251,* 266-267.

Raudenbush, S. W., Bryk, A. S., Cheong, Y. F., & Congdon, R. (2000). *HLM 5: Hierarchical linear and nonlinear modeling.* Chicago, IL: Scientific Software International.

Reynolds, A. J., & Walberg, H. J. (1992). A structural model of high school mathematics outcomes. *Journal of Educational Research, 85(3),* 150-158.

Reynolds, A. J., & Walberg, H. J. (1991). A structural model of science achievement. *Journal of Educational Psychology, 83(1),* 97-107.

Schmidt, W. H., McKnight, C. C., Cogan, L. S., Jakwerth, P. M., & Houang, R. T. (1999). *Facing the consequences: Using TIMSS for a closer look at US mathematics and science education.* Boston, MA: Kluwer Academic Publishers.

Schmidt, W. H., Wolfe, R. G., & Kifer, E. (1992). The identification and description of student growth in mathematics achievement. In L. Burstein (Ed.), *The IEA study of mathematics III: Student growth and classroom process.* New York: Pergamon Press.

Shamos, M. H. (1995). *The myth of scientific literacy.* New Brunswick, NJ: Rutgers University.

Steen, L. A. (Ed.). (1997). *Why numbers count: Quantitative literacy for tomorrow's America.* New York: College Entrance Examination Board.

TIMSS International Study Center. (1998). *TIMSS released item set for the final year of secondary school.* Chestnut Hill, MA: Boston College.

Tsai, S. L., & Walberg, H. J. (1983). Math achievement and attitude productivity in junior high school. *Journal of Educational Research, 76(5),* 267-272.

Walberg, H. J. (1981). A psychological theory of educational productivity. In F. H. Farley & N. J. Gordon (Eds.), *Psychology and education.* Chicago: National Society for the Study of Education.

Walberg, H. J. (1984). Improving the productivity of America's schools. *Educational Leadership, May,* 19-27.

Walberg, H. J. (1992). The knowledge base for educational productivity. *International Journal of Educational Reform, 1(1),* 5-15.

Wilkins, J. L. M. (1999). Demographic opportunities and school achievement. *Journal of Research in Education, 9(1),* 12-19.

Wilkins, J. L. M. (2000). Preparing for the 21st century: The status of quantitative literacy in the United States. *School Science and Mathematics, 100(8),* 405-418.

Willms, J. D. (1992). *Monitoring school performance: A guide for educators.* London: Falmer.

Young, D. J., Reynolds, A. J., & Walberg, H. J. (1996). Science achievement and educational productivity: A hierarchical linear model. *Journal of Educational Research, 89(5),* 272-278.

Chapter 19

INDICATORS OF ICT IN MATHEMATICS: STATUS AND COVARIATION WITH ACHIEVEMENT MEASURES

Willem J. Pelgrum and Tjeerd Plomp

IEA has a tradition of addressing questions about the infusion of computers (since the mid-1990s, usually referred to as Information and Communication Technologies–ICT) in education. A first attempt was made in the Second International Science Study (SISS) conducted in 1984, where one item in the school questionnaire concerned the availability of microcomputers in the participating schools. The fact that the researchers at that time only included one item about computers reflected the common opinion that computers were not yet widespread, and that it was hardly worthwhile to investigate the way they were used in science education.

By the end of the 1980s, governmental plans regarding the introduction of computers in education had been launched in numerous countries and, as a consequence, many schools were equipped with microcomputers. IEA then began to study this phenomenon from an international comparative perspective. The Computers in Education (CompEd) surveys of 1989 and 1992 provided a rich database with base line information that captured the beginning of the introduction of technology in education on a large scale. In the early 1990s, when political interest in information technology in education seemed to have considerably declined (as compared to the mid and late 1980s), the designers of TIMSS–95 realized that, given IEA's interest in this area, the inclusion of indicators of technology use in mathematics and science was of great importance. This resulted in estimates of infrastructure and indicators of the use of ICT from more than 50 countries. Not only will future historians value the existence of these data, the designers of IEA's Second International Technology in Education Study (SITES) were also pleased with these data, because they offered an opportunity to look at changes over time between 1995 and 1998 in terms of the availability of computer

D.F. Robitaille and A.E. Beaton (eds.), Secondary Analysis of the TIMSS Data, 317–330.
© 2002 *Kluwer Academic Publishers. Printed in the Netherlands.*

equipment in schools. In this chapter the terms "computers" and "ICT" will be treated as synonyms.

The aim of this chapter is to offer a description of the ICT data that were collected in TIMSS–95 and, where possible, to compare these data with similar data that were collected earlier or later. Due to time and space considerations, the focus will be on mathematics in TIMSS Population 2. During the course of data processing and writing this chapter, the first descriptive reports of TIMSS-99 became available. Where appropriate, ICT statistics from those reports are also included.

The TIMSS–95 database is unique to the extent that it combines indicators of student achievement in mathematics with indicators of the use of computers in this school subject. Therefore, as it is the first time in a large-scale international comparative assessment that ICT and achievement data were simultaneously collected, this chapter will, in addition to the descriptive part, also examine the extent to which these indicators covary in a meaningful (that is, interpretable) way.

AVAILABILITY OF COMPUTERS FOR EDUCATIONAL PURPOSES

In TIMSS–95, school principals were asked about the number of computers that were available in the school for use by teachers or students. On the basis of these responses and the available estimates of the student enrollment in schools, student-to-computer ratios were calculated. These offer a rough indication of the availability of hardware in the schools. In addition, students who participated in TIMSS were asked about the availability of computers at home.

Availability of computers at schools

Table 1 contains the student-to-computer ratios in schools in 1995. One may observe that the ratios differ quite substantially between countries. Whereas in Australia, Canada, England and Scotland there was in 1995, on average, one computer available for every 15 or fewer students, in other countries much larger groups of students had to share the available equipment.

One may expect that, as a result of decreasing prices for equipment and large-scale governmental programs for stimulating the use of ICT in education, student-to-computer ratios will rapidly improve over the years. As shown in Figure 1, which contains a comparison of student-to-computer ratios for countries that participated in TIMSS–95 and SITES 1998, these improvements were, in some countries, enormous. For instance, in Norway, the ratio improved from 55 to 9. Pelgrum (1999) concluded that there was convincing evidence from multiple sources that the comparability of the indicators from SITES and TIMSS were not at stake.

Table 1. Student-to-computer ratios in TIMSS–95: Population 2

Country	Ratio	Country	Ratio
Scotland	9	Hong Kong	39
England	11	Hungary	42
Australia	13	Germany	43
Canada	14	Belgium (Fl)*	44
United States	15	Norway	55
Israel	16	Korea	57
Denmark	17	Greece	58
New Zealand	18	Latvia (LSS)*	58
Sweden	19	Czech Republic	63
Austria	19	Slovenia	82
Singapore	20	Russian Federation	87
Kuwait	25	Spain	94
Netherlands	26	Colombia	96
Japan	27	Slovak Republic	98
Switzerland	28	Lithuania	117
Iceland	29	Portugal	140
Belgium (Fr)*	29	Thailand	206
France	29	Cyprus	339
Ireland	31	Romania	880

* The Flemish (Fl) and French (Fr) educational systems in Belgium participated separately in TIMSS. Latvia is annotated LSS for Latvian Speaking Schools only.

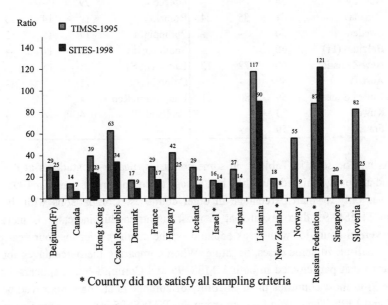

* Country did not satisfy all sampling criteria

Figure 1. Student-to-computer ratios for TIMSS–95 and SITES–98.

Accessibility of computers at students' homes

Computers are, especially since the Internet became available for the general public, increasingly becoming a tool that is available in many households. In roughly half of the countries that participated in TIMSS–95, 50 percent or more of the students had access to a computer at home (see Table 2).

Table 2. Percentages of students in the TIMSS Population 2 upper grade (mostly Grade 8) that had access to home computers in 1995 and 1999 and change (DIFF) between those years

Country	1995	1999	DIFF	Country	1995	1999	DIFF
Japan	-	52	-	Singapore	49	80	31
Scotland	90	-	-	Slovenia	47	66	19
England	89	85	-4	Spain	42	-	-
Netherlands	85	96	11	Lithuania	42	16	-26
Ireland	78	-	-	Korea	39	67	28
Iceland	77	-	-	Portugal	39	-	-
Denmark	76	-	-	Cyprus	39	58	19
Israel	76	80	4	Hong Kong	39	72	33
Australia	73	86	13	Hungary	37	50	13
Germany	71	-	-	Czech Republic	36	47	11
Belgium (Fl)	67	86	19	Russian Fed.	35	22	-13
Switzerland	66	-	-	Slovak Republic	31	41	10
Norway	64	-	-	Greece	29	-	-
Canada	61	85	24	Romania	19	14	-5
Sweden	60	-	-	Philippines	17	15	-2
Belgium (Fr)	60	-	-	South Africa	15	11	-4
New Zealand	60	72	12	Latvia (LSS)	13	15	2
Austria	59	-	-	Colombia	11	-	-
United States	59	80	21	Iran, Islam. Rep.	4	7	3
Kuwait	53	-	-	Thailand	4	8	4
France	50	-	-				

As can be seen in Table 2, in some countries, nearly all students at the lower secondary level, claimed that they had access to computers at home in 1995, while this was only a small percentage in other countries. Comparisons with the more recent TIMSS–99 data reveal that in most countries this indicator is increasing. However, in countries with weak economies, the changes over the four-year period were small or, in some cases, negative. When comparing the percentages for those countries that participated in both TIMSS–95 and CompEd–92, it appears that (see Table 3) in most countries the availability of computers at home improved between 1992 and 1995. The reason for comparing the TIMSS–95 indicator with Comped–92 (instead of using the TIMSS–99 result) is that for this comparison the greatest overlap of countries was possible.

Table 3. Percentages of students at lower secondary level indicating that a computer was available in their homes in 1992 and 1995, and the difference (DIFF) between these years

Country	1992	1995	DIFF
Austria	45	59	14
Germany	64	71	7
Greece*	31	29	-2
Japan	21	-	-
Netherlands	60	85	25
United States	48	59	11

- Data not collected.

* The sample in 1992 consisted of students in computer-using schools and, hence, the percentage of students who have access to a computer at home may be upwardly biased.

USE OF COMPUTERS FOR MATHEMATICS

On the basis of the available research literature it is not possible to advocate the use of ICT for mathematics learning, because there is no unequivocal proof of the added value of technology use. However, over the past ten years, the availability of electronic sources (software, websites, applets) that can be used in mathematics has increased substantially. Therefore, periodically examining the use of ICT in school subjects, in our case mathematics, may offer an operational indication of the extent to which educational practitioners perceive ICT as useful tool in the mathematics curriculum.

Although the availability of computers as reported above has increased considerably over years, it is not clear that ICT will also be used for learning mathematics at school or at home.

During the CompEd studies (1989 and 1992, see for instance Pelgrum & Plomp, 1993; Pelgrum, Janssen, Reinen, & Plomp 1993), it was concluded that computers were mainly an object of instruction (namely for learning *about* computers) and were hardly ever used as a tool in school subjects. In 1992, even though the overall use of computers for mathematics was relatively the highest when compared to other school subjects, most students at the lower secondary level reported that they never used computers as part of their mathematics learning. An interesting question is to what extent this situation had changed by 1995.

Table 4 contains the percentages of students in 1995 and, where available, 1999 who said that they had never used computers for learning mathematics at school. The reason for reporting the percentage of students who said that they never used computers for mathematics is that this answer category was comparable on the 1992, 1995, and 1999 questionnaires. One may observe in Table 4 that the majority of students in most countries reported that they never used computers for mathematics.

Between 1995 and 1999 only Singapore reported a considerable decrease of students who said that they never used computers for mathematics. For 1992 these percentages were for Austria, Germany (student samples from 9 out of 11 Bundeslander), Greece (only computer-using schools), Netherlands, Japan, the United States, 44, 51, 74, 79, 65, 69, respectively. From these few comparisons with 1992, there is no indication that the use of computers in mathematics increased between 1992 and 1995. Rather, during that period, there seemed to be a trend towards a decrease in use.

Table 4. Percentage of students, in 1995 and 1999 and the differences (DIFF) between these years, indicating that they never used computers for mathematics

Country	% never use			Country	% never use		
	1995	1999	DIFF		1995	1999	DIFF
Portugal	97	-	-	Greece	83	-	-
Ireland	96	-	-	Canada	82	67	-15
Colombia	95	-	-	Switzerland	82	-	-
Slovak Republic	94	95	1	Iceland	81	-	-
Belgium (Fl)	94	93	-1	Netherlands	81	80	-1
Russian Fed.	94	97	3	New Zealand	79	73	-6
Belgium (Fr)	94	-	-	Romania	78	93	15
Korea	93	83	-10	Philippines	78	80	2
Spain	93	-	-	Kuwait	78	-	-
Iran, Islamic Rep.	92	96	4	Australia	77	71	-6
Hungary	92	92	0	Japan	77	76	-1
Lithuania	92	-	-	Israel	76	67	-9
Latvia (LSS)	91	95	4	Cyprus	73	81	8
Thailand	91	85	-6	South Africa	70	-	-
Hong Kong	91	75	-16	United States	69	61	-8
Singapore	90	46	-44	Austria	62	-	-
Slovenia	89	81	-8	Sweden	61	-	-
Czech Republic	88	84	-4	Scotland	53	-	-
Norway	88	-	-	England	45	46	1
France	88	-	-	Denmark	40	-	-
Germany	84	-	-				

As one might expect, there was rather a strong covariation between student-to-computer ratios and the percentages of student who indicated that they never used computers in mathematics lessons. For instance, in the Russian Federation, Colombia, Spain, and the Slovak Republic, the student-to-computer ratios were very high (which means a low availability of computers) and, at the same time, more than 90 percent of the students had never used computers in mathematics. On the other hand, in England where the ratio was very favorable, 45 percent of the students said that they never used computers for mathematics.

Characteristics of students using computers for mathematics lessons

One of the most intriguing questions for policymakers who make decisions regarding ICT in education and who are held accountable for (sometimes huge) public investments in ICT for education is what the added value of ICT use is (as compared to instruction without ICT). Addressing this question within the framework of TIMSS–95, we need to find out, on the basis of the TIMSS data, whether there are any indications that students profit in one way or the other from the use of technology during their mathematics lessons. The question is formulated very cautiously because one should not expect that data which are collected at one moment in time will allow for strong inferences about causal relationships. To explore answers to this question, it is important to understand how technology is used in mathematics lessons and what the characteristics of the users are. An obvious implication from the results that were presented above is that analyses of TIMSS–95 data for addressing this question should be undertaken with care. One complication is that the number of students in the sample who frequently use computers was extremely small in quite a number of countries. Therefore for the analyses that are reported below, the criterion was used that only those countries where at least 150 students were found in the high-ICT use and low-ICT use groups respectively would be included. Furthermore, as 150 cases is statistically still a relatively small number (bearing in mind that these are likely to be located in only a few classrooms) we decided to treat the countries as replications of ICT use in mathematics and present the results of analyses in a non-parametric way. That is to say, if all or the majority of the analyses point in the same direction, our trust in the robustness of a finding will be strengthened.

A first relationship that has been explored is that between the use of ICT and achievement. For that purpose, the mathematics achievement of "high" ICT users has been compared with that of "low" ICT users. The "high" ICT users are students who used computers almost always or very often for mathematics and the "low" ICT users are the ones who used computers never or once in a while. It appears that "high" ICT users have much lower achievement scores than the "low" ICT users (see Table 5). In order to help interpret the achievement differences in Table 5, we have quantified this discrepancy by comparing the differences in achievement scores with the average growth in achievement from Grade 7 to Grade 8 (or in TIMSS jargon, the lower and upper grades). The difference between the average achievement score of the lower and upper grade students can be interpreted as the growth in mathematics in one year. Therefore, in Table 5, for each country, the differences between "high" and "low" ICT users is also expressed in terms of a fraction of the score difference between the upper and lower grade. So, for instance, if the difference in a country on the mathematics test between the upper and lower grade is 20 scale points and the difference between the students who used computers frequently and those who hardly used them was 40, in Table 5 this would be shown as a difference of 2 years. Such conversion should be treated with care because the TIMSS achievement scale is not of a ratio type, which would allow interpretations that the difference between 500 and 480 is the same as a difference between 370 and 350. Nevertheless, when interpreted with caution, it gives an indication of the magnitude of the observed score differences. On the basis of this calculation, one may conclude that in 10 out of the 14 countries where this calculation was possible (given the criterion of at least 150 available cases in either group), the high ICT-users were lagging roughly between one and two school years behind the low ICT-users.

This finding may at first sight, when interpreted causally, be quite surprising for those who believe that the future development of education greatly depends on the application of new information and communication technologies. However, it seems too early for overly pessimistic interpretations as some alternative hypotheses have not been investigated. For instance, a plausible hypothesis might be that the frequent use of computers was heavily dominated by a focus on drill and practice exercises for lower ability students (a popular approach for remedial purposes, see Pelgrum & Plomp, 1993). If such a hypothesis were tenable, the low scores are a result of selection bias in the group of high-ICT users.

Table 5 *Differences in achievement between groups with high and low ICT use, and upper and lower grade, and number of years high ICT use group was lagging behind (second column divided by the third)*

Country	Achievement score differences		Years behind
	High ICT– Low ICT	Upper grade–Lower	
Canada	−50	33	1.5
Cyprus	−48	28	1.7
Denmark	−23	39	0.6
Greece	−43	44	1.0
Iran, Islamic Rep.	−33	29	1.1
Japan	−8	34	0.2
New Zealand	−66	37	1.8
Philippines	−31	13	2.3
Romania	−15	27	0.6
Sweden	−65	36	1.8
Thailand	−13	25	0.5
England	−56	31	1.8
Scotland	−45	35	1.3
United States	−47	22	2.2

In order to shed light on which characteristics the frequent computer-using students differed from their non-computer-using counterparts, a number of factors were examined. These had to do with students' home backgrounds and instructional approaches during mathematics lessons.

The home background variables offer some insight into whether the high- and low-use groups differed in terms of home conditions commonly associated with achievement. If the low-use groups were found to include more students with relatively unfavorable home backgrounds, this would corroborate the tenability of the selection bias hypothesis. The variables that were selected as indicators for the home-background of students were:

- born in country,
- speak language of test at home,
- watch more than 5 hours television,
- play more than 5 hours computer games,
- have more than 100 books in the home,
- mother has university degree,
- father has university degree,
- availability of home computer,
- mother wants student to do well in mathematics.

The selection of these variables was based on evidence from TIMSS–95 that they covary, taking the whole sample as basis for the calculation, with the mathematics scores. For instance, typically, across countries, students from families where at least one of the parents had a university degree scored roughly 40 points higher than students from families where the highest educational level of the parents was primary education. Such differences are of a magnitude comparable to the increase in scores from lower to upper grade, which is equivalent to roughly the effect of one year of schooling.

For each intensity-of-computer-use group, for each of the variables from the list above, the percentage of students having that characteristic (born in country, etc.) was calculated. Next, the differences in these percentages between the high-ICT-use group and low-ICT use group were determined.

Due to space considerations, the results of these analyses are presented in condensed form as boxplots. In Figure 2 one can observe, for instance, that the difference between the high-ICT use and low-ICT use groups in terms of the percentage of students born in the country varied roughly between 5 percent and –5 percent, while the median was around zero. In terms of the percentage of students speaking the language of the test at home, the differences varied roughly between 8 percent and –20 percent. In one country, the low ICT-use group had 35 percent more students who spoke the language of the test at home. The results also indicate that students who were frequently using computers for mathematics tended, when regarding the median, not to differ very much from the "low" ICT using students on the other home background variables. However, the availability of computers at home tended to be somewhat higher in the high ICT-use group.

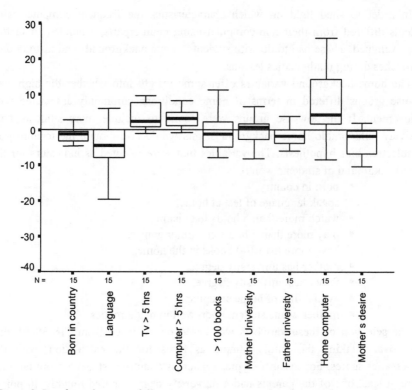

Figure 2. Boxplots of country differences in percentages of students with certain background characteristics between high-ICT users and low-ICT users (outliers excluded).

With regard to instructional approaches a list of statements from the student questionnaire was examined. These statements (for which students judged the frequency of occurrence) concerned the use of the following approaches during mathematics lessons:

- teacher shows how to do problems,
- copy notes from the board,
- have a quiz or test,
- work from worksheets on our own,
- work on projects,
- use calculators,
- use computers,
- work in pairs or small groups,
- solve with objects from everyday life,
- teacher gives homework,
- begin homework in class,
- teacher checks homework,
- check each other's homework,
- discuss completed homework.

The data regarding these statements were factor analyzed and it was discovered that student-centered approaches (with typical items, such as 5, 8, and 9 in the list above) referring to an emphasis on active student participation were distinguishable from more traditional, teacher-centered approaches (typical items were 1, 2, and 10). When examining the differences between the "high" and "low" ICT using groups, the student-centered items appeared to differentiate particularly strongly. This is illustrated in Figure 3. In all countries a higher percentage of students in the frequent use group also frequently were involved in project work, working in small groups or working on mathematical problems related to daily life.

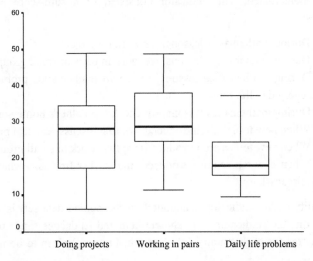

Note. Only for countries where each group contained at least 150 (unweighted) cases

Figure 3. Summary of country differences in percentages (doing projects, working in pairs and use daily life problems in mathematics) between high- and low-computer-use groups.

SUMMARY AND DISCUSSION

This chapter contained a description of indicators related to ICT and mathematics education. A major finding was that, in most countries, despite the fact that the availability of computers had increased rapidly over the years, ICT was only marginally used at the lower secondary level and that this situation (except for some countries) hardly changed between 1995 and 1999. Another interesting finding from TIMSS–95 was that students who used ICT frequently for mathematics learning had much lower achievement scores than students who hardly used or did not use ICT.

Explorations were conducted in order to determine to what extent the use of ICT co-varies with other variables that may contribute to foster our understanding of this phenomenon. These analyses showed that:

1.   Students who used computers for mathematics quite frequently did not seem to differ in terms of home background when compared with students who never or hardly used computers for mathematics learning.

2.   Frequent use of computers tended to be quite strongly associated with the existence of student-centered pedagogical approaches during mathematics lessons.

In order to examine the second point in somewhat more depth, a composite indicator was constructed. This indicator consisted of a sum-score across the following items:

- During mathematics lessons, we work on projects.
- During mathematics lessons, we work in pairs or small groups.
- During mathematics lessons, we solve mathematics problems with every day life things.
- During mathematics lessons, we check each other's homework.
- When new topics are introduced, we first discuss practical problems.
- When new topics are introduced, we first work in small groups.
- When new topics are introduced, the teacher first asks what students already know.

The reliability of this indicator (calculated on the pooled data set) is 0.72. The mean scores on the indicator for student-centered didactics and the mean mathematics scores for each country (as shown in Figure 4) seem to be negatively associated.

With reference to Figure 4, one might argue that the potential danger of this type of analyzes is what Postlethwaite (1999) called the "ecological fallacy." This implies that certain relationships, which seem to exist at the aggregated level, do not exist at non-aggregated levels, or in some cases the sign of the relationship may be even reversed. Hence, the association as presented in Figure 4 was also examined within countries, at the student level. From this analysis it appears that, at the student level as well, the association between the indicator of student-centered instructional practices and mathematics scores was negative in almost all countries. In other words, their association does not exist only at the aggregated level.

The strong association between student-centered didactics and the use of computers does fit nicely with the currently popular rhetoric regarding ICT, education, and the information society. This rhetoric has been formulated in many policy documents (European Commission, 1995; ERT, 1997; Panel on Educational Technology/PCACT/PET, 1997) which calls for the fostering of life-long learning together with the use of ICT as one of the cornerstones of the information society. In this rhetoric, a shift from a traditional pedagogical paradigm (teacher-centered, whole-class teaching, etc.) to a paradigm focusing on independent learning (doing projects, teamwork, etc.) is foreseen, and in numerous documents it is assumed that ICT can facilitate the adoption and implementation of such reform.

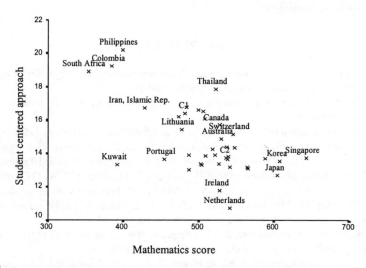

Mathematics score

Legend.
C1: Cyprus, England, Greece, Hong Kong, New Zealand, Romania, Spain, United States;
C2: Austria, Belgium-Fl, Belgium-Fr, Czech Republic, Denmark, France, Germany, Hungary, Iceland, Israel, Norway, Russian Federation, Slovak Republic, Slovenia, Spain, Sweden.

Figure 4. Plot of mean values per country for TIMSS–95 mathematics scores and an indicator of students centered didactics.

One may argue that a shift toward more learner-centered approaches in mathematics may result in having less time for the traditional focus on the reproduction of facts and "standard" problems and, hence, that scores on traditionally constructed mathematics tests may decline. Such an argument is no longer purely speculative or theoretical, since in Singapore it appeared that, "To facilitate the development of such a learner-centered environment (supported by the availability of technology and digital resources), a 10 to 30 percent reduction in curriculum content was instituted towards the end of 1998." (Teng & Yeo, 1999)

The evidence presented in this chapter seems to suggest that the use of ICT tends to take place in situations in which a somewhat higher emphasis is placed on learner-centered approaches. A tentative hypothesis about the large score difference between the high-computer-use and low-computer-use groups is that this is caused by a pedagogical approach in which less emphasis is placed on competencies such as those measured in the TIMSS–95 mathematics tests. On the other hand, one may wonder why a high emphasis on student-centered approaches typically seems to exist more in developing countries (Figure 4). Nevertheless, from a purely empirical point of view, one can maintain that cause and effect are indistinguishable on the basis of the data currently available. One may hope that TIMSS–99 may help to find further answers regarding the nature of the relationship between computer use and the learning of mathematics.

REFERENCES

ERT/European Round Table of Industrialists. (1997). *Investing in knowledge: The integration of technology in European education.* Brussels: Author.

European Commission. (1995). *Teaching and learning: Towards the learning society.* Brussels: European Union.

Evaluation and Educational Policy Analysis.Panel on Educational Technology. (1997). *Report to the President on the use of technology to strengthen K-12 education in the United States.* Washington, DC: President's Committee of Advisors on Science and Technology.

Mullis, I. V. S., Martin, M. O., Gonzalez, E. J., Gregory, K. D., Garden, R. A., O'Connor, K. M., Chrostowski, S. J., & Smith, T. A. (2000). *TIMSS 1999 International mathematics report.* Chestnut Hill, MA: International Study Center Lynch School of Education, Boston College.

Pelgrum, W. J. (1999). Infrastructure. In W. J. Pelgrum, & R. E. Anderson (Eds.), *ICT and the emerging paradigm for life-long learning.* Amsterdam: IEA.

Pelgrum, W. J., Janssen Reinen, I. A. M., & Plomp, T. J. (1993). *Schools, teachers, students and computers: A cross-national perspective.* The Hague: IEA.

Pelgrum, W. J., & Plomp, T. J. (Eds.). (1993). *The IEA study of computers in education: Implementation of an innovation in 21 education systems.* Oxford: Pergamon Press.

Postlethwaite, T. N. (1999). *International studies of educational achievement: Methodological issues. CERC Studies in Comparative Education 6.* Hong Kong: University of Hong Kong, Comparative Education Research Center.

Teng, S. W. & Yeo, H. M. (1999). Singapore. In W. J. Pelgrum & R. E. Anderson (Eds.), *ICT and the emerging paradigm for life-long learning.* Amsterdam: IEA.

Chapter 20

EXTRA-SCHOOL INSTRUCTION IN MATHEMATICS AND SCIENCE

Richard M. Wolf

One educational phenomenon that has received very little attention is extra-school instruction. Although this phenomenon varies widely in both its form and content, its prevalence cannot be ignored. Activities undertaken under the general heading of extra-school instruction can include individual tutoring in a particular school subject, group instruction in school subjects, preparation for high stakes national or regional tests, and guidance with regard to selection for the next higher level of education. It usually takes place at locations outside the school and when school is not in session.

The cost of such activity is invariably borne by parents. In Korea, for example, it is estimated that the cost of such extra school instruction in 1997 was $25 billion annually which was fully 150 percent of the government's educational budget (Asia Week, 1997). A household typically spent $1950 a year on tutoring for a secondary school student and $1500 a year for a child in elementary school. Many families are said to spend up to half their income on private teachers. In Japan, it was estimated that in 1998 a family spent about $400 a month on extra school instruction in the *Juku*, the name given to privately owned and operated extra school instruction enterprises (Personal communication, Ryo Watanabe, National Institute for Educational Research, 1998).

Some writers (Baker et al., 2001) have used the term "shadow education" to refer to this out-of-school activity. However, this seems somewhat ambiguous so in this chapter the term extra-school instruction (ESI) will be used. It is intended to denote teaching and coaching activities in mathematics and science taking place outside of the regular school structure. It excludes extra help given to students by teachers under the auspices of the school. The purpose of this study is to investigate the phenomenon of ESI to determine how extensive it is, who receives it, how much time is spent on it, and some of its correlates.

The TIMSS student questionnaire included two items about ESI. At Population 1 (Grades 3 and 4 in most countries), the following items appeared on the international version of the student questionnaire:

D.F. Robitaille and A.E. Beaton (eds.), Secondary Analysis of the TIMSS Data, 331–341.
© 2002 *Kluwer Academic Publishers. Printed in the Netherlands.*

Outside of school do you ever have <extra lessons/cramming school>*
before or after school in mathematics? ____Yes ____No

Outside of school do you ever have <extra lessons/cramming school>
before or after school in science? ____Yes ____No

*The TIMSS national center in each country replaced the text between chevrons with the
term(s) commonly used in that country.

At Population 2 (Grades 7 and 8), the following items appeared on the student
questionnaire:

During the week, how much time before or after school do you spend
taking extra lessons in mathematics?

__None __Less than one hour __1-2 hours __3-5 hours __more than 5 hours

During the week, how much time before or after school do you spend
taking extra lessons in science?

__None __Less than one hour __1-2 hours __3-5 hours __more than 5 hours

METHOD

Since the questionnaire items about ESI varied between Populations 1 and 2, it
was necessary to analyze the resulting data for the two population levels differently.
At Population 1, the percentages of students responding Yes or No were obtained
and used to answer questions about ESI. At Population 2, it was possible to obtain
percentages for each category of response and obtain an average weighted by the
percentage of students in each category of response. These could then be related to
other variables in the study. Not all countries administered the items regarding ESI
so some countries that were in the TIMSS study are not included in the following
analyses. Further, it was found that there was a translation error in the items
regarding ESI in Colombia so that country was also excluded from the analyses.

RESULTS

The basic results of the analyses are presented in Tables 1 through 4. These
tables present information for each participating country concerning the percentage
of students enrolled and not enrolled in ESI, their average scores in mathematics and
science, as well as the total mathematics score for the country. The difference
between the means of the ESI group and the non-ESI group is also presented. At
Population 2, the average number of hours of ESI is presented for the ESI group.
While two grade levels were tested at each population level, results for only the

lower grade at each population level are presented in this chapter. Results for the upper grade at each population level were quite similar to those for the lower grade.

Table 1. ESI statistics for mathematics Population 1, lower grade

Country	Total Math	%ESI	Math-ESI	% No ESI	Math-No ESI	Diff
Australia	483	12	445	88	490	-45
Austria	487	15	428	85	500	-72
Canada	469	22	448	78	479	-31
Cyprus	430	6	398	94	435	-37
Czech Republic	497	12	459	88	505	-46
England	456	20	439	80	463	-24
Greece	428	15	403	85	436	-33
Hong Kong	524	20	504	80	530	-26
Hungary	476	12	474	88	479	-5
Iceland	410	22	388	78	415	-27
Iran, Islamic Rep.	378	31	377	69	389	-12
Ireland	476	22	456	78	485	-29
Japan	538	31	539	69	537	2
Korea	561	45	558	55	565	-7
Latvia (LSS)	463	38	441	62	581	-140
Netherlands	493	21	461	79	503	-42
New Zealand	440	16	418	84	449	-31
Norway	421	16	415	84	424	-9
Portugal	425	31	406	69	437	-31
Scotland	458	10	441	90	460	-19
Singapore	552	53	543	47	561	-18
Slovenia	488	45	468	55	504	-36
Thailand	444	36	437	64	448	-11
United States	480	25	454	75	492	-38
Mean	470	24	450	76	482	-32

The ESI participation rates are notable. The average participation rate in mathematics for Population 1 is 24 percent, while the average participation rate in mathematics for Population 2 is 35 percent. In science, the average participation rate for Population 1 is 13 percent, while the participation rate for Population 2 is 26 percent. The average participation rates for Population 2 are about 10 percent higher than for Population 1 in both mathematics and science. Also, the participation rates for mathematics are about 10 percent higher than for science at both population levels.

The average participation rates, while noteworthy, mask the fact that the variability in participation rates across countries is high. In mathematics at Population 1, the participation rates range from 6 percent in Cyprus to 53 in Singapore. In science at Population 1, the participation rates range from 3 percent in Iceland to 38 in Singapore. In mathematics at Population 2, the range is from 13 percent in Germany and Norway to 79 in Latvia. In science at Population 2, the range is from 6 percent in Austria, Germany, and France to 78 in the Slovak Republic. Clearly, there is enormous variability in participation rates in both subjects at both population levels.

Table 2. ESI statistics for science Population 1, lower grade

Country	Total Sci.	%ESI	Sci-ESI	% No ESI	Sci-No ESI	Diff
Australia	510	4	457	96	514	-57
Austria	505	7	435	93	512	-77
Canada	490	13	452	87	499	-47
Cyprus	415	3	386	97	418	-32
Czech Republic	494	6	456	94	499	-43
England	499	8	456	92	507	-51
Greece	446	12	426	88	452	-26
Hong Kong	482	13	472	87	485	-13
Hungary	464	4	428	96	467	-39
Iceland	435	3	412	97	440	-28
Iran, Islamic Rep.	356	22	346	78	373	-27
Ireland	479	10	459	90	486	-27
Japan	522	9	506	91	524	-18
Korea	553	26	544	74	558	-14
Latvia (LSS)	465	24	448	76	480	-32
Netherlands	499	5	469	95	502	-33
New Zealand	473	8	469	92	481	-12
Norway	450	7	420	93	455	-35
Portugal	423	27	405	73	433	-28
Scotland	484	7	477	93	486	-9
Singapore	488	38	479	62	493	-14
Slovenia	487	17	472	83	492	-20
Thailand	433	18	431	82	432	-1
United States	511	16	480	84	523	-43
Mean	473	13	449	87	480	-30

The next question that was addressed was who participates in ESI. This was approached in terms of gender and achievement levels. The percentage of boys and girls participating in ESI was tabulated for every country at each population level for both mathematics and science. The results are fairly easily summarized. In most cases, the difference between the percentage of boys and girls enrolled in ESI was negligible. The only instances in which the percentage difference between participation rates for boys and girls was greater than 10 percent in mathematics were Iran at Population 1, where the difference was 11 percent higher for boys; Cyprus at Population 2, where the difference was 20 percent higher for boys; and the Czech Republic at Population 2, where the difference was 11 percent higher for boys. In science, at Population 2 in the Slovak Republic the participation rate was about 12 percent higher for boys. All of the above are exceptions to the general finding of negligible differences in ESI participation rates between boys and girls.

Table 3. ESI statistics for mathematics Population 2, lower grade

Country	Total Math	% ESI	Math-ESI	Hours	% No ESI	Math-No ESI	Diff
Australia	498	19	472	1.1	81	506	-34
Austria	509	18	491	1.4	82	517	-26
Belgium (Fl)	558	32	532	1.4	68	571	-39
Belgium (Fr)	507	32	489	1	68	518	-29
Canada	494	28	482	0.9	72	500	-18
Cyprus	446	44	425	1.6	56	468	-43
Czech Republic	523	55	514	1	45	537	-23
England	476	15	453	1.1	85	487	-34
France	492	16	470	1.3	84	500	-30
Germany	484	13	462	1.3	87	491	-29
Hong Kong	564	55	567	1	45	564	3
Iran, Islamic Rep.	401	43	398	1.7	57	410	-12
Ireland	500	23	482	0.8	77	510	-28
Japan	571	60	566	1.8	40	580	-14
Korea	577	50	600	2.2	50	556	44
Latvia (LSS)	462	79	467	1.4	21	457	10
Lithuania	428	40	425	1.2	60	437	-12
Netherlands	516	19	476	0.8	81	527	-51
New Zealand	472	21	457	0.9	79	481	-24
Norway	461	13	421	1.3	87	468	-47
Portugal	423	27	413	1.1	73	429	-16
Romania	454	48	461	1.3	52	453	8
Russian Fed.	501	54	498	1.2	46	511	-13
Scotland	463	21	453	0.8	79	469	-16
Singapore	601	33	595	2.6	67	604	-9
Slovak Republic	508	77	506	1.1	23	515	-9
Slovenia	498	58	493	2	42	507	-14
Spain	448	28	429	1.3	72	458	-29
Sweden	477	16	438	1.1	84	487	-49
Switzerland	506	24	483	0.9	76	515	-32
United States	476	32	466	0.9	68	483	-17
Mean	493	35	480	1.3	65	500	-20

When one examines the differences in achievement levels between students who participate in ESI and those who do not, there are considerable and generally consistent results. The last column of information in Tables 1 to 4 shows the average difference in test performance between students enrolled in ESI and students not enrolled in ESI. The average difference at Population 1 in both mathematics and science is around 30 points, while the average difference at Population 2 in both mathematics and science is around 20 points. The negative values of these numbers indicate that students who were not enrolled in ESI performed better on the tests than students enrolled in ESI. This may indicate that ESI is not used for enrichment but rather for remediation since it is the lower performing students who tend to be enrolled. Since the tests were standardized to have a mean of 500 and a standard deviation of 100 points, a difference of 30 points denotes three tenths of a standard deviation while a difference of 20 points denotes two tenths of a standard deviation.

Table 4. ESI statistics for science Population 2, lower grade

Country	Total Sci	% ESI	Sci-ESI	Hrs.	% No ESi	Sci-No ESI	Diff
Australia	504	9	479	0.9	91	509	-30
Austria	519	6	482	1.4	94	523	-41
Belgium (Fl)	529	24	510	1.1	76	535	-25
Belgium (Fr)	442	26	430	1.1	74	447	-17
Canada	499	15	476	0.8	85	505	-29
Cyprus	420	17	396	1.4	83	430	-34
Czech Republic	533	60	529	1.3	40	541	-12
England	512	10	469	1.1	90	523	-54
France	451	6	434	1	94	455	-21
Germany	499	6	457	0.8	94	506	-49
Hong Kong	495	43	500	0.8	57	495	5
Hungary	516	37	512	1.3	63	523	-11
Iran, Islamic Rep.	436	31	426	1.5	69	449	-23
Ireland	495	18	480	0.8	82	504	-24
Japan	531	40	520	1.3	60	538	-18
Korea	535	25	549	1.7	75	531	18
Latvia (LSS)	435	72	435	1.1	28	443	-8
Lithuania	403	29	403	1.2	71	408	-5
Netherlands	517	11	477	0.9	89	524	-47
New Zealand	481	11	454	1	89	489	-35
Norway	483	8	452	0.9	92	488	-36
Portugal	428	14	408	1.1	86	433	-25
Romania	452	48	453	1.1	52	457	-4
Russian Fed.	484	39	486	1.1	61	489	-3
Scotland	468	17	448	0.8	83	477	-29
Singapore	545	17	544	2.5	83	545	-1
Slovak Republic	510	78	508	1.2	22	518	-10
Slovenia	530	37	543	1.9	63	525	18
Spain	477	20	454	1.2	80	485	-31
Sweden	488	12	447	0.9	88	497	-50
Switzerland	484	13	464	0.9	87	489	-25
United States	508	23	472	1.2	77	518	-46
Mean	488	26	472	1.2	74	494	-22

There are a few rather interesting findings in the differences between the two groups (ESI and non-ESI). In Latvia, the non-ESI group scored 140 points higher in mathematics at Population 1. This was the largest difference between the two groups at any population level in either subject. There are also several instances where students in the ESI group scored higher than students in the non-ESI group.

Another way of studying the relationship between students participating in ESI and those not participating is to examine the correlations between the extent of participation in ESI and test performance. Accordingly, correlations were computed between ESI participation and test scores. Since the question regarding ESI at Population 1 was asked simply as Yes/No item, the correlations tell little more than that the means between the groups are different. However, at Population 2, the question was asked in terms of the amount of participation in ESI. Thus, it was possible to obtain the correlation between the amount of participation in ESI and the total scores on each test for each country. At Population 2, the correlations between the amount of ESI and mathematics achievement were, with one exception, negative with the average correlation around -0.10. The sole exception was in Korea where the correlation between the amount of ESI and mathematics achievement was +0.26. Similarly, the correlations between the amount of ESI and science achievement at Population 2 were negative with the average correlation also about -0.10. Again, the sole exception was in Korea where the correlation was +0.13. What this signifies is that, within countries, there is a slight negative relationship between the amount of ESI and achievement. This indicates that ESI tends to be used for remediation rather than enrichment and those students who are in need of remediation tend to be enrolled in ESI.

When one studies the results across countries, a different picture emerges. At Population 2, the results between the percentage of students enrolled in ESI and the average scores on the mathematics and science tests for those students show a positive relationship between the two variables. These are:

Test	Correlation between %ESI and Test Score at Population 2
Mathematics	+0.37
Science	+0.27

This is a very different picture. It describes a clear positive relationship between achievement in mathematics for those students and the percentage of students enrolled in ESI. It stands in contrast to the within country correlations which were small and almost always negative, averaging around –0.01 for both mathematics and science. In science, there was a small positive correlation with science achievement (+0.14). This finding of small negative correlations within countries and substantial positive correlations across countries is an example of an ecological fallacy. It is not easily interpreted. Since education takes place within a nation, it seems that greater weight should be given to the within country correlations.

At Population 1, similar results were obtained:

Test	Correlation between %ESI and Test Score at Population 1
Mathematics	+0.37
Science	+0.14

At this level, all correlations were positive and moderate in size. They stand in contrast to the generally small negative correlations within countries. They are rather consistent within each subject. They do, however, reflect the considerable variation within countries in participation rates in ESI, as do the correlations at Population 2. They should also probably be given less consideration than the within country correlations.

The final set of analyses conducted on participation in ESI and achievement related the average number of hours of ESI to test achievement across countries. This analysis could only be carried out at Population 2 since hours of participation were not obtained at Population 1. The correlations between average hours of participation in ESI and test performance in each subject were as follows:

Test	Correlation between Time in ESI and Test Score at Pop. 2
Mathematics	+0.44
Science	−0.27

The results for mathematics show a moderate relationship between hours of participation and ESI, while in science the results are just the opposite. The correlation is negative (−0.27).

The results presented above are quite provocative. ESI is a common phenomenon in all the participating countries. No country showed an absence of ESI. In a number of countries, the proportion of students participating in ESI was well above 50 percent and, in some cases, close to 80 percent. Furthermore, since the questions about ESI were asked only for mathematics and science, we have no idea what the answers for other subjects, e.g., mother tongue, might be. The total amount of time spent in ESI could be considerable.

This is a phenomenon that cannot be ignored. ESI requires a real commitment on the part of both parents and students since it is not publicly financed. In Korea, for example, parents invest heavily in ESI, as do parents in Japan. Students have to give up what otherwise would be free time to attend ESI. How committed students are to ESI is a matter for future research. How much students benefit from ESI cannot be determined from the TIMSS data. The generally lower test performance of students in ESI (about two to three tenths of a standard deviation) indicates that ESI is used primarily for remediation rather than enrichment. There are negligible differences between participation rates for boys and girls in ESI in virtually all countries, and the correlations between amount of ESI received and test scores are generally small and negative in most countries at Population 2.

One has to go to other sources to find additional information regarding ESI. The article in Asia Week (1997) furnishes some basic information regarding the costs of ESI in Korea. At Population 2, Korean students enrolled in ESI scored substantially higher in both mathematics and science than students not enrolled in ESI. The differences between the ESI and non-ESI groups were sizable, ranging from 16 to 44 points in the two subjects. This is reflected in the correlations between amount of participation in ESI and achievement. The correlations ranged from +0.11 (mathematics) to +0.30 (science). This stands in marked contrast to the generally low and negative correlations between ESI participation and achievement in virtually all other countries. It would seem that ESI is used for enrichment in Korea.

There are also three studies carried out in Japan that provide some useful information on students' views about the *Juku*, the Japanese institution that provides ESI. One study was conducted by the Japanese Ministry of Education (1993), one by the National Institute for Research Advancement (undated), and the third by the Japanese National Parents and Teachers Association (1987). The National Institute for Research Advancement study contains statistics on attendance at the *Juku* and reasons for attending it. The sampling for the study consisted of parents who were generally from higher income groups and lived in urban areas. It does contain a good deal of background material on the *Juku*, including its history. Parents who were questioned gave three major reasons for sending their children to the *Juku*. These were to prepare for high stakes entrance examinations to higher education, to review material covered in the regular curriculum, and to provide children with a place to stay after school. Unfortunately, there is virtually no response data in this study regarding students' views and opinions about the *Juku*.

The PTA (1987) study questioned parents, *Juku* teachers, and students regarding the value of *Juku* among other things. Two-thirds of the parents and teachers felt that there was excessive use of the *Juku*. When the *Juku* teachers were questioned, two - thirds reported that they used the *Juku* to go over weak points in the students' understanding of the subject being studied and 62 percent indicated that students could not get ahead without *Juku* attendance. Students reported that *Juku* helped them learn their school subjects, improve their grades and improve their learning. On the negative side, parents reported that the *Juku* was expensive and that it reduced children's free time. Students also reported the loss of free time as a negative (72 percent) and 69 percent reported that they got tired attending the *Juku* after school.

The Ministry of Education study (1993) contained some interesting results regarding students' views of the *Juku*. Parents reported that the *Juku* provided considerable individual direction regarding learning school subjects and promoted interest in learning. This sentiment was echoed by students. In fact, students reported that the *Juku* did more to help them learn their subjects than the regular school because of the individualized attention given them. However, there was a feeling of excessive use of the *Juku*. This is consistent with the results of the PTA study. However, unlike Korea, the correlations between ESI and achievement are virtually zero, ranging from −0.05 to +0.05.

Unfortunately, there is no additional information available on ESI from any of the other participating countries. It is hoped that future IEA studies will inquire more deeply into ESI. However, a few conclusions can be drawn from the information that is available. First, ESI is a worldwide phenomenon. None of the participating countries reported an absence of ESI. The percentage of students taking part in ESI varied considerably across countries with anywhere from 3 to 82 percent of students enrolled in ESI in a particular subject at a grade level. On average, larger percentages of students at Population 2 participated in ESI than at Population 1. Also, enrollment in ESI was greater in mathematics than in science at all grade levels. What is not known is how much total time students spent in ESI since the question was not asked for all subjects but just for mathematics and science. It would seem useful if future studies could inquire about total time spent in ESI and what subjects were studied. Finally, it is interesting to note that there are virtually no gender differences in ESI participation.

The findings from the internal Japanese studies strongly suggests that if one wishes to understand the educational process in a country, restricting investigations to the regular school classroom is insufficient to gaining such understanding. Since a very high percentage of students in a number of countries——notably Korea, Japan, Latvia, Romania, and the Slovak Republic—are enrolled in ESI, efforts will need to be made to capture how much time is spent in ESI and what happens there. This is not to suggest that separate studies of ESI have to be undertaken, but at least some questions should be asked of students regarding their ESI participation. Studies of classrooms, even videotape studies, are apt to be misleading since they omit what students such as those in Japan see as a critical part of their education. In fact, Japanese students report that the *Juku* plays a central role in their education and, in fact, is often viewed as more central to their learning than their regular classroom instruction.

One of the rather noteworthy findings of this study is that, in almost every country, ESI is used more for remediation than enrichment. This is reflected in the generally higher scores for students in the non-ESI group than in the ESI group for both mathematics and science. This is reflected in the generally negative numbers in the last column of Tables 1-4 in this report. Clearly, more research is needed about this educational phenomenon.

REFERENCES

Baker, D., Akiba, M., LeTendre, G., & Wiseman, A. (2001). Shadow education, mass schooling, and cross-national mathematics achievement. *Educational Evaluation and Policy Analysis, 23*(1), 1–18.

Banning Tutors. (1997, May 2). *Asia Week, 23*(17), p.20.

Ministry of Education. (1993). *Report of researches into the state of Juku.* Tokyo, Japan: Author.

National Institute for Research Advancement. (n.d.). *Japanese education as seen through its cram Schools.* (NIRA Research Report No. 950073). Tokyo: Author.

National Parents and Teachers Association. (1987). *Questionnaire research regarding Juku: Report of results middle school third year, elementary school sixth year.* Tokyo: Author.

Chapter 21

TEACHERS' SOURCES AND USES OF ASSESSMENT INFORMATION

Thomas Kellaghan and George F. Madaus

Over the last two decades, assessment has assumed an increasingly important role in efforts to improve the quality of education throughout the world (Kellaghan & Greaney, 2001). Most resources have been invested in assessment involving external agents, probably because such assessment is relatively inexpensive (compared, for example, to the professional development of teachers); it can be externally mandated; it can be established and implemented in a short period of time; and results can be made visible (Linn, 2000). Insofar as external assessments aspire to affect student learning, however, they can only hope to do so indirectly. Classroom-based assessment, on the other hand, when integrated into the teaching and learning process, has as its primary and immediate objective the facilitation of student learning and indeed some evidence is available, mostly from the United States and Great Britain, that its impact can be positive (Black & Wiliam, 1998; Crooks, 1988).

We know little about the assessment practices of teachers even in countries with long traditions of educational research (see Frary, Cross, & Weber, 1993; McMillan, 2001; Plake, Impara, & Fager, 1993; Stiggins & Conklin, 1992; Stiggins, Conklin, & Bridgeford, 1986). Further, we would expect those practices to vary from one education system to another, depending on the conditions in which they are implemented, such as the size of classes, grouping and tracking practices, teachers' educational backgrounds, systems of classroom management, the impact of external testing systems, teacher-student relations, teachers' conceptions of the learning process, degree of individualization of instruction, and general approaches to the regulation of learning (see, e.g., Perrenoud, 1998). Little systematic information is available on the nature or extent of variation in teachers' assessment practices, either between or within education systems. TIMSS may go some way towards filling this information gap since, in addition to having students take tests in mathematics and science, teachers were asked in a questionnaire to respond to questions relating to contexts of learning and instruction (Martin et al., 1999), including questions about assessment. In the study described in this paper, we examined teachers' responses to

D.F. Robitaille and A.E. Beaton (eds.), Secondary Analysis of the TIMSS Data, 343–356.
© 2002 Kluwer Academic Publishers. Printed in the Netherlands.

questions regarding the weight they assigned to various sources of information in assessing students' work and the uses they made of assessment. Analyses were confined to data relating to mathematics for teachers and students of the upper grade of Population 2 (Grade 8 in most countries) in 38 education systems and addressed the following issues:

1. differences between education systems in the weight given by teachers to seven sources of information in assessing the work of their students,
2. differences between education systems in the use of assessment by teachers for six purposes,
3. relationships between students' mathematics achievement and the weight given by teachers to seven sources of information across and within education systems,
4. relationships between students' mathematics achievement and six uses by teachers of information across and within education systems.

VARIABLES AND ANALYSES

Teachers were asked to indicate the weight (none, little, quite a lot, a great deal) they gave in assessing the work of their students in mathematics class to seven sources of information: standardized tests developed outside the school; teacher-made short answer or essay tests that required students to describe or explain their reasoning; teacher-made multiple choice, true-false, and matching tests; how well students did on homework assignments; how well students did on projects or practical exercises; observation of students; and responses of students in classes.

Teachers were also asked how often (none, little, quite a lot, a great deal) they used assessment information gathered from students to provide student grades or marks; to provide feedback to students; to diagnose students' learning problems; to report to parents; to assign students to courses, streams or sets; and to plan for future lessons.

The first plausible value of Rasch mathematics test scores was used as an index of students' mathematical achievement.

Analyses

To describe the assessment practices of teachers, responses to all items dealing with sources of assessment information and use of assessment information were collapsed into two categories. The first was formed by combining the "Quite a lot" and "A great deal" responses; the second by combining the "None" and "Little" responses. For each of the 38 education systems, the percentage of teachers falling into each of these two constructed categories was computed. In reporting results on sources of information, we refer to the "Quite a lot/A great deal" category as high weight and the "None/Little" category as low weight. In the case of uses of information, "Quite a lot/A great deal" is described as a lot and "None/Little" as a little.

To examine relationships between teachers' assessment practices and student achievement, responses to the seven items relating to sources of information and to the six items relating to uses of information were assigned the following values: 1 (none); 2 (little); 3 (quite a lot); and 4 (a great deal). The mean mathematics achievement score of all students associated with individual teachers was calculated, and correlated with teachers' assessment scores. Correlations between aggregated country mean scores and aggregated mean scores for teachers' assessment practices, as well as within-country correlations between individual teachers' assessment practices and the mean scores of students associated with a teacher, were calculated.

We can speak in a rather loose way of "class" means in describing student achievement and, in some education systems, the student achievement mean is a class mean. However, because of variation in the way education systems are organized and differences between countries in sampling procedures, this is not always the case. Thus, "class" is more correctly interpreted as the group of students associated with a particular teacher.

In considering the results of analyses the reader should bear in mind that, because of time constraints, standard errors that took into account sample design effects were not available for the questionnaire data. This means that tests of statistical significance are strictly speaking only applicable in the case of the between-country correlations between teachers' assessment practices and students' mathematics scores.

ASSESSING STUDENTS' WORK IN MATHEMATICS

Use of Standardized Tests Produced Outside the School

When asked how much weight they gave to standardized tests produced outside the school in assessing the work of students in their mathematics class, the average across all education systems (excluding Singapore and the Russian Federation, which did not administer this question) falling in the high-weight category was 30 (Table 1, Column 1). Percentages ranged from the atypical values of 77 (Israel, in which the number of participating teachers was small) and 75 (Slovak Republic) to 0 in Germany and 4 in Austria. Approximately half the teachers in Denmark, Latvia, Romania, Slovenia, and Sweden indicated they gave a high weight to externally produced standardized tests. These high- and relatively high-weight education systems contrasted sharply with 13 systems in which at least 8 in 10 teachers indicated they gave low weight to such test information (Australia, Austria, Belgium (Fl), Belgium (Fr), Canada, Colombia, Germany, Japan, Lithuania, New Zealand, Portugal, Spain, and the United States). It is interesting to note that in all but six countries (Denmark, Israel, Latvia, Slovak Republic, Slovenia, and Sweden), more

Table 1. Percents of teachers attaching "Quite A Lot/A Great Deal" of weight to seven sources of information in assessing the work of their students in mathematics

	n	Externally produced tests	Teacher open-ended tests	Teacher multiple-choice tests	Home-work	Projects/ practical exercises	Student obser-vation	Student response in class
Australia	307	8	42	24	26	29	37	34
Austria	216	4	29	1	47	23	97	81
Belgium (Fl)*	154	10	94	11	15	16	50	55
Belgium (Fr)*	120	6	85	16	35	6	47	58
Canada	388	16	49	18	44	32	43	41
Colombia	143	16	81	55	90	77	88	94
Cyprus	102	40	71	56	96	67	88	100
Czech Rep.	149	43	100	19	14	29	74	96
Denmark	128	54	75	21	66	74	97	91
England	133	36	32	7	68	48	71	66
France	338	23	83	25	28	16	49	54
Germany	134	0	55	7	18	40	74	81
Greece	172	32	92	44	58	45	87	99
Hong Kong	85	32	40	40	74	12	68	74
Hungary	150	34	71	24	43	90	69	87
Iceland	136	45	42	9	92	53	73	68
Iran, Islamic Rep.	191	22	88	24	60	14	45	86
Ireland	131	35	26	25	75	37	76	86
Israel	60	77	29	64	61	70	54	46
Japan	151	16	54	20	44	34	68	71
Korea	147	36	54	32	24	20	31	62
Kuwait	64	30	78	77	62	32	61	88
Latvia (LSS)*	140	52	61	33	79	62	83	100
Lithuania	146	10	31	11	34	16	24	83
Netherlands	95	29	99	31	30	14	36	42
New Zealand	148	14	52	20	34	36	52	46
Norway	148	27	100	3	25	15	55	59
Portugal	142	14	69	16	79	61	89	97
Romania	164	48	90	51	81	37	78	97
Russian Fed.	179	NA	100	54	64	52	97	NA
Singapore	137	NA	30	6	72	37	61	70
Slovak Rep.	146	75	97	24	35	36	89	99
Slovenia	113	56	76	22	59	44	70	73
Spain	154	5	92	23	75	42	90	95
Sweden	255	59	90	19	50	53	87	79
Switzerland	264	28	77	6	13	14	47	54
Thailand	147	22	52	71	75	21	51	66
United States	292	20	51	26	57	35	44	45
Mean		30	67	27	53	38	66	74

Note: The percentage of teachers who choose the "none" or "a little" options may be obtained by subtracting the figures in the table from 100.

*The Flemish (Fl) and French (Fr) educational systems in Belgium participated separately in TIMSS. Latvia is annotated LSS for Latvian Speaking Schools only.

than half the teachers indicated they gave low weight to externally produced standardized test information, while in 20 countries the percentage falling in this category was 70 or higher.

It would seem that there is a strong disregard among teachers in most countries for standardized, externally produced test results. This may be a function of the timing of test administration. It may be that if tests are given late in the academic year, the results come too late for teachers to act on them. In this situation, the tests may, from the teachers' point of view, be summative rather than formative (Bloom, Madaus, & Hastings, 1981). However, it may also be that teachers give more credence to their own informal sources in assessing students than to the information provided by formal tests developed outside the school (Kellaghan, Madaus, & Airasian, 1982).

Weight Given to Teacher-Made Short-Answer or Essay Tests

The mean percent responses across education systems of teachers giving high weight to their own essay/short-answer (supply-type) tests was 67, and the range 26 (Ireland) to 100 (Norway) (Table 1, Column 2). In 10 education systems, 90 percent or more of teachers gave high weight to this source, while in 10 others between 70 and 90 percent of teachers fell into this category. On the other hand, 60 percent or more of teachers in Austria, England, Hong Kong, Ireland, Israel, Lithuania, and Singapore gave low weight to this type of information. It seems strange that teachers would disregard information from their own supply-type tests, especially mathematics teachers, whom one would expect to give tests in which students produced short answers (a number, an equation, or a few words). A possible explanation is that teachers focused on the word "essay" which modified the noun "test" rather than on the "short-answer" adjective when responding, and considered "essay" tests irrelevant to mathematics.

Weight Given to Teacher-Made Multiple-Choice, True-False, and Matching Tests

The mean percent of teachers across education systems that indicated that they gave high weight to teacher-made multiple-choice, true-false, or matching (selection-type) tests was low (27), and the percentage that gave low weight correspondingly high (73) (Table 1, Column 3). At one extreme, 77 percent of teachers gave high weight to this source of information (Kuwait); at the other, only 1 percent did (Austria). Three out of 4 teachers in 25 education systems said that they gave low weight to teacher-made multiple-choice, true-false, and matching tests in their assessments. These findings may reflect the fact that mathematics teachers are more likely to set short-answer quizzes and tests than multiple-choice ones. In seven countries, however, a majority of teachers gave considerable weight to the results of their multiple-choice tests (Colombia, Cyprus, Israel, Kuwait, Romania, Russia, and Thailand).

Weight Given to How Well Students Do on Homework Assignments

When asked how much weight they gave to student performance on homework assignments, 53 percent of teachers overall fell into the high-weight category (Table 1, Column 4). The percentages ranged from 13 (Switzerland) to 96 (Cyprus). More than half the teachers in 21 education systems fell into this category. In 11 systems, the figure was more than 70 percent of teachers. At the other extreme, in nine systems, 70 percent or more of teachers indicated that they gave low weight to this source of information. These conflicting data on the importance of homework may reflect differences between countries in the culture of schools and the beliefs of teachers. It may be that in some countries homework is seen as a learning or practice experience for students rather than as an evaluative exercise. In other contexts, teachers may regard homework as an exercise to be graded.

Weight Given to How Well Students Do on Projects or Practical Exercises

The mean percent of teachers across all education systems that accorded high weight to projects or practical exercises in the use of assessment information was 38, and for the low-weight category it was 62 (Table 1, Column 5). The percentage in the former category ranged from 6 (Belgium (Fr)) to 90 (Hungary). However, a majority of teachers fall into this category in only 10 education systems. In the other 28 systems, less than half of teachers gave high weight to projects or practical exercises in their assessment of students. Such differences in approach toward this source of information may reflect differences across countries in curriculum practices as much as in modalities of assessment; in some of these, projects and practical exercises may feature while in others they may not.

Weight Given to Observations of Students

Teachers continually observe their students, from which they derive a complex mosaic of cognitive, affective, and social information about individual students and about the class as a social group (Jackson, 1968; Stiggins et al., 1986). One would expect that most teachers would give considerable weight to such information. The overall mean percent of teachers who gave high weight to their own observations of students was 66, and for individual countries ranged from 24 (Lithuania) to 97 (Austria) (Table 1, Column 6). While the mean percent ranked third among the sources of information used by teachers, the variation is sufficient to indicate that the assumption that most teachers give considerable weight to the observation of students in making assessments is not uniformly supported. True, in 11 education systems, 8 or more teachers out of 10 fell in the high-weight category, and 50 to 80 percent fell in the category in 16 systems. However, a majority of teachers in 11 education systems said they gave low weight to their observation of students.

Weight Given to Responses of Students in Class

Student responses in class, like observations of students, are a rich source of information for teachers, and so we might expect considerable weight to be given them in assessing students. The data with few exceptions confirm this expectation. The mean percent of teachers across 37 education systems (the Russian Federation had no data for this item) in the high-weight category was 74 (Table 1, Column 7), and the range 34 (Australia) to 100 (Cyprus, Latvia). In 31 systems, 50 percent or more of teachers indicated they would give high weight to student responses. In 10 systems, 9 or more teachers out of 10 responded this way, while in seven others, the figure was between 8 and 9 out of 10. A majority of teachers fall in the low-weight category for this item in only six education systems (Australia, Canada, Israel, Netherlands, New Zealand, and the United States).

USES OF ASSESSMENT FOR SIX PURPOSES

Use of Assessment Information for Student Grades

The mean percent of teachers across the 37 education systems that used assessment information a lot in grading students is 78 (data for Austria were not available). Percentages ranged from a low of 14 (Israel) to a high of 100 (Cyprus) (Table 2, Column 1). In 11 education systems, more than 90 percent of teachers indicated that they used assessment information a lot. However, little use was made of assessment information for assigning grades in Denmark (26 percent), where formal grading of students is not a common feature of educational practice.

Use of Assessment Information to Provide Feedback to Students

The mean percent of teachers across education systems that indicated that they used assessment information a lot to provide feedback to students was 80 (Table 2, Column 2). Percentages ranged from 14 (Israel) to 97 (Russian Federation). In 22 systems, more than 8 out of 10 teachers used assessment data a lot for this purpose. However, in a number of systems, the percentage of teachers that made little use of assessment information for feedback was relatively high: Israel (86), Korea (58), Lithuania (48), and Japan (40). Japan and Korea, it will be noted, were two of the highest-scoring countries on the TIMSS mathematics test. It may be that the interpretation of the terms assessment and feedback by teachers in these countries differed from the interpretation in countries in which teachers reported using assessment information a lot to provide feedback to students.

Table 2. Percent of teachers using assessment information "Quite A Lot/ A Great Deal" for six purposes, By education system

	n	Provide student grades/ marks	Provide feedback to students	Diagnose students' learning problems	Report to parents	Assign students to courses, streams, sets	Plan future lessons
Australia	307	86	89	76	77	55	73
Austria	216	-*	72	75	39	17	53
Belgium (Fl)	154	70	78	88	80	84	54
Belgium (Fr)	120	92	81	92	61	-*	89
Canada	388	87	92	84	79	52	79
Colombia	143	68	90	92	53	37	94
Cyprus	102	100	93	96	96	60	91
Czech Rep.	149	94	93	100	67	38	98
Denmark	128	26	85	85	54	68	85
France	133	91	91	84	81	78	85
England	338	89	93	90	61	36	91
Germany	134	84	86	89	48	28	86
Greece	172	97	88	90	89	41	77
Hong Kong	85	72	82	81	13	13	74
Hungary	150	58	71	95	81	83	79
Iceland	136	84	71	82	78	10	91
Iran, Isl. Rep.	191	83	71	81	63	62	79
Ireland	131	72	83	84	76	54	85
Israel	60	14	14	20	27	36	7
Japan	151	73	60	66	9	29	58
Korea	147	39	42	65	10	3	56
Kuwait	64	70	75	81	53	66	83
Latvia (LSS)	140	97	69	96	39	42	95
Lithuania	146	78	52	54	54	45	78
Netherlands	95	86	68	65	57	68	50
New Zealand	148	87	87	81	86	45	76
Norway	148	69	77	47	31	57	82
Portugal	142	92	80	95	64	43	90
Romania	164	94	90	94	75	78	95
Russian Fed.	179	90	97	98	25	90	98
Singapore	137	71	87	88	39	31	76
Slovak Rep.	146	74	79	90	68	12	78
Slovenia	113	73	97	95	76	40	92
Spain	154	95	93	90	86	72	92
Sweden	255	73	91	85	53	32	93
Switzerland	264	85	92	88	47	23	80
Thailand	147	65	77	84	41	72	86
United States	292	96	91	80	82	30	86
Mean		78	80	82	58	47	79

Note: The percent of teachers who chose the "none" or "a little" options may be obtained by subtracting the figures in the table from 100.

- * Data not available.

Use of Assessment Information to Diagnose Students' Learning Problems

The mean percent of teachers across education systems that indicated that they used assessment information a lot to diagnose students' learning problems was high (82) (Table 2, Column 3). The range however, was considerable: from 20 percent (for Israel, which it should be acknowledged was again atypical) to 100 percent (Czech Republic). Apart from Israel, the lowest percentage was 47 (Norway). In 14 systems, 90 percent or more of teachers used assessment information for diagnostic purposes a lot. The mean for little use at 18 percent was the lowest for items dealing with teachers' use of assessment information. In six systems (Israel, Japan, Korea, Lithuania, Netherlands, and Norway), one-third or more of teachers made little use of assessment information for diagnostic purposes.

Use of Assessment Information to Report to Parents

Overall mean percents of teachers that used assessment information to report to parents a lot (58) and a little (42) were not greatly dissimilar across education systems (Table 2, Column 4). Percentages for high use, however, ranged from 9 (Japan) to 96 (Cyprus). In 14 systems, a large majority (at least three-quarters) used assessment information a lot to report to parents. These education systems contrasted sharply with nine in which 60 percent or more of teachers reported low use. Four of these were Asian: Hong Kong, Japan, Korea, and Singapore. The remaining five were Austria, Israel, Latvia, Norway, and the Russian Federation. These data may reflect differences between countries in parent-teacher relationships and in the role that parents play in the formal educational system.

Use of Assessment Information to Assign Students to Courses or Streams

When asked how often they used assessment information to assign students to courses or streams, teachers split fairly evenly. However, for the first time the mean country percentage of teachers in the category reporting little use (53) exceeded the percentage in the group reporting high use (47) (Table 2, Column 5). The individual country mean for high use ran from a low of 3 percent (Korea) to a high of 90 percent (Russian Federation). At least 7 out of 10 teachers in seven education systems reported using assessment information for this purpose a lot. At the other extreme, in nine education systems 7 or more teachers out 10 said that they made little or no use of assessment information in assigning students to courses, streams, or sets. The data probably say more about differences between countries in grouping practices, and in how students are assigned to groups when they exist, than they do about differences in assessment practices.

Use of Assessment Information to Plan for Future Lessons

In most countries, a large proportion of teachers used assessment information to plan for future lessons. The mean percentage of teachers across all education systems that used assessment information for this purpose a lot was 79 (Table 2, Column 6). With the again notable exception of Israel, at least half the teachers in all education systems used assessment information a lot to plan lessons.

RELATIONS BETWEEN STUDENTS' MATHEMATICS ACHIEVEMENT AND WEIGHT GIVEN BY TEACHERS TO SOURCES OF INFORMATION

Between-Education System Correlations

The correlation between the mean mathematics scores for education systems and the mean weights given to information by teachers in assessing the work of students in each country was statistically significant at the 0.05 level for three variables: teacher-made multiple-choice tests (–0.33), students' performance on homework (–0.41), and students' responses in class (–0.36). All the correlations are negative.

Within-Education System Correlations

We have already noted that standard errors that took into account sample design effects were not available for within-education system correlations between class mean mathematics scores and individual class teachers' weights given to sources of assessment information. To get some impression of the extent to which within-education system correlations are statistically significant, a design effect of two was assumed for the teacher variables. On this assumption, the effective sample size in each country would be half the actual size. Thus, sample sizes were halved in calculations. Of 188 correlations involving the weight given to five sources of information, only seven reached significance at the 0.01 level.

RELATIONS BETWEEN STUDENTS' MATHEMATICS ACHIEVEMENT AND TEACHERS' USE OF INFORMATION

Between-Education Systems Correlations

The correlation between country mean mathematics scores and the country mean frequency for teachers' use of information was statistically significant for three variables: reporting to parents (–0.38), assigning students to courses, streams, sets (–0.33), and planning for lessons (–0.35). It will be noted that all the correlations are negative.

Within-Education System Correlations.

Within-country correlations for class mean mathematics scores and individual class teachers' uses of assessment information using half the number of teachers in calculations yielded nothing by way of significance. Only 2 of 301 correlations were significant at the 0.01 level.

CONCLUSION

In general, the TIMSS questionnaire was not very successful in identifying sources to which teachers attached a lot of weight in assessing the work of their students in mathematics. The most frequently cited source was student response in class, mentioned by 74 percent of teachers. About two-thirds attached a similar weight to teacher-made open-ended tests (67 percent) and observation of students (66 percent), and about half (53 percent) to homework assignments. An overall majority of teachers attached little or no weight to other sources.

These figures mask considerable variation between education systems in the proportions of teachers according weight to various sources. For example, while over 90 percent of teachers in 10 education systems said that they gave high weight to student responses in class in the assessment of their students, less than half the teachers in six systems did. Again, 90 percent or more teachers in 10 systems, compared to less than half in 10 systems, accorded a high weight to teacher open-ended tests.

There is also evidence of variation within education systems. While the figures we have just cited, in which more than 90 percent of teachers in an education system accorded a high weight to sources in assessing students, indicate a high degree of uniformity in practice within a country, in 10 education systems between 40 and 60 percent of teachers said that they gave considerable weight to student responses and teachers' open-ended tests, reflecting considerable variation within these systems in the weight teachers accorded sources in assessing students.

There is evidence that, in a few education systems, teachers gave consistently higher or lower weights to the sources in the TIMSS questionnaire than teachers in most systems. Thus Colombian teachers accorded relatively high weights to homework assignments, projects or practical exercises, observation of students, and student responses in class. In Cyprus, Denmark, Greece, Romania, and the Slovak Republic, large percentages of teachers also gave high weight to three or more sources. On the other hand, teachers in Australia, Lithuania, the Netherlands, and Switzerland consistently gave low weights to sources. In general, teachers in most education systems did not indicate that they gave high weight to the sources listed in the questionnaire, which may indicate that the questionnaire failed to identify the sources that they actually used.

The TIMSS questionnaire was, on the whole, more successful in identifying teachers' uses of assessment than it was in identifying the weights teachers gave to the sources they used in their assessment practices. A relatively high level of use was recorded for diagnosis of student learning problems (82 percent), for providing feedback to students (80 percent), for planning lessons (79 percent), and for providing student grades or marks (78 percent). More than 80 percent of teachers in 30 education systems used assessment information a lot to diagnose learning problems; and similar percentages in 22 systems used such information a lot to provide feedback to students, and in 21 systems to plan lessons.

Not all teachers, however, used assessment information a lot for the purposes listed in the questionnaire. This was particularly the case for assigning students to courses or streams, and for reporting to parents, for which purposes teachers in a number of countries said that they did not use assessment information. Even in the case of planning future lessons, for which teachers frequently used assessment information, teachers in one country, Israel, did not.

While there can be little variation in an education system in which a large majority of teachers indicate that they use assessment information for a particular purpose a lot, there are indications that practice in the use of assessment information is far from uniform in some systems. Divergence within systems was most frequent in the use of assessment to assign students to courses or streams; to report to parents; and to provide feedback to students.

Findings on relationships between teachers' assessment practices and students' achievements in mathematics are difficult to interpret. Between-education system correlations, when significant, are all negative in the case of both the weight teachers give to sources of assessment information and their uses of information. It was equally surprising that within-country analyses yielded practically nothing by way of significance, particularly in the correlations between students' achievements and teachers' uses of assessment to diagnose students' learning difficulties and to provide feedback. In some education systems, this may be a function of lack of variance in teachers' assessment practices.

It is clear that our analyses present a complex, and even perplexing, picture of assessment practices in the education systems that participated in TIMSS. There certainly is evidence of the variation in conditions that Husén (1973) had envisaged that international studies of student achievement would reveal, but not only are the variations not readily interpretable, in some cases they seem counterintuitive. We do not have enough information to say why a large majority of teachers in some countries, but only a small percentage in others, give weight to a particular source of information or use assessment information in making decisions for a variety of purposes. In seeking explanations, we have to acknowledge that the instrumentation in TIMSS may have been inadequate to describe important practices of teachers. Only a small amount of information was obtained on teachers' assessment practices in the questionnaire.

At a more basic level, it may not make sense to attempt to extract and interpret the significance of particular aspects of teachers' assessment practice without considering the contexts in which they are used. Our analyses, however, did not relate assessment practices to characteristics of systems, such as evaluation policies and practices (see Robitaille, 1997), or to factors in the teaching-learning context (e.g., systems of classroom management, general approaches to the regulation of learning) that might be relevant to an interpretation of their use, or lack of use. Furthermore, not all items might have been equally appropriate in all education systems. For example, "grading" or the use of standardized tests may not be a feature of all systems, while in some systems, students may not be assigned to separate courses or streams, or, if they are, assignment may not be on the basis of teacher assessment. Such differences could well have led to differences and difficulties in teachers' interpretation of the items. Finally, teachers may not be very aware or conscious of the components of their assessment practices, in which case the items could have evoked unreliable responses.

Our findings might be regarded as having been more successful in raising questions than in providing answers. Certainly, the variation in assessment practices revealed in analyses, both between and within countries, and the unexpected findings on the relationships between assessment practices and student achievement invite further investigation. They also, however, point to many difficulties in exploring the "one big educational laboratory" that Husén (1973) had envisaged as providing information through international studies that would improve policymakers' understanding of the relationship between school achievement and the conditions in which students learn.

REFERENCES

Black, P., & Wiliam, D. (1998). Assessment and classroom learning. *Assessment in Education, 5,* 7-74.

Bloom, B. S., Madaus, G. F., & Hastings, T. (1981). *Evaluation to improve student learning.* New York: McGraw-Hill.

Crooks, T. J. (1988). The impact of classroom evaluation practices on students. *Review of Educational Research, 58,* 438-481.

Frary, R. B., Cross, L. H., & Weber, L. J. (1993). Testing and grading practices and opinions of secondary teachers of academic subjects: Implications for instruction in measurement. *Educational Measurement: Issues and Practice, 12*(3), 23-30.

Husén, T. (1973). Foreword. In L. C. Comber & J. P. Keeves (Eds.), *Science achievement in nineteen countries.* New York: Wiley.

Jackson, P. W. (1968). *Life in classrooms.* New York: Holt, Rinehart & Winston.

Kellaghan, T., & Greaney, V. (2001). The globalisation of assessment in the 20th century. *Assessment in Education, 8,* 87-102

Kellaghan, T., Madaus, G. F., & Airasian, P. W. (1982). *The effects of standardized testing.* Boston: Kluwer-Nijhoff.

Linn, R. L. (2000). Assessments and accountability. *Educational Researcher, 29*(2), 4-16.

Martin, M. O., Mullis, I.V.S., Gonzalez, E. J., Smith, T. A., & Kelly, D. L. (1999). *School contexts for learning and instruction.* Chestnut Hill MA: TIMSS International Study Center, Boston College.

McMillan, J. H. (2001). Secondary teachers' classroom assessment and grading practices. *Educational Measurement: Issues and Practice, 20*(1), 20-32.

Perrenoud, P. (1998). From formative evaluation to a controlled regulation of learning processes. Towards a wider conceptual field. *Assessment in Education, 5,* 85-102.

Plake, B. S., Impara, J. C., & Fager, J. J. (1993). Assessment competencies of teachers: A national survey. *Educational Measurement: Issues and Practice, 12*(4), 10-25.

Robitaille, D. (Ed.). (1997). *National contexts for mathematics and science education. An encyclopedia of the education systems participating in TIMSS.* Vancouver, Canada: Pacific Educational Press.

Stiggins, R. J., & Conklin, N. F. (1992). *In teachers' hands: Investigating the practices of classroom assessment.* Albany NY: State University of New York.

Stiggins, R. J., Conklin, N. F., & Bridgeford, N. J. (1986). Classroom assessment: A key to effective education. *Educational Measurement: Issues and Practices, 5*(2), 5-17.

5 FOCUS ON METHODOLOGY

Chapter 22

Extending the Application of Multilevel Modeling to Data from TIMSS

Laura M. O'Dwyer

Innovations in educational research, whether methodological or applied, are generally directed toward a greater understanding of how schools, classrooms, and teachers affect student outcomes. It is generally accepted that attempts to reform educational policy or practice should be guided by the most appropriate, efficient, and powerful analytical techniques available. Policy decisions regarding the allocation of funding or the implementation of reform are often made on the basis of results from the statistical modeling of student or school outcomes. In the last decade, with advances in software capabilities, multilevel modeling techniques have become increasingly available to researchers. The application of multilevel modeling has the potential to lead to a greater understanding of how formal education is affected by schools, classrooms, teachers, and changes that occur over time.[1]

The sample design employed in any research study dictates the complexity of the multilevel model that can be formulated. More complex sample designs allow more levels to be modeled, thus yielding more useful and more accurate information about the contexts that affect outcomes of interest. The TIMSS sampling procedure required that participating countries sample at least one classroom per target grade in each school. Population 2 refers to the two adjacent grades with the largest proportion of 13-year-olds. For most countries these were Grades 7 and 8. Sampling adjacent grades ensured that the data collection process was efficient and that the sampling burden was kept to a minimum (Martin & Kelly, 1996). The majority of countries adhered to the sampling procedure and sampled only one classroom per grade in each school. This design limits the number of levels that can be modeled, in that only a two-level model, in which the between-classroom and among-school variances are confounded, can be formulated for the majority of countries. To facilitate more in-depth modeling, Australia, Cyprus, Sweden, and the United States chose to sample two classrooms per grade from each school (Martin & Kelly, 1996). In this paper, these countries are referred to as validation countries.

While the two-level model is useful, it cannot accurately partition the total variance in the outcome into its within-classroom, between-classroom, and among-

D.F. Robitaille and A.E. Beaton (eds.), Secondary Analysis of the TIMSS Data, 359–373.
© 2002 *Kluwer Academic Publishers. Printed in the Netherlands.*

school components. In this way, the two-level model cannot be used to assess the effects of context variables at the classroom level independently of the school level. A three-level model is a more appropriate method of modeling student outcomes since it allows individual or student effects to be modeled in the same way as with the two-level model, while also allowing classroom effects to be modeled separately from the school effects. The aim of many secondary analyses of educational data, particularly with data as rich and extensive as TIMSS, is to guide policy decisions relating to formal education. It would seem appropriate that, where feasible, analysts employing multilevel regression techniques should use the most powerful and accurate models of student outcomes available to them, namely the three-level model. This research was conducted to extend the application of multilevel modeling beyond the two levels dictated by the one-classroom-per-grade sample design used in the majority of TIMSS countries. The method outlined in this research is referred to as the pseudo-classroom procedure. The student outcome modeled throughout this research is the first plausible value for mathematics achievement.[2]

The Pseudo-Classroom Procedure

Conceptually, the pseudo-classroom procedure involves the creation of a second or pseudo-classroom at the grade of interest by adjusting student achievement at the adjacent grade. The adjustment is the addition (or subtraction) of the mean difference between the lower and upper adjacent grades for the entire sample to (or from) the achievement scores of the students at the adjacent grade. When the mean difference is added or subtracted, mean achievement across the adjacent grades is roughly equivalent, but the variance of the adjusted classroom's achievement remains unchanged. In this way, a one-classroom-per-grade sample design at adjacent grades can be used to simulate a two-classroom-per-grade design and thus allow a three-level model to be formulated.

The four validation countries provided a unique opportunity to validate the use of the pseudo-classroom procedure for simulating a two-classroom-per-grade sample design. In these countries it was possible to compare the three-level variance estimates when two actual classrooms are modeled, to the three-level variance estimates when one classroom was randomly replaced by a pseudo-classroom from an adjacent grade.

A further aim of this research was to extend the application of the procedure to those countries that sampled only one classroom per grade in each school. Given the complexity and diversity of the participating education systems, one would not expect the pseudo-classroom procedure to be valid in all countries. For example, differences in levels of tracking or a transition between primary and secondary education between adjacent grades might prohibit the valid application of the pseudo-classroom procedure. In order to predict the countries in which adjacent grades are sufficiently similar in terms of their variance distributions to be combined into a single sample, the two-level variance structures for each of the four validation countries at the adjacent grades were compared. This was an attempt to assess whether differences in tracking levels or the structure of the education system are manifested as differences in the two-level variance structure across the two grades.[3]

The final aim of this research was to examine whether the pseudo-classroom procedure could be used to model the covariance structure when predictors were added at each level. To examine this question, mathematics achievement in the United States was modeled at Grade 8 using context variables from the student background questionnaire. The three-level regression coefficients, their associated significance levels, and the percent of variance explained when two actual Grade 8 classrooms were modeled were compared to those resulting when one actual Grade 8 classroom was randomly replaced by a pseudo-classroom.

RESULTS

The results are divided into three sections. First, to explore whether the valid use of the pseudo-classroom procedure could be predicted, the two-level unconditional variance structures at the adjacent grades in the four validation countries were compared. Second, to examine whether the pseudo-classroom procedure could be used to simulate a two-classroom-per-grade sample design, the three-level unconditional variance estimates in the four validation countries were examined. Finally, the pseudo-classroom procedure is discussed in terms of modeled variance at Grade 8 in the United States.

Two-level, Unconditional Variance Structure in Four Validation Countries

Table 1 presents the within-classroom and between-classroom/school two-level, unconditional variance estimates for the United States, Cyprus, Sweden, and Australia.[4] Table 1 shows that the percent of variances within classrooms and between classrooms or schools for both Grades 7 and 8 are approximately equivalent in Cyprus, Sweden, and the United States. Although the differences between the structures were more pronounced than in either the United States or Cyprus, the two-level variance structures in Sweden were roughly equivalent across the two grades. In Australia, the two-level variance structure was markedly different across the adjacent grades; at the lower grade, 33 percent of the variance lay between classrooms or schools, compared to 46 percent at the upper grade. To explore these observed differences in the two-level, unconditional variance structures across adjacent grades in Australia, the education systems in each of the four countries were examined.

In Australia, the grade level sampled by TIMSS at Population 2 was not consistent across all eight states. This complex sample design was due in part to the fact that some states make the transition from primary to secondary education between the target grades. In New South Wales, Victoria, Tasmania, and the Australian Capital Territory, Population 2 was made up of Grades 7 and 8, while in Queensland, South Australia, Western Australia, and the Northern Territory, Grades 8 and 9 made up Population 2 (Robitaille, 1997). The problem is further compounded by the fact that there were differences in the levels of tracking across states as well as across grades. The complexity of the Australian sample affected the unconditional, two-level variance structure across the adjacent grades and may prohibit the valid use of the pseudo-classroom procedure.

Table 1. Two-level variance estimates for lower and upper grades in four validation countries

		Variance Component	Variance Estimates	Percent of Variance
United States	Lower Grade	Between Classroom/School	3365	44%
		Within Classroom	4341	56%
	Upper Grade	Between Classroom/School	3786	45%
		Within Classroom	4565	55%
Cyprus	Lower Grade	Between Classroom/School	574	8%
		Within Classroom	6206	92%
	Upper Grade	Between Classroom/School	631	8%
		Within Classroom	7074	92%
Sweden	Lower Grade	Between Classroom/School	2345	31%
		Within Classroom	5244	69%
	Upper Grade	Between Classroom/School	3063	36%
		Within Classroom	5535	64%
Australia	Lower Grade	Between Classroom/School	2738	33%
		Within Classroom	5465	67%
	Upper Grade	Between Classroom/School	4394	46%
		Within Classroom	5156	54%

Students in Sweden, Cyprus, and the United States did not experience a break in the education system between the target TIMSS grades. In Cyprus, students in both sampled grades were part of the *gymnasium* or junior secondary program and all students were taught in self-contained, mixed ability classrooms (Robitaille, 1997). In the United States, students in the adjacent grades were either enrolled in junior high or in middle school. Beyond Grade 6, students in the United States are typically taught mathematics by specialist teachers.

The similarity of the two-level, unconditional variance structure in Cyprus, Sweden, and the United States predicts that the pseudo-classroom procedure can simulate a two-classroom-per-grade sample design very closely. The differences between the upper and lower grade variance structures in Australia imply that the pseudo-classroom procedure will be unable to simulate a two-classroom per grade sample design in that country.

Three-level, Unconditional Variance Structure in Validation Countries

To examine whether the two-level variance structure can be used to predict the validity of the application of the pseudo-classroom procedure, the three-level unconditional model was used to partition the total variance in mathematics achievement into its within- and between-classroom, and among-school components. In the four validation countries, the three-level variance structure when two real classrooms were modeled (the real sample) will be compared to the variance structure when a pseudo-classroom randomly replaced one real classroom (the pseudo-sample). Should the percent of variance at the student, classroom, and school levels be roughly equivalent for the real and pseudo-samples, then the pseudo-classroom procedure may be considered a tenable solution to the limiting nature of the TIMSS sample design.

Figures 1 through 4 contain the three-level unconditional variance structure for the real and pseudo-samples for Cyprus, the United States, Sweden, and Australia, respectively.

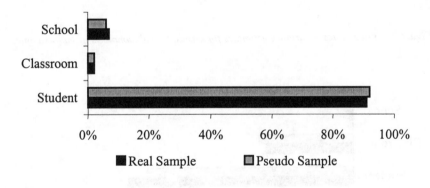

Figure 1. Cyprus —Variance estimates for Grade 7, real sample versus pseudo-sample.

Figure 1 shows that, for Cyprus, the variance at each of the levels does not differ substantially from the real sample to the pseudo-sample: approximately 92 percent of the variance lies within classrooms, 2 percent between classrooms, and 6 percent among schools.

Figure 2 shows that, for the United States, the within-classroom variance is approximately 57 percent, the between-classroom variance 30 percent, and the among-school variance 13 percent for both samples.

Though not large, the difference between the variance estimates for the real and pseudo-samples was larger in Sweden than in either the United States or Cyprus, particularly at the individual level. Figure 3 shows that, in Sweden, 72 percent compared to 66 percent of the variance lay within classrooms, for the real sample and pseudo-sample, respectively. Twenty-eight percent of the variance lay between classrooms in the real sample compared to 30 percent in the pseudo-sample. Among schools, the variances were 1 percent and 3 percent for the real and pseudo-samples, respectively.

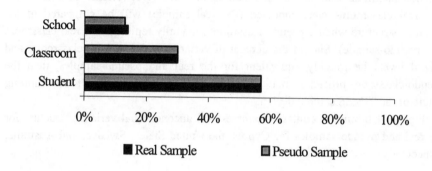

Figure 2. United States—Variance estimates for Grade 8, real-sample vs. pseudo-sample.

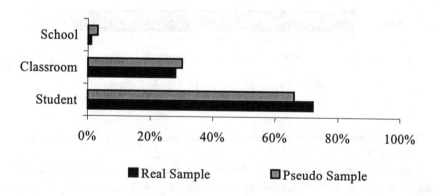

Figure 3. Sweden—Variance estimates for Grade 7, real-sample versus pseudo-sample.

Figure 4 shows large differences between the estimates in Australia compared to those in the other countries, with 55 percent compared to 60 percent of the variance lying within classrooms for the real- and pseudo-samples, respectively. Between classrooms, the variance estimate was 41 percent for the real sample and 19 percent for the pseudo-sample. Among schools, the estimates were 4 percent and 20 percent for the real and pseudo-samples, respectively.

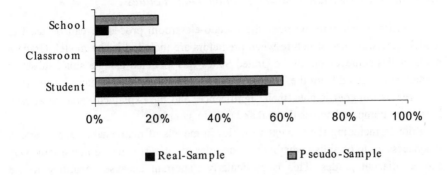

Figure 4. Australia—Variance estimates for upper grade, real sample versus pseudo-sample.

Table 1 showed that the two-level variance structures in Sweden, Cyprus, and the United States are very similar, particularly in Cyprus and the United States. The two-level, unconditional results in Australia predicted that the pseudo-classroom procedure would be unable to simulate a two-classroom-per-grade sample design. This was supported by the lack of similarity between the three-level variance structure in the real and pseudo-samples. The three-level results in the validation countries suggest that the degree to which the two-level variance structures are equivalent across the adjacent grades can be used to predict the accuracy with which the pseudo-classroom procedure can simulate a two-classroom-per-grade sample design. The more similar the two-level variance structure, the more accurately the pseudo-classroom procedure will simulate a two-classroom-per-grade sample design. This finding has implications for the valid application of the pseudo-classroom procedure beyond these countries.

Due to the complex nature of the variance structure, and the lack of independence between the lower and upper grades (both grades are in the same school), it is not possible to use a valid significance test to examine the equivalence of the two-level variance estimates across the adjacent grades. This research suggests that there is a continuum along which future analysts must evaluate the equivalence of the two-level estimates prior to interpreting the three-level estimates as accurate. Ultimately, the more similar the two-level variance estimates the more confident analysts can be that the pseudo-classroom procedure will accurately simulate a two-classroom-per-grade sample design.

Although it is not possible to predict fully how the pseudo-classroom procedure will operate in those TIMSS countries that sampled only one classroom per grade, it is useful to examine the two-level variance structure in each of the participating countries for both Populations 1 and 2. Tables containing this information are available on the internet at www.csteep.bc.edu/research/odwyer/t95levelvar.html (see for future analyses with regard to which countries are suitable for the application of the pseudo-classroom procedure).

Modeling the Covariance Structure in the United States

In order to examine whether the pseudo-classroom procedure can be used to model the covariance structure when predictors are included in the model, Grade 8 mathematics achievement in the United States was modeled using a small subset of predictors extracted from the TIMSS student background questionnaire.[5] The three-level regression coefficients, their significance, and the explained variance at each level were compared across the real and pseudo-samples.

Prior to including the context variables in models of mathematics achievement, it is necessary to establish whether the variables of interest behaved in the same way across adjacent grades. This is particularly important because according to the pseudo-classroom procedure, the adjacent grades were combined into a single sample. Differences between the upper and lower grades with respect to some predictors may occur for a variety of reasons, including the natural maturation process that occurs during adolescence, or perhaps a heightened awareness of the importance of school-based learning.

Establishing whether the adjacent grades were sufficiently similar in terms of the predictors of interest was particularly important for the valid application of the pseudo-classroom procedure. To address this issue the mean difference between the adjacent grades with respect to each predictor to be included in the multilevel model was examined. When calculating the mean difference between predictor scores for the grades, it was essential to take into account the complex sample design employed in TIMSS. In order to calculate the correct mean difference between the grades and the associated standard error in the entire sample, it was necessary to compute the mean difference between the adjacent grades within each school. Subsequently, the Jackknife procedure was used to compute the weighted mean difference and its standard error for the entire sample.[6]

Table 2 contains the mean difference, the standard error of the mean difference, and the associated significance for a small group of student home background predictors. According to the values in Table 2, the mean difference between the adjacent grades for the presence of both parents in the home composite was significantly different from zero. This implies that this composite may behave differently, in terms of regression coefficients, their significance, or the percent of variance explained by the predictors across the real and pseudo-samples. The significance of the difference between the lower and upper grades for the presence of both parents in the home composite was possibly due to noise in the sample rather than actual differences between the adjacent grades. Given that a second classroom was simulated, some extraneous noise is to be expected.

Prior to including predictors in a three-level model, it is necessary to formally present the results of the three-level, unconditional model of mathematics achievement that was first presented for the United States in Figure 2.

Table 2. United States jackknife mean difference between upper and lower grades for home background variables

	Jackknife Mean Difference between adjacent grades (SE)	Significance
Home Background Composite[1]	0.05 (0.08)	$p > 0.05$
Presence of Computer in the Home[2]	-0.02 (0.02)	$p > 0.05$
Presence of Both Parents in the Home[3]	0.04 (0.02)	$p < 0.05$

[1] The Home Background index was computed by calculating the mean of three variables: number of books in the home, mother's education, and father's education.
[2] The presence of computer in the home variable was reverse-coded. The higher end of the scale indicates the presence of a computer.
[3] The presence of both parents in the home composite was computed by combining the reverse-coded presence of father in home and presence of mother in home variables.

Tables 3 and 4 contain the unconditional three-level results for mathematics achievement in the real sample and the pseudo-samples in the United States.[7] Approximately 57 percent of the variance lay within classrooms in the two samples, with 31 percent lying between classrooms in the real sample and 29 percent in the pseudo-sample. The smallest percentage of variance lay among schools, at 13 percent in the real sample and 14 percent in the pseudo-sample.

Table 3. United States unconditional model for real sample

Fixed Effect		Coefficient	s.e.	t-ratio	
Average School Mean γ_{000}		492.57	4.11	119	
Random Effect	Variance Component	df	χ^2	p-value	Variance Decomposition (% by Level)
Student (Level 1) e_{ijk}	4650				56%
Classroom (Level 2) r_{0jk}	2618	150	869	0.000	31%
School (Level 3) u_{00k}	1093	149	263	0.000	13%

Table 4. United States unconditional model for pseudo-sample

Fixed Effect		Coefficient	s.e.	t ratio	
Average School Mean γ_{000}		491.11	4.05	121	
Random Effect	Variance Component	df	χ^2	p-value	Variance Decomposition (% by Level)
Student (Level 1) e_{ijk}	4546				57%
Classroom (Level 2) r_{0jk}	2326	150	789	0.000	29%
School (Level 3) u_{00k}	1162	149	284	0.000	14%

The variation between classrooms was statistically significant in the two samples: $X^2 = 869$ with 150 df ($p < 0.001$) in the real-sample and $X^2 = 789$ with 150 df ($p < 0.001$) in the pseudo-sample. The variance among schools was also statistically significant in both samples; $X^2 = 263$ with 149 df ($p < 0.001$) in the real-sample and $X^2 = 284$ with 149 df ($p < 0.001$) in the pseudo-sample.

In the three-level contextual model, student mathematics achievement was predicted at Level 1 according to the equation:

$$Y_{ijk} = \pi_{ojk} + \pi_{1jk}(HBI_{ijk}) + \pi_{2jk}(Computer_{ijk}) + \pi_{3jk}(BOTHPARENTS_{ijk}) + e_{ijk} \quad (1)$$

The intercept for classroom j, in school k was represented by π_{0jk}, and the Level 1 regression coefficients indicate the magnitude and direction of the relationships between the predictors and the outcome. At the first macro level, the classroom in this case, the Level 1 intercept was modeled by student level observations aggregated to classroom level:

$$\pi_{0jk} = \beta_{00k} + \beta_{01k}(ClHBI_{0jk}) + \beta_{02k}(ClComputer_{0jk}) + \beta_{03k}(ClBOTHPARE\ NTS_{0jk}) + r_{0jk} \quad (2)$$

At Level 3, the classroom level intercept was modeled by the predictors aggregated to the school level. The regression equation became:

$$\beta_{00k} = \gamma_{000} + \gamma_{001}(SchHBI_{00k}) + \gamma_{002}(SchComputer_{00k}) + \gamma_{003}(SchBOTHPARENTS_{00k}) + \mu_{00k} \quad (3)$$

The multilevel model results for the real and pseudo-samples are shown in Tables 5 and 6, respectively. The results show that all three home background predictors are positively related to mathematics achievement at the student, classroom, and school levels in the two samples. The largest effects were between classrooms and among schools, with only small to modest effects within classrooms. Within classrooms in both samples, the presence of a computer in the home variable had the strongest effect. The coefficients in the real and pseudo-samples were roughly equivalent, approximately 7 in the two samples. In both samples, the presence of a computer in the home variable was significant within classrooms for at least $p < 0.01$.

Within classrooms, the home background composite had a smaller effect, with coefficients of approximately 5 in the real and pseudo-samples. At the student level the dichotomous presence of the both parents in the home composite behaved differently in the two samples. This was predicted by the significance of the mean difference between the predictor scores for the lower and upper grades in Table 2. In the real sample, the coefficient was approximately 5 points, compared to 0.68 in the pseudo-sample. The effect of the predictor was significant at $p < 0.01$ in the real sample and non-significant in the pseudo-sample.

Table 5. United States three-level model results for real sample

Variable	Coefficient	s.e.	t-ratio	p-value
Within Classroom				
Intercept	-169.00	55.44	-3.05	0.003
Home Background Composite	5.08	0.92	5.51	0.000
Presence of Computer in Home	6.81	2.11	3.23	0.002
Presence of Both Parents in Home	4.79	1.83	2.61	0.009
Between Classroom				
Home Background Composite	56.21	10.08	5.58	0.000
Presence of Computer in Home	93.29	26.51	3.52	0.001
Presence of Both Parents in Home	89.08	23.48	3.80	0.000
Among School				
Home Background Composite	35.51	8.96	3.96	0.000
Presence of Computer in Home	59.86	25.47	2.35	0.019
Presence of Both Parents in Home	118.53	17.17	6.90	0.000

Random Effect	Variance Component	df	χ^2	p-value
Student (Level 1) e_{ijk}	4595			
Classroom (Level 2) r_{0jk}	1067	147	949	0.000
School (Level 3) u_{00k}	222	146	200	0.002

The effects of predictors at the classroom level were noticeably larger in both samples and all three predictors were significant at the classroom level. Again, the presence of a computer in the home variable had the largest coefficient in the two samples, 93 in the real sample and 109 in the pseudo-sample. Between classrooms, the presence of both parents in the home coefficient was roughly 90 in the real sample and roughly 100 in the pseudo-sample.

Among schools, the presence of both parents in the home composite had the strongest effect in both samples, approximately 119 in the real sample and 110 in the pseudo-sample. In both samples, the presence of a computer in the home variable appeared to exert a stronger influence among schools than the home background composite. The contribution to the within- and between-classroom variance, and among-school variance was calculated by comparing the model in Tables 5 and 6 to the fully unconditional models in Tables 3 and 4. Tables 7 and 8 contain the total variance and level specific variance accounted for by the model in the real and pseudo-samples, respectively.

The three variables included in the model account for only about 1 percent of within-classroom variance in the real and pseudo-samples; within classrooms, the model does little to explain student achievement. Given that the predictors used in the model were all home background variables, this was a reasonable percentage. According to the TIMSS sample, once students are in the classroom, the type of home they came from no longer mattered.

Table 6: United States three-level model results for pseudo-sample

Variable	Coefficient	s.e.	t-ratio	p-value
Within Classroom				
Intercept	-135.72	50.14	-2.71	0.007
Home Background Composite	5.05	0.91	5.57	0.000
Presence of Computer in Home	7.39	2.08	3.55	0.001
Presence of Both Parents in Home	0.68	1.83	0.37	0.708
Between Classroom				
Home Background Composite	42.85	9.28	4.62	0.000
Presence of Computer in Home	109.22	26.02	4.20	0.000
Presence of Both Parents in Home	101.35	23.48	4.32	0.000
Among School				
Home Background Composite	39.34	8.65	4.55	0.000
Presence of Computer in Home	46.35	23.77	1.95	0.051
Presence of Both Parents in Home	110.04	15.57	7.07	0.000

Random Effect	Variance Component	df	χ^2	p-value
Student (Level 1) e_{ijk}	4502			
Classroom (Level 2) r_{0jk}	1006	147	961	0.000
School (Level 3) u_{00k}	195	146	196	0.004

At the macro levels, the predictors accounted for an increased percentage of the between-classroom and among-school variance. At the classroom level, the predictors accounted for approximately 59 percent and 57 percent of the between classroom variance in the real and pseudo- samples, respectively. This indicates that the relationship between the variables in the model and achievement was stronger at the classroom level than at the individual level.

At the school level, the predictors in the model accounted for approximately 80 and 83 percent of the among-school variance in the real and pseudo-samples, respectively. This indicates again that, the relationship between the predictors and achievement was stronger among schools than at the individual level.

Table 7. United States explained variance in real sample

	Within-Classroom	Between-Classroom	Among-School
Total Variance	4650.0	2618.0	1093.0
Percent of Variance at each Level	56.0	31.0	13.0
Variance Predicted by Variables	4595.0	1067.0	222.0
Percent of Variance Predicted by Variables	1.2	59.2	79.7
Percent of Total Variance Predicted by Variables	0.5	18.6	10.4

Table 8. *United States explained variance in pseudo-sample*

	Within-Classroom	Between-Classroom	Among-School
Total Variance	4546.0	2326.0	1162.0
Percent of Variance at each Level	57.0	29.0	14.0
Variance Predicted by Variables	4502.0	1006.0	195.0
Percent of Variance Predicted by Variables	1.0	56.7	83.2
Percent of Total Variance Predicted by Variables	0.5	16.4	12.0

In terms of total variance, the within-classroom predictors accounted for approximately half a percentage point of the total variance in both samples. Between classrooms, the three predictors accounted for 18.6 percent of the total variance in the real sample and 16.4 percent of the total variance in the pseudo-sample. Among schools the predictors accounted for about 10 and 12 percent of the total variance in mathematics achievement in the real and pseudo-samples, respectively. In both the real and pseudo-samples, the model accounted for approximately 28 percent of the total variance in mathematics achievement in the United States.

The results in Tables 7 and 8 indicate that the predictors accounted for roughly the same percentages of variance, both in terms of total variance and level-specific variance in the two samples. Although the regression coefficients and their significance levels are different in the two samples for the presence of both parents in the home composite, interestingly this had little effect on the percentage of variance explained by the model in the two samples.

From a methodological viewpoint, the results from this model were very positive. It appeared that the magnitude of the variance accounted for by the model at each of the levels was roughly the same both samples. Although there was some variation in the size of the coefficients, the general patterns in their magnitudes were the same across the samples. Some variation and noise are to be expected in the pseudo-sample model given that the second classroom is created to simulate a two-classroom-per-grade design. This does not, however, appear to have affected the percentage of variance accounted for by the predictors in the model. This finding has implications for future applications of the pseudo-classroom procedure.

CONCLUSION AND DISCUSSION

The results presented in this chapter indicate that the pseudo-classroom procedure has the potential to be a tenable solution to the confounding of the between-classroom and among-school variance in the two-level regression model. In the validation countries, the three-level unconditional variance estimates in the real and pseudo-samples indicated both that the two-level variance structure can be used to predict country suitability, and that the pseudo-classroom procedure can be used to simulate a two-classroom-per-grade sample design. Beyond the four validation countries, this research also showed that the two-level variance structure predicted that the pseudo-classroom procedure can be applied in the majority of countries participating at Population 1 and Population 2.

A simple model of mathematics achievement, formulated at Grade 8 in the United States, showed that the general pattern in the magnitude of the coefficients remained constant across the real and pseudo-sample models. The results also indicated that both in terms of level specific variance and total variance, the percentage of variance explained remained reasonably constant for both samples.

With regard to future applications and interpretations of the pseudo-classroom procedure, this research suggests that researchers should evaluate the similarity of the two-level, unconditional variance structure prior to implementing the procedure. Without a valid test of the equivalence of the two-level variance structure, the correct application of the pseudo-classroom procedure does not rest with a binary decision. Rather, this research suggests there is a continuum along which researchers must evaluate the similarity of the two-level variance structures prior to interpreting the three-level variance estimates as accurate. In addition, the simple contextual, three-level model of mathematics achievement in the United States suggests that researchers should interpret the relative rather than the absolute effects of the predictors in the solution.

The results presented in this study are very encouraging. They suggest that, with attention to the similarity of the two-level variance structure and the interpretation of the relative effects of the predictors, the pseudo-classroom procedure can be used effectively to extend the application of multilevel modeling beyond that which is currently permitted by the TIMSS sample design. The results from this study have implications for future analyses of TIMSS 1995, Population 1 and 2 achievement data, as well as for future IEA studies. With increased interest in and availability of multilevel modeling software, there is no doubt that future analysts of TIMSS data will rely increasingly on multilevel regression techniques. The high quality of the TIMSS design, administration, scoring, and reporting, as well as its enormity, all ensure that researchers will be analyzing its findings for years to come. The correct application of the pseudo-classroom procedure introduced in this research will aid future researchers in formulating more powerful and accurate models of student outcomes.

NOTES

[1] For a more complete discussion of the advantages of multilevel modeling, see Bryk & Raudenbush (1997), Goldstein (1995), Kreft and DeLeeuw (1998).

[2] Although it is more accurate to use all five plausible values, the software used in this research does not allow modeling of all five across three levels. In general, using one plausible value will give consistent estimates of the population parameters, but will underestimate the uncertainty in the parameter estimate (Beaton & Gonzalez, 1995). Using appropriate software, future research will focus on modeling all five plausible values, nested within-individuals, at a fourth level.

[3] Standard tests for the homogeneity of variance are not suitable for examining whether the adjacent grades are sufficiently similar in terms of tracking and school structure. This is due to the fact that the within-classroom and between-classroom/school variance in each of the schools within a country contribute to the overall variance. In this way, differences between the variances at the adjacent grades will be masked in any standard examination of the homogeneity of variance.

[4] For a complete discussion of multilevel models see Bryk & Raudenbush (1992), Goldstein (1995), and Kreft and deLeeuw (1998). The multilevel results were produced using HLM software by Bryk, Raudenbush, and Congdon (1996).

[5] The purpose of the three-level contextual model is methodological and so it is necessarily simpler than one used to make substantive conclusions regarding student outcomes.

[6] For a complete discussion of the use of the Jackknife procedure in TIMSS see Gonzalez & Smith (1997).

[7] The results presented in this table may be slightly different to those presented in Figure 2. This is due to the slight reduction in sample size because of the listwise deletion of missing predictor values and the convergence procedure used by the HLM software.

REFERENCES

Beaton, A. E. & Gonzalez, E. J. (1995). *NAEP primer*. Center for the Study of Testing, Evaluation, and Educational Policy, Boston College. Chestnut Hill: MA.

Bryk, A. S., & Raudenbush, S. W. (1992). *Hierarchical linear models: Applications and data analysis methods*. Newbury Park, CA: Sage

Bryk, A. S., Raudenbush, S. W., & Congdon, R. (1996) *Hierarchical Linear and Nonlinear Modeling with the HLM/2 and HLM/3 programs*. Chicago, IL: Scientific Software International, Inc.

Goldstein, H. (1995). *Multilevel statistical models*. London: Edward Arnold.

Gonzalez, E. J., & Smith, T. A. (1997). *User guide for the TIMSS International Database*. Chestnut Hill, MA: Center for the Study of Testing, Evaluation, and Educational Policy, Boston College.

Kreft, I., & de Leeuw, J. (1998). *Introducing Multilevel Modeling*. Thousand Oaks, CA: SAGE Publications, Inc.

Martin, M. O., & Kelly, D. L . (Eds.). (1997). *Third International Mathematics and Science Study: Technical report volume II: Implementation and analysis*. Chestnut Hill, MA: Center for the Study of Testing, Evaluation, and Educational Policy, Boston College.

Raudenbush, S. W., & Bryk, A. S. (1997). Hierarchical linear modeling. In J.P. Keeves (Ed.), *Educational research, methodology, and measurement: An international handbook* (2nd edition). New York, NY: Elsevier Science, Inc.

Robitaille, D. F. (Ed.). (1997). *National contexts for mathematics and science education. An encyclopedia of the education systems participating in TIMSS*. Vancouver, Canada: Pacific Educational Press.

Chapter 23

APPLICATION OF THE SCALE ANCHORING METHOD TO INTERPRET THE TIMSS ACHIEVEMENT SCALES[1]

Dana L. Kelly

Angoff (1984), writing about test scores and interpreting test scores, said that the test user

> needs to understand the meaning of the score itself and what it represents. . . . And, clearly, meaning is essential, for without meaning the score itself is useless; and, without a meaningful score to transmit the value of the test performance, the test ceases to be a measuring instrument and becomes merely a practice exercise for the student on a collection of items. (pp. vii-viii)

Angoff points out a critical issue in educational measurement, that of needing to provide an appropriate scale structure for achievement data and of communicating what a scale score means. The construction of educational and psychological scales enables us to represent unobservable mental characteristics, attitudes, or abilities in a quantitative manner and to compare individuals and groups with respect to the measured characteristics. In educational measurement, scales are used to place estimates of student achievement along a continuum for the purposes of understanding the achievement in a normative or criterion manner, or both. While we are familiar with measurement scales such as the Fahrenheit temperature scale or the yardstick, people are less familiar with scales of mental measurement, even those cales which we are exposed to throughout our educational experiences, such as the IQ scale or the College Board SAT scale. The challenge for measurement specialists is not only to derive scales by which to quantify student achievement, but also to describe the scales so that meaningful interpretations of student performance can be made.

The primary means of reporting the TIMSS achievement data was through item response theory (IRT) scale scores. TIMSS reported, for each grade, national

D.F. Robitaille and A.E. Beaton (eds.), Secondary Analysis of the TIMSS Data, 375–390.
© 2002 *Kluwer Academic Publishers. Printed in the Netherlands.*

average scale scores for mathematics and science and presented within-country distributions of achievement and average achievement by gender in the scale score metric. To provide a means of interpreting the scale scores in terms of content knowledge and conceptual understandings, sample items were used to illustrate performance at different points on the scale. One of the other main displays of achievement presented the percentage of students reaching empirically derived international benchmarks on the achievement scales.

To describe the international benchmarks of performance in terms of what students reaching those points know and can do, the TIMSS International Study Center conducted a scale anchoring analysis. Scale anchoring is a way of attaching meaning to a scale by describing what students know and can do at different points on the scale. In scale anchoring, several points along a scale are selected as anchor points and the items that students scoring at each anchor point can answer correctly (with a specified probability) are identified and grouped together by anchor point. Subject-matter experts review the items that "anchor" at each level and identify the content knowledge and conceptual understandings represented by each item. The item descriptions are then summarized to yield a general description, illustrated by sample items, of what students scoring at the anchor points are likely to know and be able to do.

The four scales for the primary and middle school assessments were anchored in this study:

1. Population 1 (third and fourth grades) mathematics;

2. Population 2 (seventh and eighth grades) mathematics;

3. Population 1 (third and fourth grades) science; and

4. Population 2 (seventh and eighth grades) science.

The scales were analyzed at four scale points corresponding to the 25th, 50th, 75th, and 90th international percentiles for fourth and eighth grades. The rationale for the scale anchoring analysis and the procedures used are described in this chapter.

INTERPRETING ACHIEVEMENT SCALES THROUGH SCALE ANCHORING

For any achievement test, determining how examinees perform in relation to other examinees is relatively straightforward. Regardless of how a test was constructed or what it measures, it is possible to determine a student's or a group's relative standing along the continuum of achievement. Determining how examinees perform in terms of the content they know or the conceptual understandings they have is not as straightforward. Over the years, several approaches have been suggested for providing information about what a particular scale score means in terms of content knowledge and conceptual understanding, including behavioral scaling (Tucker, 1955, cited in Carroll, 1993), scale books (Ebel, 1962), and item mapping.

Bock, Mislevy, and Woodson (1982) introduced the concept of "content referencing" as a means of describing IRT scales and recommended a procedure whereby items are placed on the scale at particular response probability points to illustrate the kinds of skills and understandings students reaching those points have. This is essentially "item mapping," which has been widely used to interpret scales of large-scale assessments including the U.S. Young Adult Literacy Survey, National Adult Literacy Survey (NALS), National Assessment of Educational Progress (NAEP), and TIMSS.

An extension of what Bock et al. (1982) call "content-referencing" and what NAEP, NALS, and TIMSS call item mapping is the scale anchoring method (Beaton & Allen, 1992). In scale anchoring, several points along a scale are selected and the items that "anchor" at those points are identified. An item is said to anchor at a selected point if it is an item that most of the students scoring at that anchor point can do but the majority of students scoring at the next lowest level cannot. For example, an item would anchor at a particular point if examinees scoring at that scale point had at least a 65 percent probability of answering correctly, but students at the next lowest anchor point had less than a 50 percent probability of answering correctly. In item mapping, the items that students reaching a particular scale score have a certain (and high) probability of answering correctly are identified and then (in some cases) reviewed to generalize to proficiencies typical of levels on the scale. In scale anchoring, however, an additional criterion is employed to identify items that discriminate between adjacent anchor points on the scale. After identifying anchor items, the items are grouped together by anchor level and examined by subject-matter experts who summarize the knowledge and skills that students scoring at successively higher anchor points exhibit, beyond those shown at the next lowest level.

Scale anchoring can provide both normative and content-referenced information about student proficiency on the measured construct (Beaton & Allen, 1992). Normative information is possible insofar as students (or a larger unit such as schools or countries) are assigned scores on a scale, allowing comparisons of their performance against other students (or the larger unit). The use of scale anchoring enables a content-referenced interpretation in the sense that the scores can be interpreted in terms of content knowledge and skills. Scale anchoring can be used for any continuous scale, not just IRT scales. Any scale for which performance can be considered hierarchical and for which individuals receive scores on the items can be anchored.

Scale anchoring was developed and first used by NAEP in the 1980s to synthesize the achievement results and communicate them in an easily understood manner (Mullis, Applebee, & Langer, 1985). The method was used until 1992 as the primary means of interpreting the NAEP mathematics, science, reading, civics, geography, and history scales. The original purpose of the scale anchoring method was to provide succinct interpretations of what students scoring at different points on the NAEP scales know and can do in the assessed subject. Generally, four or five empirically-derived scale points (standard deviations) were used as "anchor" points and proficiency at each level was described in one paragraph.

TIMSS SCALE ANCHORING ANALYSIS

The scale anchoring method pioneered by NAEP was adapted and used to interpret the TIMSS Population 1 (third and fourth grades) and Population 2 (seventh and eighth grades) mathematics and science achievement scales. The analysis was a two-part effort. First, analyses were conducted on the achievement data for each scale to identify items that students scoring at four international benchmarks (anchor points) could correctly answer with a high probability. These "anchor items" were assembled in binders, and then grouped by international benchmark and content area. Second, subject-matter experts examined the anchor items and determined what students correctly answering each item know and can do (or what they do to answer an item correctly) and summarized the item-level information in detailed descriptions of performance at each international benchmark.

Anchor Points

Four empirically-derived international benchmarks were used as anchor points in the analysis: the 25th, 50th, 75th, and 90th international percentiles for the fourth and eighth grades. TIMSS reported the percentages of students reaching the 50th, 75th, and 90th international percentiles for mathematics and science in the international reports (Beaton, Martin, Mullis, Gonzalez, Smith, & Kelly, 1996; Beaton, Mullis, Martin, Gonzalez, Kelly, & Smith, 1996; Martin, Mullis, Beaton, Gonzalez, Smith, & Kelly, 1997; and Mullis, Martin, Beaton, Gonzalez, Kelly, & Smith, 1998). The international percentiles for fourth and eighth grades were used as anchor points in this study to yield a content-referenced view of the mathematics and science achievement of the top half, the top quarter, and the top 10 percent of fourth- and eighth-grade students internationally. The 25th international percentile also is used as an anchor point to provide information about achievement at the lower end of the scales. The percentiles for the fourth and eighth grades were used (as opposed to third and seventh grades) because in many countries the fourth and eighth grades represent important transition points in the education system.

The international percentiles were computed for each grade using the combined data from the countries that participated. For the fourth grade, student scale scores from 26 countries were used; and, for the eighth grade, student scale scores from 41 countries were used. Student scores were weighted to contribute to the computation of the percentiles proportional to the size of the student population in a country. The scale scores representing the international percentiles for the fourth and eighth grades are shown in Table 1.

Table 1. TIMSS international percentiles for fourth and eighth grades*

	Percentile			
	25th	50th	75th	90th
Fourth Grade Mathematics (26 countries)	464	535	601	658
Eighth Grade Mathematics (41 countries)	425	509	587	656
Fourth Grade Science (26 countries)	466	541	607	660
Eighth Grade Science (41 countries)	451	522	592	655

*Fourth and eighth grades in most countries. International percentiles were computed from the combined data from the participating countries. Achievement for each subject and grade was estimated separately and thus scale scores across subjects and grades are not comparable.

Given that it is unlikely that very many examinees will score exactly at one of the anchor points, it was necessary to include in the analysis students scoring in a range around each anchor point. A range of plus or minus 5 scale points was used in this analysis. This yields a homogeneous group of students (in terms of achievement) and also a sufficient pool of respondents for each anchor point, yet the adjacent anchor levels are far enough away from each other to differentiate performance. The ranges around the international percentiles and the number of observations within each range are shown in Table 2.

Table 2. Range around each anchor point and numbers of observations within ranges

	Percentile			
	25th	50th	75th	90th
Population 1 Mathematics				
Range	459-469	530-540	596-606	653-663
Number of Observations	6430	6529	4458	2393
Population 2 Mathematics				
Range	420-430	504-514	582-592	651-661
Number of Observations	9153	10330	7250	3759
Population 1 Science				
Range	461-471	536-546	602-612	655-665
Number of Observations	6300	6775	4345	2093
Population 2 Science				
Range	446-456	517-527	587-597	650-660
Number of Observations	9970	10873	7512	3633

Note: Achievement for each subject and grade was estimated separately and thus scale scores across subjects and grades are not comparable.

Item Percent Correct at Each Anchor Level

For each item, the percent of students in the range around each anchor point that answered the item correctly was computed. To compute these percentages, students were weighted to contribute proportionally to the size of the student population in a country. Most of the TIMSS items were scored dichotomously, i.e., right or wrong. For these items, the percent of students at each anchor point who answered each item correctly was computed. Some of the open-ended items were scored on a partial-credit basis (one, two, or three points). These items were transformed into a series of dichotomously-scored items and the percentages were computed for each. For the Population 1 scales, the percent of third and fourth grade students across the 26 countries correctly answering each item (or in the case of open-ended items, the percentage of students earning at least one point, at least two points, etc.) was computed at each of the four anchor levels. Similarly, for the Population 2 scales and for each anchor level, the percent of seventh and eighth grade students across the 41 countries correctly answering each item (or receiving partial credit) on the scale was computed.

Identifying Anchor Items

Anchor items are the items that reflect the content knowledge and conceptual understandings that students at different scale points demonstrate with a high degree of probability. To identify the items that anchored at each of the benchmarks, response probability criteria were selected and applied.

The first step was to determine a reasonable response probability. A response probability of 50 percent would provide a set of items at each anchor point which students were equally likely to answer correctly or incorrectly. A response probability of 80 percent would provide a set of items at each anchor point which students were very likely to have answered correctly (at least 80 percent of the time), but might result in a set of fairly easy items. To strike a balance between these two extremes, yet still provide an interpretation of "mastery," a response probability of 65 percent was used. This response probability criterion was used by TIMSS to map items onto the mathematics and science achievement scales and has been used by NAEP to identify anchor items.

In scale anchoring, the anchor items for each level are intended to be those that differentiate between adjacent anchor points. To meet this goal, the criteria for identifying the items must take into consideration performance at more than one anchor point. Therefore, in addition to a criterion for the percentage of students at a particular anchor point correctly answering an item, it is necessary to use a criterion for the percentage of students scoring at the next lowest anchor point correctly answering an item. The criterion of less than 50 percent was used because with this response probability, students are more likely to answer the item incorrectly than to answer the item correctly. This criterion has been used by NAEP as well.

To summarize, the criteria used to identify items that "anchored" were as follows:

For the 25th percentile, an item anchored if
- at least 65 percent of students scoring in the range around the anchor point answered the item correctly (Because the 25th percentile was the lowest level, items were not identified in terms of performance at a lower level.)

For the 50th percentile, an item anchors if
- at least 65 percent of students scoring in the range answer the item correctly and
- less than 50 percent of students at the 25th percentile answer the item correctly

For the 75th percentile, an item anchors if
- at least 65 percent of students scoring in the range answer the item correctly and
- less than 50 percent of students at the 50th percentile answer the item correctly

For the 90th percentile, an item anchors if
- at least 65 percent of students scoring in the range answer the item correctly and
- less than 50 percent of students at the 75th percentile answer the item correctly

To supplement the pool of anchor items, an additional yet slightly less stringent set of criteria also was used. Items that at least 60 percent of the students scoring at an anchor point and less than 50 percent of the students at the next-lowest anchor point answered correctly were considered to "almost anchor." To further supplement the pool of items, items that met only the criterion that at least 60 percent of students answered correctly (regardless of the performance of students at the next lowest level) were also identified.

The three categories of items are mutually exclusive. That is, an item could anchor, almost anchor, or meet the 60 percent criterion, and could do so for only one anchor level. All remaining items were those that were too difficult for students at the 90th percentile.

The criteria were applied to identify, for each scale, the items that anchored, almost anchored, or met only the 60 percent criterion at each anchor point. The percent correct for each item was examined to determine at which anchor level an item anchored. For example, the results shown in Figure 1 were obtained for Population 2 mathematics Item L16. This item was determined to anchor at the 75th percentile. Students scoring at the 75th percentile had a 73 percent probability of answering the item correctly. Students scoring at the 50th percentile were less likely (42 percent probability) to answer the item correctly.

Find x if $10x - 15 = 5x + 20$

P2-L16
                    Answer: _____

Anchor Point	25th Percentile	50th Percentile	75th Percentile	90th Percentile
Percent Correct	15	42	73	84

Figure 1. Sample output for items.

Tables 3 through 6 present the number of items anchoring, almost anchoring, or meeting only the 60 percent criterion at each anchor level for each scale.

Table 3. Number of items anchoring at each international benchmark: Mathematics Population 1

	Anchors	Almost Anchors	60 percent criterion	Total
25th Percentile	23	6	--	29
50th Percentile	13	4	18	35
75th Percentile	9	3	16	28
90th Percentile	0	3	12	15
Too difficult for 90th				6
				113

Table 4. Number of items anchoring at each international benchmark: Mathematics Population 2

	Anchors	Almost Anchors	60 percent criterion	Total
25th Percentile	15	5	--	20
50th Percentile	17	12	23	52
75th Percentile	25	7	24	56
90th Percentile	11	2	14	27
Too difficult for 90th				7
				162

Table 5. Number of items anchoring at each international benchmark: Science Population 1

	Anchors	Almost Anchors	60 percent criterion	Total
25th Percentile	28	7	--	35
50th Percentile	9	5	12	26
75th Percentile	5	3	10	18
90th Percentile	2	2	15	19
Too difficult for 90th				7
				105

Table 6. Number of items anchoring at each international benchmark: Science Population 2

	Anchors	Almost Anchors	60 percent criterion	Total
25th Percentile	29	7	--	36
50th Percentile	6	8	18	32
75th Percentile	4	4	29	37
90th Percentile	2	1	21	24
Too difficult for 90th				17
				146

Expert Panel Review of Items

Two panels of mathematics and science educators from the TIMSS countries were assembled to examine the items and develop descriptions of performance at the international benchmarks. The mathematics panel comprised ten individuals and the science panel comprised nine individuals with extensive experience in their subject areas and a thorough knowledge of the TIMSS curriculum frameworks and achievement tests.

The panelists' assignment consisted of three tasks: (1) work through the items one by one and arrive at a short description of what students answering each item correctly know and can do, or what they do to answer the item correctly; (2) based on the items anchoring at each international benchmark, draft a detailed description of the knowledge, understandings, and skills demonstrated by students; and (3) select TIMSS sample items to support and illustrate the benchmark descriptions.

In preparation for review by the panels, the items were assembled in binders. Four binders, one per scale, were prepared. Each binder had four sections, corresponding to the four international benchmarks. Within each section, the items were sorted by content area and then by the anchoring criteria they met: items that "anchor," followed by items that "almost anchor," followed by items that "meet only the 60 percent criterion." The following information was included for each item: its TIMSS content area and performance expectation categories, answer key, percent correct at each anchor level, overall international percent correct by grade, and item difficulty. For open-ended items, the scoring guides were included.

Item Descriptions

The first step in developing the descriptions was to review the items one by one and identify what students answering the item correctly know or are able to do. For example, as shown in Figure 2, an eighth-grade algebra item that anchored at the 75th percentile was described as "solves linear equation with variables on both sides of the equation." Figure 3 presents an eighth-grade science item from the life science category that anchored at the 75th percentile; it was described as "recognizes the relationship between hibernation and decreased energy consumption."

L 16

Find x if $10x - 15 = 5x + 20$

Answer:_____

_____Percent Correct_____

25th	50th	75th	90th	Overall (8th gr.)
15	42	73	84	46

Solves linear equation with variables on both sides of the equation

Figure 2. Sample item description for mathematics.

P04

What happens when an animal hibernates?

A. There is no life in any of its parts.
B. It stops breathing.
C. Its temperature is higher than when it is active.
D. It is absorbing energy for use when it is active.
E. It is using less energy than when it is active.

_____Percent Correct_____

25th	50th	75th	90th	Overall (8th gr.)
48	67	71	85	56

Recognizes the relationship between hibernation and decreased energy consumption

Figure 3. Sample item description for science item.

Descriptions of Achievement at Each International Benchmark

Based on the descriptions of the items anchoring at each international benchmark, the panels developed detailed descriptions of student performance. These describe students' mathematics and science understandings and the sorts of tasks they can successfully do. In addition to detailed descriptions of performance at each benchmark, the panelists identified the major accomplishments and hallmarks of each level. To illustrate, Figure 4 shows the summary descriptions of performance at the four international benchmarks on the eighth-grade mathematics scale. Descriptions of the international benchmarks, and example items, for all four scales are available in Kelly, Mullisand Martin (2000) and Smith, Martin, Mullis and Kelly (2000).

25th Percentile

Understand different representations of fractions—verbal and decimal; add and subtract decimals with the same number of decimal places; read, locate, and compare data in charts and graphs; calculate average of whole numbers

50th Percentile

Use understanding of rounding in problem situations; perform basic operations with familiar fractions; understand place value of decimal numbers; understand measurement in several settings; locate data in charts and graphs to solve word problems; know and use simple properties of geometric figures to solve problems; identify algebraic expressions and solve equations with one variable

75th Percentile

Order, relate, multiply, and divide fractions and decimals; relate area and perimeter; understand simple probability; use knowledge of geometric properties to solve problems; identify algebraic expressions and solve equations with two variables

90th Percentile

Organize information in problem-solving situations; solve time-distance-rate problems involving conversion of measures within a system; apply relationships – fractions and decimals, ratios, properties of geometric figures, and algebraic rules – to solve problems; solve word problems involving the percentage of increase

Figure 4. Performance at international benchmarks of the TIMSS eighth-grade mathematics achievement scale.

In providing descriptions of performance at the benchmarks, this analysis has revealed the major factors that distinguish performance across the scales. In mathematics, three factors appear to distinguish performance at the four anchor levels: the mathematical operation required, the complexity of the numbers or number system, and the problem situation. For example, fourth grade students scoring at the lower end of the Population 1 scale demonstrate facility with whole numbers in simple problem situations, while their peers scoring at higher levels on the scale can use basic operations on whole numbers and solve multi-step word problems. Students scoring at the 90th percentile demonstrate an understanding of fractions and decimals and can perform simple division. Eighth grade students scoring at the lower end of the Population 2 scale demonstrate an understanding of fractions and decimals and perform basic operations on decimal numbers, while students scoring at higher levels can perform basic operations on fractions, locate and use data in charts and graphs to solve problems, and have a grasp of beginning algebra. Students scoring at the 90th percentile demonstrate that they can "bring things together." They organize information in problem solving situations and apply relationships to solve problems.

In science, several general factors appear to distinguish performance across the anchor levels: terminology and context, facts vs. principles, and the ability to communicate knowledge. For example, fourth grade students scoring at the lower end of the scale demonstrate knowledge of basic facts expressed in non-technical language and everyday contexts, and demonstrate little in the way of communication skills. Higher on the scale, students demonstrate knowledge of some scientific principles in addition to facts, understand scientific terms, and can communicate their knowledge through short responses. Eighth grade students scoring at the lower end of the Population 2 scale demonstrate knowledge of basic facts expressed in non-technical language and communicate knowledge by providing a single fact. Higher on the scale, students demonstrate an understanding of scientific principles and can communicate their knowledge through increasingly longer responses. Another factor that seems to influence achievement on TIMSS in science is the amount of information presented in the items and what is required of students to answer them. Lower on the scale, the items tend to be brief and to the point. Items anchoring higher on the scale include more text and require students to extract information from a contextualized situation. Of course, at both populations, knowledge of science content plays an important role in how students perform, although it is more difficult to make a general statement about what differentiates students across the levels in terms of subject matter content.

INTERPRETING THE BENCHMARK DESCRIPTIONS

The scale anchoring analysis has provided useful descriptions of student performance at different scale points. There are, however, some important issues to keep in mind when interpreting the descriptions of performance. First, performance on the TIMSS scales is cumulative. That is, students reaching the 50th percentile typically demonstrate the knowledge and understandings characterizing the 50th percentile as well as the competencies of students at the 25th percentile. Students reaching the 75th percentile typically demonstrate the knowledge and understandings characterizing the 75th percentile as well as the 25th and 50th percentiles, and so on. Second, the method used to identify items and build descriptions of performance relies on the probability that students with certain scores correctly answer the items. It is important to recognize that some students scoring below an international benchmark may correctly answer items anchoring at that point, particularly when the student's score is very close to the anchor point. Similarly, students scoring above a benchmark may incorrectly answer an item anchor at that point. A related point is that the anchor level descriptions should be interpreted as characterizations of what students reaching each point are *likely* to know and be able to do. Although the anchor level descriptions include phrases such as "Students at this level can..." and "Students at this level demonstrate...," interpretations must be made with caution. It must be recognized that "can" has been defined in terms of having a high probability of success. Moreover, it is not appropriate to interpret an anchor level description in terms of what students below an anchor point cannot do. For example, for the 90th percentile international benchmark it is not appropriate to say that the 90 percent of students below that scale point do not demonstrate any of the competencies typical of that anchor point.

Finally, the descriptions of performance are limited to the content of the TIMSS tests. Although the tests included a wide range of mathematics and science topics and were designed to elicit a range of performance skills from students, there are undoubtedly some concepts that students reaching a benchmark have mastered but that were not included in the TIMSS tests.

CONCLUSION

The utility and validity of assessment results depend heavily on the way results are reported. Every assessment program has a professional and ethical responsibility to report results that are easily understood by different audiences, provide meaningful and useful information for decision-makers, and minimize misinterpretations. This is necessary to ensure that the interpretations made by various audiences and the decisions that may result from those interpretations are appropriate. To accomplish this, it is necessary to give meaning to the scores upon which assessment results are based. Providing "content-referenced" interpretations of the TIMSS achievement scale scores improves communication of the TIMSS achievement results and the validity of the interpretations and use of these results.

The scale anchoring analysis provides a greater understanding of what different scores on the TIMSS scales mean. Together with the percentages of students reaching each international benchmark, the profiles of performance provide countries with an understanding of their fourth and eighth graders' strengths and weaknesses in mathematics and science, and how this compares with other countries. For example, in Singapore, the highest-performing country, 45 percent of eighth graders reached the 90th international percentile (scale score of 656) in mathematics (Beaton, Mullis, Martin, Gonzalez, Kelly, & Smith, 1996). As was shown in Figure 4, students reaching this benchmark that demonstrate they can organize information in problem solving situations, solve time-distance-rate problems involving conversion of measures within a system, apply relationships to solve problems, and solve word problems involving the percent of increase. Nearly all (94 percent) of the eighth graders in Singapore reached the 50th international percentile, which was characterized by students' ability to use rounding, perform basic operations with familiar fractions, understand place value of decimal numbers, understand measurement in several settings, locate data in charts and graphs to solve word problems, know and use simple properties of geometric figures to solve problems and identify algebraic expressions and solve equations with one variable. In contrast, in Sweden, which performed around the international average, just 5 percent of eighth grade students reached the 90th percentile and about half (53 percent) reached the 50th percentile. Educators and policymakers in Sweden can use the benchmark descriptions and information about how many of their students and students in high-performing countries, such as Singapore, reached the different levels, to better understand how well their students performed on TIMSS relative to other countries and in terms of content understanding.

The benchmark descriptions can be used by countries to examine their standards and curriculum to determine if they are "world class." For example, the TIMSS International Study Center analyzed two U.S. national curriculum standards documents: the National Council of Teachers of Mathematics' *Principles and Standards for School Mathematics* (2000) and the National Research Council's *National Science Education Standards* (1995) in light of the international benchmark descriptions (Kelly, Mullis, & Martin, 2000; Smith, Martin, Mullis, & Kelly, 2000). Descriptions of performance at the benchmarks were compared with the expectations for student learning put forth in the standards.

The scale anchoring analysis also laid the groundwork for interpreting achievement on TIMSS-Repeat, conducted in 1999. A scale anchoring analysis based on the TIMSS 1995 analysis was conducted on the TIMSS 1999 scales and used as the basis for reporting achievement in the international reports (Mullis, Martin, Gonzalez, Gregory, Garden, O'Connor, Chrostowski, & Smith, 2000; Martin, Mullis, Gonzalez, Gregory, Smith, Chrostowski, Garden, & O'Connor, 2000).

The close scrutiny that the items underwent during the panel review highlighted some issues that can be used to inform future TIMSS item development. For example, panelists noted some content areas or topics for which more items were included in the test than was probably necessary. Also, panelists noted a few items did not contribute substantially to the measurement of the construct. Future TIMSS assessments, including TIMSS 2003 which is currently under development, will need to be updated to address current and future priorities in mathematics and science education, and the lessons learned through this scale anchoring study can further inform item development and test coverage.

NOTES

[1]This paper is based upon work supported by the National Science Foundation under Grant No. REC-9815001. The author acknowledges Ina V.S. Mullis and Albert E. Beaton, who served as advisor and reader, respectively, for the dissertation (Kelly, 1999) upon which this chapter is based, as well as the contribution of Eugenio. J. Gonzalez.

REFERENCES

Angoff, W. H. (1984). *Scales, norms, and equivalent scores*. Princeton, NJ: Educational Testing Service.

Beaton, A. E., & Allen, N. L. (1992). Interpreting scales through scale anchoring. *Journal of Educational Statistics, 17*, 191-204.

Beaton, A. E., Martin, M. O., Mullis, I. V. S., Gonzalez, E. J., Smith, T. A., & Kelly, D. L. (1996a). *Science achievement in the middle school years: IEA's Third International Mathematics and Science Study (TIMSS)*. Chestnut Hill, MA: Boston College.

Beaton, A. E., Mullis, I. V. S., Martin, M. O., Gonzalez, E. J., Kelly, D. L., & Smith, T. A. (1996b). *Mathematics achievement in the middle school years: IEA's Third International Mathematics and Science Study (TIMSS)*. Chestnut Hill, MA: Boston College.

Bock, R. D., Mislevy, R., & Woodson, C. (1982). The next stage of educational assessment. *Educational Researcher, 2*, 4-11,16.

Ebel, R. L. (1962). Content standard test scores. *Educational and Psychological Measurement, 22* (1), 15-25.

Kelly, D. L. (1999). Interpreting the Third International Mathematics and Science Study (TIMSS) achievement scales using scale anchoring (Doctoral dissertation, Boston College, 1999) *Dissertation Abstracts International, 60-03A*.

Kelly, D. L., Mullis, I. V. S., & Martin, M. O. (2000). *Profiles of student achievement in mathematics at the TIMSS international benchmarks: U.S. performance and standards in an international context*. Chestnut Hill, MA: International Study Center, Boston College.

Martin, M. O., Mullis, I. V. S., Gonzalez, E. J., Gregory, K. D., Smith, T. A., Chrostowski, S. J., Garden, R. A., & O'Connor, K. M. (2000). *TIMSS 1999 international mathematics report: Findings from IEA's repeat of the Third International Mathematics and Science Study at the eighth grade*. Chestnut Hill, MA: International Study Center, Boston College.

Martin, M. O., Mullis, I. V. S., Beaton, A. E., Gonzalez, E. J., Smith, T. A., & Kelly, D. L. (1997). *Science achievement in the primary school years: IEA's Third International Mathematics and Science Study (TIMSS)*. Chestnut Hill, MA: Boston College.

Mullis, I. V. S., Martin, M. O. Gonzalez, E. J., Gregory, K. D., Garden, R. A., O'Connor, K. M., Chrostowski, S. J., & Smith, T.A. (2000). *TIMSS 1999 international science report: Findings from IEA's repeat of the Third International Mathematics and Science Study at the eighth grade*. Chestnut Hill, MA: International Study Center, Boston College.

Mullis, I. V. S., Martin, M. O., Beaton, A. E., Gonzalez, E. J., Kelly, D. L., & Smith, T. A. (1997a). *Mathematics achievement in the primary school years: IEA's Third International Mathematics and Science Study (TIMSS)*. Chestnut Hill, MA: Boston College.

Mullis, I. V. S., Martin, M. O., Beaton, A. E., Gonzalez, E. J., Kelly, D. L., & Smith, T. A. (1998). *Mathematics and science achievement in the final year of secondary school: IEA's Third International Mathematics and Science Study (TIMSS)*. Chestnut Hill, MA: Boston College.

National Council of Teachers of Mathematics (2000). *Principles and standards for school mathematics: Discussion draft*. Reston, VA: Author.

National Research Council (1995). *National science education standards*. Washington, DC: National Academy Press.

Smith, T. A., Martin, M. O., Mullis, I. V. S., & Kelly, D. L. (2000). *Profiles of student achievement in science at the TIMSS international benchmarks: U.S. performance and standards in an international context*. Chestnut Hill, MA: International Study Center, Boston College.

Tucker, L. R. (1955). Some experiments in developing a behaviourally determined scale of vocabulary. Paper presented at the meeting of the American Psychological Association, San Francisco.

Chapter 24

EFFECTS OF ADAPTATIONS ON COMPARABILITY OF TEST ITEMS AND TEST SCORES

Kadriye Ercikan and Tanya McCreith

A basic concern of international assessments is whether students' performance on test items is comparable when items are adapted to different languages. There are several basic differences between languages that might make it difficult to create items that function similarly across languages. Among these are the variations in the frequency of word use and word difficulty. For example, words that may be commonplace and "easy" in one language may not be equally so in another language. Another adaptation problem occurs when grammatical forms either do not have equivalents, or else have many of them in another language. These differences between languages create difficulties in developing items that have equivalent (a) meanings and functions of words, sentences, and passages; (b) content of the items, and (c) skills measured by the items. The degree to which item features are changed during adaptation will affect the extent to which the equivalence of items is maintained.

There has been a good deal of research on test item adaptation and translation, and a large number of rules for checking test adaptations have been documented (Hambleton, 1993; Ercikan, 1998). Even if all guidelines for item adaptation have been followed carefully and content and linguistic experts have checked the results, translated items may not be equivalent to their original versions in terms of the cognitive processes evoked in examinees. This chapter focuses on understanding how adaptations might be affecting the equivalence of items and tests in different languages. In particular, it focuses on effects of adaptations on the comparability of English and French versions of the TIMSS items.

The purpose of this chapter is to identify items that functioned differently in the English and French versions of TIMSS, and to determine the extent to which differential functioning is due to adaptations. Differential functioning of items can happen even when the items are administered in only one language. In fact the term, "differential item functioning" (DIF), is typically used in relation to differential functioning of items for different ethnic and gender groups and has been used to examine item bias.

D.F. Robitaille and A.E. Beaton (eds.), Secondary Analysis of the TIMSS Data, 391–405.
© 2002 *Kluwer Academic Publishers. Printed in the Netherlands.*

There are many statistical methods for identifying DIF, and they are all designed to identify items that result in differing psychometric properties for the comparison groups, such as males and females. Several researchers have investigated potential causes of DIF for gender and ethnic groups. These include item content, and additional skills such as reading (Scheunemann & Geritz, 1990; Schmitt & Dorans, 1990). In a review, O'Neil and McPeek (1993) summarized the results of empirical research on the relation between DIF and characteristics of test items. Their results indicated that all cases of DIF were related to additional knowledge required for these items that was not intended to be measured and included in scoring. Related to this, another way of interpreting DIF is explained by Ackerman (1992) as, "If two different groups of examinees have different underlying multidimensional ability distributions and the test items are capable of discriminating among levels of abilities on these multiple dimensions, then any unidimensional scoring scheme has the potential to produce bias."

In the case of international assessments, differential item functioning does not necessarily imply bias against one group or another. Multidimensional abilities can differ from one country to another for a variety of reasons such as cultural, linguistic, and curricular differences. Cultural differences can influence intrinsic interest in and familiarity with the content of items. Curriculum-related differences can result in varying degrees of student exposure to the domain of items depending upon the student's country. The order in which topics are introduced and subsequent instruction provided to students could be different between comparison groups. For example, if one of the groups has not covered algebra when the assessment took place, then this group would be expected to do more poorly relative to their performance on the rest of the test. This in turn may create differential functioning for this group on algebra items. In other words, certain topics or subjects will undoubtedly have differential coverage in different countries and differential coverage can lead to differential response patterns and different difficulty levels, independent of any problems due to adaptation. When the tests are assumed to be unidimensional and these differences are not taken into account, the results can give rise to DIF. Typically, DIF procedures are applied to test items to identify differential functioning for groups (gender or ethnic) who are taking the same form of the test. The main difference between these applications and international assessments is that, in the latter, the examinees are responding to different versions of the items in that one set of items has been translated into another language. Therefore, there is an additional factor that might impact DIF.

One of the challenges for examining differential item functioning is interpretation of DIF and its sources. The interpretation of DIF in this chapter employs three strategies. First, judgmental reviews by multiple bilingual translators of all items are conducted. Second, DIF analyses are conducted in three comparisons and consistency of DIF across these comparisons is analyzed. Third, analyses are conducted to determine whether DIF could be due to curricular differences. Other major factors that might affect DIF, such as cultural differences, are not examined and the interpretations of DIF are not exhaustive. Rather, an evidence gathering approach is taken where analyses are conducted to gather evidences that may provide support for or against the following hypotheses:

1. DIF is due to adaptation effects

Two types of evidence were used to support this hypothesis. The first was identification of differences in meaning between translated versions of items by judgmental review. Second was the replication of DIF for the items in more than one comparison. In order to increase the likelihood of identifying adaptation related DIF, this study focuses on adaptation effects in one country, namely Canada, which is expected to minimize effects due to cultural and curricular differences. Canada, a bilingual country in French and English, administers TIMSS assessments in both of these languages. The study uses other comparisons of French and English versions of TIMSS to determine whether DIF in fact is due to adaptation effects.

2. DIF is due to curricular differences

Clustering of DIF items that are all in favor or against the same group in a topic area, such as algebra, was interpreted as evidence supporting the hypothesis that there was an association between DIF and that topic area. In other words, DIF may be due to curricular differences in relation to a topic.

Even though we may not be able to identify sources of DIF for all items, the existence of DIF points to possible differences between groups that may jeopardize the comparability of scores. The last part of the chapter discusses the degree to which the comparability of scores is affected by DIF.

METHOD

TIMSS was conducted in 1995 and included a survey of 13-year-old students' (Population 2) mathematics and science achievement in 45 countries. This chapter focuses on the tests that were administered in Canada (in English and in French) England, France, and the United States. The selection of these countries allowed for investigation of the equivalence of English and French versions of items (i.e., England vs. France, and the United States vs. France), as well as comparing English and French versions when cultural differences are expected to be minimized (i.e., English and French administrations in Canada). The sample size, and numbers of items are presented in Table 1.

Identification of Differential Item Functioning

DIF was identified using the Linn and Harnisch (LH) (1981) procedure. This procedure computes the observed and expected mean response (expected and observed p-values) and the difference between them (observed minus predicted, p_{diff}) for each item by deciles of the specified group. The expected p-values are calculated based on modeling of item responses using item response theory. Based on the difference between the expected and observed p-values a Z-statistic is calculated for each decile and an average Z-statistic for the item is computed for identifying the degree of DIF. The level of DIF is determined by the set of rules presented in Table 2.

Table 1. Sample sizes and number of items used in comparisons

Country	N	Number of items	
		Mathematics	Science
Canada		156	140
English speaking	5,445		
French speaking	2,925		
England – France		156	139
England	3,572		
France	6,002		
United States – France		154	139
United States	10,945		
France	6,002		

Table 2. Statistical rules for identifying three levels of DIF

DIF Level	Rule	Implication
Level 1	$|Z| < 2.58$	No DIF
Level 2	$|Z| \geq 2.58$ and $|p_{diff}| < 0.10$	DIF
Level 3	$|Z| > 2.58$ and $|p_{diff}| \geq 0.10$	Serious DIF

Judgmental Reviews

The French and English versions of items administered in Canada were examined by four translators bilingual in French and English for potential adaptation problems. If any problems were identified, the two versions of the items were evaluated by each translator who would rate the degree to which there were differences in meaning due to adaptations. Items were assigned ratings between 0 and 3. A rating of 0 indicated there were no differences in meaning in the adaptation of the items. Ratings between 1 and 3 were assigned to items in which there was an adaptation problem identified between the two versions, indicating the degree of impact to the meaning of the item. For example, a rating of 1 was assigned when there were minimal differences in meaning due to adaptations. Items were assigned a rating of 2 if there were clear differences between the two versions, but they were not expected to lead to differences in examinee performance for the two groups. A rating of 3 was assigned if the adaptation problems were expected to lead to differences in comparability of performances for the two groups. Translators were also used to interpret differences identified by statistical DIF procedures.

Analyses

Analyses were intended to determine (1) whether there was differential functioning between English and French versions of items, (2) whether differential functioning could be attributed to curricular or language differences, and (3) the extent to which differential item functioning was expected to affect comparability of scores. Pair wise DIF analyses were conducted separately for the Canadian English and Canadian French versions of the items, for England and France, and for the United States and France. These analyses provided information about the degree of DIF in the English and French versions of items. If DIF was replicated in two or more comparisons, it provided support for an interpretation that the differences might be due to adaptations rather than cultural or curricular differences. If, on the other hand, DIF occurred in the same direction (e.g., all in favor of French-speaking students) clustered by topic this provided support for interpretation that DIF might be due to curricular differences.

RESULTS

TIMSS provided large sample sizes and large numbers of items for examining the hypotheses stated earlier (see Table 1). Items were combined across test booklets and analyses were conducted separately for mathematics and science. The results of the analyses examining degree of DIF in the English and French versions of the items, and evidence supporting whether DIF was due to adaptation effects or curricular effects are presented below.

DIF in English and French versions of Items

The results of the DIF detection procedure are summarized in Tables 3 and 4, for mathematics and science items, respectively. Out of 156 items, 22 were identified as DIF in the English and French versions of the mathematics portion of TIMSS that were administered in Canada. There were larger numbers of items identified as DIF when the French items administered in France were compared to the English ones administered in England and the United States: 61 of 156 items and 91 of 154 items, respectively. Forty-one percent of the mathematics DIF items functioned in favor of the French-speaking examinees in the Canadian comparison group, 43 percent in the England-France comparison, and 53 percent in the United States-France comparison.

Table 3. Number of mathematics DIF items

	Pro-English		Pro-French	
Comparisons	Level 2	Level 3	Level 2	Level 3
Canada English - French	11	2	8	1
England - France	32	3	21	5
U. S. - France	33	10	35	13

Table 4. Number of science DIF items

Comparisons	Pro-English		Pro-French	
	Level 2	Level 3	Level 2	Level 3
Canada English - French	16	2	30	4
England - France	24	1	26	3
U. S. – France	28	11	32	20

More items were identified as DIF in the science portion than in mathematics in two of the comparisons. There were 52 DIF science items out of 140 items in the Canadian comparison, 54 out of 139 in the England-France comparison, and 91 out of 139 in the U. S.-France comparison. Sixty-five percent of the science DIF items functioned in favor of the French-speaking group in the Canadian comparison, 54 percent in the England-France comparison, and 57 percent in the United States-France comparison.

Evidence Supporting Adaptation Effects as Explanations for DIF

Examining the degree to which DIF may be due to adaptation effects involved three processes. These were a judgmental review of all items by bilingual translators, an interpretation of DIF items by bilingual translators, and an analysis to determine whether DIF due to adaptation differences was replicated in all three comparisons. Reviews of all the items were conducted prior to an interpretation of DIF by the bilingual translators, in order to examine the degree to which adaptation-related problems were associated with differential functioning of the items.

Judgmental Reviews of All Items

Canadian English and French versions of items were used in this process. The translators identified adaptation-related problems in 64 out of 296 items (156 mathematics, 140 science). Twenty-two (approximately 34 percent) of the 64 items were identified as differentially functioning in the Canadian English and French comparisons. Samples of adaptation problems that were expected to lead to differences in examinee performance for the two groups (rating of 3) by at least one translator are displayed in Table 5. Seven of the 13 items that were identified by the bilingual translators as having serious adaptation problems (rating of 2) also displayed DIF as identified by the statistical procedures. The adaptation-related differences between the two versions of items included differences in the level of specificity of the vocabulary, the clarity of the statement of the question, the difficulty of the vocabulary in the test items, the clues that were expected to guide examinee thinking processes, and the length of the item in terms of number of words.

Table 5. Samples of adaptation problems with a judgmental review rating of 3 by at least one translator

Item DIF level (Favored version)	English Version	French Adaptation	Problem
2 (English)	Commonly used vocabulary	French words not as common	English version clearer and easier to understand
	Stem and distractors use the same vocabulary	Stem and distractors use different vocabulary to refer to the same thing	
2 (French)	A son	*Un enfant*	English version more specific
2 (French)	Has a side of 10cm	*A une largeur de 10cm*	French version more descriptive-tell you exactly which side
2 (English)	Both the stem and the distractors use simple vocabulary	French version has more difficult vocabulary	English version clearer and easier to understand
2 (French)	Oxygen equipment	*Bouteilles d'oxygène pour respirer*	French version provides a clue about the utility of oxygen equipment
2 (French)	Most organic material	*La plus grande partie de matière organique*	French version does not have shading in first layer which helps distinguish it from the other layers
3 (English)	Active: "he found that"	Passive: "it was stated that"	French version is more difficult
No DIF	Which species have been on earth for the shortest amount of time?	*Laquelle des espèces suivantes est apparue le plus récemment sur la Terre*	Difference in meaning and emphasis
No DIF	Complete the Grade 8 row in the pictograph	*Complèter le pictogramme*	English version specifies the row
No DIF	Use up Give off	*Consomment* *Rejettent*	French vocabulary is more sophisticated

Judgmental Reviews for Interpreting DIF

The results of the DIF analyses and their corresponding judgmental review ratings are displayed in Tables 6 and 7 for mathematics and science, respectively. Sixteen of the 22 mathematics DIF items had judgmental review ratings of 1 or lower, indicating minor to no differences between French and English versions in the Canadian assessments. Thirty-three of the 52 science DIF items had judgmental review ratings of 1 or lower. Additional judgmental reviews of the items identified as showing Level 3 DIF in the analyses were conducted to further examine potential causes of DIF. These reviews resulted in the identification of adaptation-related problems in two out of three Level 3 mathematics DIF items and two out of six Level 3 DIF science items. Almost all of these DIF items had previously been identified as displaying adaptation problems by the bilingual translators.

Table 6. Mathematics DIF items and judgmental review ratings

Judgmental Review Rating	Differential Item Functioning Level	
	2	3
0 - 1	15	1
2	4	2
3	0	0

Table 7. Science DIF items and judgmental review ratings

Judgmental Review Rating	Differential Item Functioning Level	
	2	3
0 - 1	29	4
2	10	0
3	7	2

Replication of DIF Items in Multiple Comparisons

The focus of these analyses was the replication of DIF items in the Canadian comparisons in the other two comparisons, namely, England-France and U. S.-France comparisons. In particular, replication of DIF with adaptation-related interpretation was used as additional evidence for supporting this interpretation.

Mathematics DIF Items in the Three Comparisons

The numbers of mathematics items identified as DIF in two or more comparisons are presented in Table 8. In mathematics, out of 13 DIF items in favor of the English-speaking group in the Canadian comparison, two were also identified as having DIF favoring the English group in the England-France and U. S.-France comparisons. These items had no adaptation-related interpretations.

Out of nine mathematics items in favor of the French-speaking group in the Canadian comparison, three were identified to be in favor of the French-speakers in the England-France comparison and 6 were in favor of the French-speakers in the U. S.-France comparison. Two items were identified as DIF in all three comparisons. One of the items identified as DIF in favor of the French-speaking group in all three comparisons used the word "pace" referring to steps one takes to measure the length of a room. In French, this item used the word "step," instead of "pace." A simpler term in French could have made the item easier in the French version.

Table 8. Number of mathematics DIF items in three comparisons

		Canada English-French				England- France				U. S.- France			
		Pro-English		Pro-French		Pro-English		Pro-French		Pro-English		Pro-French	
		L2	L3	L2	L3	L2	L3	L2	L3	L2	L3	L2	L3
Canada													
Pro-English	Level 2	11				4		1		4	1	1	2
	Level 3		2			2					1		
Pro-French	Level 2			8		1		2	1			2	3
	Level 3				1								1

Science DIF Items in the Three Comparisons

The numbers of science items identified as DIF in two or more comparisons are presented in Table 9. In science, out of 18 DIF items in favor of the English-speaking group in the Canadian comparison, seven were also identified as having DIF favoring English-speakers in the England-France comparison, six of which were also identified as favoring the English-speaking group in the U. S.-France comparison. Therefore, there were six items identified as favoring the English-speaking groups in all three comparisons. One of these items had a subtle but crucial difference between the two versions. In English, the stem of the item states that, "Whenever scientists carefully measure <u>any</u> quantity many times, they expect that…".The correct answer is that, "most of the measurements will be close but not exactly the same." In French, the stem of this item could be translated as, "When scientists measure the <u>same</u> quantity many times, they expect that…" In this version, the description of the scientific measurement process is different. In addition, the word "same" could lead examinees to think that this quantity was known and the answer should be that the scientists should get the same amount every time. This item was identified as DIF in favor of the English-speaking group in the Canadian and U. S.-France comparisons.

Out of 34 science items in favor of the French-speaking group in the Canadian comparison, 13 were in favor of French-speakers in the England-France comparison and 17 were in favor of the French-speakers in the U. S.-France comparison. Ten of these items were common across the three comparisons. One of these items was related to the reflection of a candle in a mirror. In the English versions, the word "reflection" is used to refer to the candle's reflection in the mirror, whereas in the

French versions the word *l'image* is used. The word "reflection" is expected to direct the examinee's thought processes more directly.

Table 9. Number of science DIF items in three comparisons

		Canada English-French				England-France				U.S.-France			
		Pro-English		Pro-French		Pro-English		Pro-French		Pro-English		Pro-French	
		L2	L3	L2	L3	L2	L3	L2	L3	L2	L3	L2	L3
Canada													
Pro-English	Level 2	16				7		1	1	7	3	2	1
	Level 3		2							1			
Pro-French	Level 2			30		3		10	2	4	2	11	5
	Level 3				4			1					1

Evidence Supporting Curricular Differences as Explanations for DIF

This section summarizes the degree to which DIF may be due to curricular differences by examining the relative distribution of DIF items by topic area for mathematics and science separately (see Tables 10 and 11). Clustering of large proportions of DIF items by topic area in favor or against a particular group was interpreted as DIF due to curricular differences.

Mathematics DIF items by Topic

There were six topic areas in mathematics: algebra; data representation, analysis, and probability; fractions and number sense; geometry; measurement; and proportionality. Curriculum-related interpretations of mathematics DIF are summarized by comparison below.

English-French Canada Comparison

Five of the mathematics topics had less than 15 percent of the items identified as DIF. However 26 percent, 6 in number, of the geometry items were identified as DIF. Five of the six, 83 percent, were in favor of the French-speaking group, which indicates that DIF in these items may be due to curricular differences.

England-France Comparison

Twenty-eight percent of the algebra items were identified as DIF and all of these items were in favor of the English group. There is clear evidence from this analysis that there were curricular differences between England and France on algebra, favoring England.

Table 10. The relative distribution of mathematics DIF items by topic area

Topic Comparison	Level 2		Level 3	
	Pro English	Pro French	Pro English	Pro French
Algebra (29 Items)				
Canada English - French	2	1	1	
England – France	7		1	
U. S. – France	8	4	4	
Data Representation, Analysis & Probability (20 Items)				
Canada English - French	1			
England – France	5	2		
U. S. – France	5	4		
Fractions & Number Sense (52 Items)				
Canada English - French	4	2	1	
England – France	12	10	2	4
U. S. – France	13	3	5	2
Geometry (23 Items)				
Canada English - French	1	5		
England – France	2	4		1
U. S. – France	2	12		6
Measurement (21 Items)				
Canada English - French	3			1
England – France	3	3		
U. S. – France	1	11		4
Proportionality (12 Items)				
Canada English - French				
England – France	3	2		
U. S. – France	4	1		1

U. S.-France Comparison

There were many more DIF items in the U. S.-France comparison than in either of the other two comparisons. Fifty-five percent of the algebra items were identified as DIF, and 75 percent of those differences were in favor of the U. S. Forty-four percent of the fractions and number sense items were identified as DIF; 78 percent, in favor of the U. S. Eighty-seven percent of the geometry items were identified as DIF; 78 percent, in favor of the France group. Seventy-six percent of the measurement items were identified as DIF; 94 percent, in favor of France. Fifty percent of the proportionality items were identified as DIF; 66 percent, in favor of the U. S. In this case, algebra and fractions and number sense items seemed to be in favor of the U. S., whereas, geometry and measurement items were in favor of France.

Table 11. The relative distribution of science DIF items by topic area

Topic Comparison	Level 2		Level 3	
	Pro English	Pro French	Pro English	Pro French
Chemistry (21 Items)				
Canada English - French	1	3		
England - France	2	5		
U. S. - France	3	5	3	3
Earth Science (23 Items)				
Canada English - French	1	8		2
England - France	5	5		1
U. S. - France	6	3	1	4
Environmental Issues and Nature of Science (16 Items)				
Canada English - French	3	5	1	
England - France	5	1		1
U. S. - France	3	2	2	
Life Science (41 Items)				
Canada English - French	5	5	1	1
England - France	4	10	1	1
U. S. - France	9	12	2	7
Physics (40 Items)				
Canada English - French	6	9		1
England - France	7	5		
U. S. - France	7	10	3	6

Science DIF items by Topic

There were five topic areas in science: chemistry, earth science, environmental issues and the nature of science, life science, and physics. Similar to the interpretation of mathematics DIF items, when large proportions of DIF items were in favor of one group, DIF was interpreted as due to curricular differences. Interpretation of curricular differences are summarized by comparison as follows.

English-French Canada Comparison

In two of the topic areas, most of the science DIF items were in favor of the French-speaking group. Seventy-five percent of the chemistry DIF items, and 90 percent of the earth science DIF items were in favor of the French-speaking group. DIF items were more evenly distributed on the other three topic areas. The strongest evidence for curricular differences between English and French speaking groups was on earth science items. Nine out of ten earth science items, 90 percent, were in favor of the French-speaking group.

England-France Comparison

In two out of five of the topics, chemistry and life science, DIF was mostly in favor of the French group. Thirty three percent of the chemistry items (7 out of 21) were identified as DIF; 71 percent, in favor of the French group. Thirty-nine percent of the life science items were identified as DIF; 69 percent, in favor of the French group. There was also strong evidence of curricular differences in favor of the English group on environmental issues and the nature of science. Forty-four percent (7 out of 16) of items on this topic were identified as DIF. Of these DIF items 71 percent were in favor of the English group. The DIF items were more evenly distributed between England and France on the two other topics.

U. S.-France Comparison

There was strong evidence for curricular differences in life science and physics items in favor of the French group. Seventy-three percent of the life science items were identified as DIF; 63 percent, in favor of the French group. Sixty-five percent of the physics items were identified as DIF; 62 percent, in favor of the French group.

SUMMARY AND DISCUSSION

There were several observations made during the judgmental reviews of this study that linked different aspects of adaptations of the tests to DIF. First was that, even though the differences may seem small in adapted versions of a test, these differences could lead to high levels of DIF. On the other hand, large differences in meaning did not necessarily lead to high levels of DIF. The association between the judgmental review ratings and DIF levels was moderate. Approximately half of the items that were identified by the bilingual translators as having serious adaptation problems displayed DIF as identified by the statistical procedure. Approximately 22 percent of the mathematics and 40 percent of the science DIF items had adaptation-related explanations.

There were four problems in adaptations that were interpreted as leading to DIF by the judgmental reviews. First, they stressed the importance of key words in the adaptation process. The key words could be important terms that provided information, or guided the thinking processes of examinees. Exclusion or inappropriate adaptation of these key words can lead to confusion or make the item easier or more difficult. Second, adaptation problems were related to the commonness of vocabulary in a given context. Third, problems were related to the look and formatting of the item. These aspects of an item are not related to adaptation but are relevant to consider when the intention of the adaptation is to create parallel forms of an item. Lastly, there needs to be an additional focus in adaptations of items and tests about the kind of information provided to examinees, how this information guides examinees' thinking processes, and whether these aspects are equivalent in the two versions of the items.

Examination of replication of DIF in three comparisons resulted in supporting adaptation-related interpretation of DIF in the Canadian comparisons. In mathematics, two of the four DIF items that were interpreted as adaptation-related were replicated in all three comparisons. The other two items were replicated in two of the three comparisons. In science, five of the 14 DIF items that were interpreted to be related to adaptation differences were replicated in all three comparisons, another four were replicated in two of the three comparisons. The numbers of DIF items identified with adaptation-related interpretations in the Canadian comparisons and that were replicated in other comparisons are presented in Table 12.

Table 12. Number of DIF items in the Canadian comparison
with adaptation and curricular interpretations

Mathematics DIF Items				Science DIF items			
Adaptation	*Curriculum*	*Uninterpreted*	*Total*	*Adaptation*	*Curric.*	*Uninterp.*	*Total*
4 (2,2) [1]	4	15	23	14 (5,4) [1]	7	31	52

[1] The first number in parentheses refers to the number of items replicated in all three comparisons, the second refers to the number of DIF items replicated in two comparisons.

As expected, clustering of DIF items by topic provided curricular differences as interpretation for DIF only for small numbers of the DIF items in the Canadian comparisons: approximately 17 percent of the mathematics DIF items and 13 percent of the science. However, larger proportions of DIF items were interpreted to be associated with curricular differences in the England-France and U. S.-France comparisons. In the England-France comparison, 24 percent of the mathematics DIF items and 54 percent of the science DIF items were interpreted to be associated with curricular differences. In the U. S.-France comparison, 58 percent of the mathematics and 36 percent of the science DIF items were interpreted to be associated with curricular differences.

There were large portions of DIF items that could not be attributed to adaptation related or curricular differences. These were 65 percent of the mathematics DIF items and 60 percent of the science DIF items (see Table 12). Several factors that were not considered in this chapter could be potential explanations for DIF. Among these include differences in instruction methods, cultural differences, and limitations in the definitions of topics. Even though we may not be able to disentangle all possible explanations for DIF, it is important to understand the effect of DIF on comparability of scores. One way of examining such an effect is by computing group scores when items identified as DIF are not included. Group scores were calculated as average *p*-value (difficulty estimate) for all the items considered. These scores were compared to those when the scores were recalculated after excluding the DIF items in each comparison. The group mean scores with and without DIF items are displayed on Table 13.

As can be seen, the differences in group scores range between 0 to 2 percentage points, and such differences are not large enough to reverse the position of a group mean in each comparison. However, these differences can impact the ranking of a group in relation to other country means

Table 13. Group mean scores (in percent) when DIF items are included versus not included

Comparisons	Mathematics		Science	
	With DIF	*Without DIF*	*With DIF*	*Without DIF*
Canada, English	51	52	54	54
Canada, French	52	52	49	48
England	51	50	58	60
France	59	57	53	54
United States	50	48	55	57
France	59	57	53	53

DIF can impact comparability of scores in more complex ways. For example, curricular differences or adaptation-related differences are not expected to impact all examinee performance uniformly. In other words, some students' performance might be more impacted by ambiguity in a test item, or exclusion of a topic from curriculum. Therefore, such differences might impact comparability of scores for some students, such as lower ability students, to a larger degree.

REFERENCES

Ackerman, T. A. (1992). A didactic explanation of item bias, item impact, and item validity from a multidimensional perspective. *Journal of Educational Measurement, 29,* 67-91.

Ercikan, K. (1998). Translation effects in international assessments. *International Journal of Educational Research, 29,* 543-553.

Hambleton, R. K. (1993). Translating achievement tests for use in cross-cultural studies. *European Journal of Psychological Assessment, 9,* 57-68.

Linn, R. L., & Harnisch, D. L. (1981). Interactions between item content and group measurement on achievement test items. *Journal of Educational Measurement, 18,* 109-118.

O'Neil, K. A., & McPeek, W. M. (1993). Item and test characteristics that are associated with differential item functioning. In P. W. Holland, & H. Wainer (Eds.), *Differential item functioning* (pp. 255-276). Hillsdale, NJ: Erlbaum.

Scheuneman, J. D., & Geritz, K. (1990). Using differential item functioning procedures to explore sources of item difficulty and group performance characteristics. *Journal of Educational Measurement, 27,* 9-131.

Schmitt, A. P., & Dorans, N. J. (1990). Differential item functioning for minority examinees on the SAT. *Journal of Educational Measurement, 27,* 67-81.

6 CONCLUSION

Chapter 25

A LOOK BACK AT TIMSS: WHAT HAVE WE LEARNED ABOUT INTERNATIONAL STUDIES?

Albert E. Beaton and David F. Robitaille

Over the several years that have elapsed since the TIMSS data were collected in the mid-1990s, a large number of reports dealing with various aspects of the study have been published. These include reports from the International Study Center at Boston College, national reports published in each of the participating countries, several books published by private sector publishers, a large number of research papers published in academic journals, as well as many university-based theses and dissertations. In addition, articles focusing on the TIMSS findings have been published in many prestigious newspapers and magazines around the world.

One of the major questions facing school systems around the world is what kinds of curricula and what kinds of instructional practices seem most likely to provide the best educational experience for a nation's children. A particular strength of the TIMSS design, and of most IEA studies, is that students and their achievement can be linked to the instructional practices of teachers and to the content of the official curriculum. The design of the study makes it possible to investigate linkages among the intended curriculum, the implemented curriculum, and the attained curriculum.

A great deal of what has been published to date about TIMSS has focused on the achievement results, and on the rankings of the participating countries in "league tables" that have been published in a wide variety of publications. Much less has been written and published so far about the more important—at least from an educational perspective—questions relating to ways and means of improving the teaching and learning experience in school mathematics and science for students in all countries.

Some of this relative lack of attention to what are undoubtedly the most important questions arising from the TIMSS findings can be attributed to the mass media's rather short attention span, and to their need for uncomplicated explanations that fit easily into prescribed formats. (Who won? What can we do? Should we increase/decrease class size/hours of instruction? ...)

However, researchers and organizations that fund international research in education share this responsibility, and everyone involved needs to consider how the

D.F. Robitaille and A.E. Beaton (eds.), Secondary Analysis of the TIMSS Data, 409–417.
© 2002 Kluwer Academic Publishers. Printed in the Netherlands.

state of the art can be improved. It seems, for example, easier to obtain funding to collect new data than to carry out more in-depth analyses of the data we already have. But unless funding is made available for those more in-depth, secondary analyses of the data from these studies—analyses such as those described in the papers included in this volume, for example—the current state of affairs is unlikely to change.

The idea behind this book has been to showcase the kind of research that has been and is being done with the TIMSS data by researchers from the TIMSS countries around the world. The papers published in this volume are excellent examples of TIMSS-based research. The many authors—those published here as well as elsewhere—from the participating countries have utilized the data from the study to illuminate important questions concerning the teaching and learning of mathematics and science, as well as questions about promising new methodological approaches for this kind of research. They serve as exemplars of the kind of research that needs to be done with these data, and they should stimulate others to investigate these kinds of questions in greater depth.

The kind of research that can be done with the TIMSS data is highly dependent upon the design employed in the study, and that design reflects developments in international assessments and assessment in general over many years. The advantages and disadvantages of past IEA assessments were reviewed and many suggestions were adapted and implemented in the TIMSS design. The psychometric technology of the National Assessment of Educational Progress (NAEP) in the United States was examined and some of its features adapted for international assessment. In short, the TIMSS design that was implemented seemed the best design that was feasible at the time it was agreed upon. If a more effective design had been available within the constraints of time and available budget, it would have been used.

This goal of this chapter is not to present a detailed description of the TIMSS design and its implementation. Chapter 2 contains a brief summary of the main elements, and more details are discussed in three TIMSS technical reports: Volume I–Design and Development; Volume II–Implementation and Analysis, Primary and Middle School Years; and Volume III–Implementation and Analysis, Final Year of Secondary School (Martin & Kelly, 1996; Martin & Kelly, 1997; Martin & Kelly, 1998). These reports are available from the TIMSS website at www.timss.org.

The goal of this chapter is to highlight certain aspects of the TIMSS design and to comment on some of the strengths and weaknesses of those aspects. TIMSS was, in many respects, a highly innovative study, and our hope is that those who undertake such studies in the future will benefit from the advances made in TIMSS. We also have a number of suggestions to make to those future researchers about issues that need to be addressed in future studies. Although we believe that the TIMSS design, as implemented, was about as good as it could be, given the many constraints in model building, we will discuss the rationale for some features and suggest some possible improvements.

IMPORTANCE OF TRAINING AND DIFFERENT PERSPECTIVES

It is important to realize that all of the countries participating in a given study will not agree on all of the details of that study. Countries participate in these studies for very different reasons, and they may come from rather different traditions with respect to achievement testing. The available staffing, funding, and timing also help define the boundaries for a study; and these may vary significantly from country to country. A successful study must look for agreement among the participants at every step of the way, including the specification of the test frameworks, the achievement items, the analytic procedures, and the way that the final results are to be presented. At each step, some countries will prefer different approaches, and they may consider the test or a particular procedure unfair. If the integrity of the comparisons is to be maintained, the participating countries cannot be allowed to vary from the international plan and so no country can expect to participate without some compromises. The aim for the assessment must be that it is equally fair, even if not perfect, to all. The final design is very much a series of compromises among the participants, compromises that are made in the context of considerations of scientific integrity identified by the technical staff associated with the study.

The successful implementation of TIMSS required that the many participating countries come together and agree upon the design and procedures. There were many debates among the participants. What made the whole system work was a series of training sessions including test administration, quality assurance, translation, and many others. Without this emphasis on training, there would have been many national variations and serious problems with comparability. The differing perspectives of some of the participating countries are demonstrated in Chapters 4. 5, 6, 9, 10, 11, 12, 13.

POPULATION DEFINITIONS

The TIMSS design included three populations covering elementary, middle, and secondary schools. Populations 1 and 2 included two included adjacent grades that included the largest proportions of students of a specific age at the time of testing. Population 1 had a target age of 9-year olds with the result that Grades 3 and 4 were sampled for most countries. Population 2 had a target age of 13 with the result that Grades 7 and 8 were assessed in the middle school grades.

This adjacent-level design feature proved very useful in several ways. First, the average growth between adjacent grades over all countries provided a useful benchmark for comparing the performance of different countries within a single grade. Sampling adjacent grades also made possible the estimation of student performance at the selected age levels, and these estimates are reasonably accurate.

The sampling of adjacent grades also made possible the partitioning of the variance between classrooms within a school (see Chapters 14 and 22) for many countries that sampled only one classroom within a school. This feature is well worth including in future studies.

Population 3, as has been noted elsewhere, was somewhat more problematic. Many participating countries had strong interests in how well their students performed at the end of secondary school, but their interests differed. Some wanted to assess the general mathematics and science literacy of their students in order to find out what they had learned in their schooling, and whether they had the skills needed for productive performance in the 21st century. Others were interested more specifically in how the students who took advanced courses in mathematics and physics performed when compared to other countries. Depending on their interests, the countries could assess any or all of the three overlapping subpopulations: all students in their graduating classes, those taking advanced mathematics courses, or those taking physics courses.

The question of whether or not Population 3 was worth pursuing remains to be answered for many people. Participating countries differ in both the number of years of schooling involved in primary and secondary education and the content of their mathematics and science curricula. The ages of the different Population 3 sub-populations differed substantially across countries. Although most countries have 12 years of primary and secondary education, some have more years and some have fewer. Moreover, substantial proportions of students may drop out before completing their education. Some countries require mathematics at the upper levels of secondary school for all students, and others do not. The proportions of students taking advanced mathematics and physics vary significantly from country to country. International comparisons of performance are affected by all these differences. One might say that comparing countries at this level is not fair; the playing field is not level.

And yet, Population 3 provided important information about educational systems around the world. TIMSS showed how much the approach to education differed in different nations. For example, all academic students in Norway take a general science course and two courses in physics; and Norwegian students did very well in physics. Students who did not have an opportunity to learn advanced subjects were not proficient in them. The Population 3 results show that countries that want high proficiency in mathematics and physics must devote a great deal of time and effort toward attaining that goal. Although assessing Population 3 was difficult, we believe that it was well worth the effort (see Chapter 18).

CURRICULUM ANALYSES

Both the development of the achievement tests and the analysis of the resulting data depend on an examination of the curricula in the participating countries. In IEA terminology, this means an analysis of the intended curriculum, the implemented curriculum, and the attained curriculum. Before the test development work for TIMSS was finalized, differences among the intended curricula for the participating countries were studied. Formal national curricula, textbooks, and other indicators of curricular intentions were investigated. The results are presented in Schmidt, et al (1997).

An in-depth analysis of the curricula of all the nations participating in a given study is essential to developing relevant tests. The TIMSS curriculum analysis was groundbreaking in its technology; but additional research is needed to improve the conceptualization of the curriculum analysis methodology so that the data are more readily intelligible and useable. Chapters 7, 8, 20, and 21 in this volume address curriculum issues.

TEST DEVELOPMENT

The aim of the test development was to have items that had maximum validity across participating countries and tested as wide a range of school mathematics and science curriculum as possible. After the test specifications were discussed, modified, and approved, test items for the paper-and-pencil component of the test were submitted from the participating countries for review, selection, and pre-testing. These included several types of items: multiple-choice, short answer, and extended response items. Professional test developers also helped in the item development. All of the items were pre-tested in several countries, and the final set of items was selected from among them.

A performance assessment instrument was also developed for use in TIMSS: originally as a required, core component of the study, but finally as an optional component for those countries that were particularly interested in participating in this form of assessment. In this part of the assessment, students were given "hands-on" problems to solve using various instruments and materials. Making the problems equivalent across countries was difficult because of differences across countries with respect to even everday materials. For example, the composition of tap water may vary substantially from one country to another. The amount of time needed and the cost of carrying out the performance assessment were high; and, as a result, many countries elected not to participate in this aspect of the study.

The test booklets were constructed using a modified form of Balanced Incomplete Block (BIB) spiralling design. The BIB design makes it possible to investigate the correlations among a large number of test items and thus to investigate the dimensionality of the mathematics and science tests. The test booklets in Populations 1 and 2 contained both mathematics and science items. Future studies might consider having only one subject in a test booklet, thus enhancing the dimensionality studies of each subject area (see Chapter 15). The price to be paid for having students assigned a test in only one subject area would be a reduced sample size, since only half of the students would be tested in each subject area.

TRANSLATION

The translation of the TIMSS tests and questionnaires from English into over 30 languages was done with great care. However, there is no way to be absolutely sure that all translations had exactly the same meaning and the same level of language difficulty. To address this problem, the items were independently translated and the results compared. Further, persons familiar with English and the target language as well as the subject matter independently checked the translations. Item analyses were also done do establish whether or not the items were behaving in the same way in different countries. Although this is the state of the art and was carefully checked, we suggest that further research be done into the translation process and its effect on international comparisons. Translation issues are addressed in Chapter 24.

SCORING

Scoring constructed response items has problems similar to those of translation. Each country appointed a head scorer who was familiar with English and with the language of the test to direct the scoring of those items. TIMSS used a double-digit scoring system that coded the type of student response as well its correctness (see Chapter 3). The head scorer trained the scorers in his or her country. TIMSS approached the issue of comparability of scores across countries by extensive training of those who were responsible for the scoring and provided materials to help those persons to train scorers in their own countries. The national representatives ran various reliability checks on the scoring and the International Study Center did further statistical checks to assure that the items responses were comparable across countries (Martin & Mullis, 1996). As with translation, more research needs to be done on scoring papers written in different languages in order to look for improvements in this methodology.

SAMPLING

The TIMSS sampling plan was complex. The basic sampling unit was schools, which were selected with probability proportional to estimated size. In some countries, such as the United States, the initial sampling unit was counties or metropolitan areas and schools were selected from these. Within schools, for Populations 1 and 2, the classrooms in which students studied mathematics were listed and either one or two classrooms were randomly selected. All students in these classrooms were assessed. For Population 3, students in their last year of secondary school were listed along with whether or not they were studying advanced mathematics or physics. Students were randomly selected from this pool and assigned TIMSS booklets according to what they were studying.

The selection of intact classrooms for Populations 1 and 2 facilitates the study of the relationship between teacher and classroom factors on the one hand, and student performance on the other. Sampling intact classrooms is also administratively easier for schools to handle. On the other hand, selecting intact classrooms generally produces a more homogeneous sample of students than simple random sampling. This means that the design effect is larger and the standard errors of population estimates will be larger.

The sample selection was monitored by the TIMSS sampling coordinator who assured that the selected sample would produce unbiased estimators of population parameters. Participation rates were also monitored. After the sample was selected and the sampling weights computed, sample estimates were compared to known population parameters in order to evaluate the implementation of the sampling procedures.

TEST ADMINISTRATION AND QUALITY CONTROL

In order for the achievement results to be comparable, the TIMSS tests had to be administered under standardized conditions. The main way of assuring standard conditions was through a number of training sessions that were held for National Research Coordinators who, in turn, trained the actual personnel who administered the tests in their respective countries.

An important part of any study is the quality assurance efforts. The TIMSS quality assurance system involved selecting and training a person in each country to interview the National Research Coordinator, review the implementation of procedures, and inspect a random selection of assessment sessions. The results of these quality assurance activities are summarized in Martin and Mullis (1996). Unfortunately, the funding for quality assurance became available only a few weeks before the data collection period and so, in a number of cases, the National Research Coordinators were asked to nominate a quality assurance monitor. Earlier funding would have made the identification of more independent monitors possible.

SCALING

The mathematics and science achievement results were scaled to summarize the data for publication and further research. Using item response theory (IRT), the Rasch one-parameter model was used. This method produced a single scale for mathematics and another scale for science. When a finer description of the subject matter was necessary, average-percent-correct methodology was used. In Population 2, mathematics was divided into several subtests: fractions and number sense, geometry, algebra, data representation, measurement, and proportionality. Science was divided into earth science, life science, physics, chemistry, and environmental. To improve parameter estimates, plausible value methodology was used.

Alternatively, the scaling might have been done using the three-parameter logistic mode. This model could be expected to fit the data better than the Rasch model, especially since multiple-choice items were involved. The Rasch model was used for TIMSS 1995 because of the availability of the required expertise and software. The TIMSS 1999 data were scaled using the three-parameter logistic model, and the TIMSS 1995 data have been rescaled using the same model for comparability. Arguments can be made for either method, but our leaning is toward the three-parameter logistic model.

ANALYSIS

The TIMSS participants agreed that the first results should be published in a straightforward and timely manner, thus making the basic results available in good time and facilitating more detailed, secondary analyses of the data. The first analyses of the TIMSS data were planned well in advance. Before the data were available, dummy tables were developed and explained to the National Research Coordinators. After various revisions were made to their satisfaction, the data were processed and the tables prepared.

Special concern was addressed to estimating the standard errors of all published statistics. The usual formulas for standard errors are appropriate for simple random samples, whereas the TIMSS sampling procedure was multi-stage with intact classes sampled. Therefore, the jackknife method was used to compute the error estimates.

A separate analysis was performed to address the issues of the fairness of the test in different countries. This analysis is called the Test Curriculum Matching Analysis (TCMA) and the results of that analysis have been published in the appendices of various reports. The general idea of the analysis was to allow each country to reject items that were not relevant to the curriculum in that country. Students in that country were then scored only on the items deemed appropriate for that country; and, all other countries were scored on the same items. The result is a square table that shows how each country scored on its own items and on the items selected by other countries. Results show little difference in international standings as a result of the selection of items.

To make the TIMSS scales more meaningful, the scales were anchored to describe what students at particular levels knew and could do. The results are described in Chapter 23.

The TIMSS data have been used to investigate issues such as gender differences, the use of computers, and the relationships with background variables (Chapters 16, 17, and 18) across nations and also to investigate the effect of outside tutoring (Chapter 20). The use of assessment information is by teachers has also been investigated (Chapter 21).

FUTURE STUDIES

It is our opinion that the TIMSS study has produced valuable information that has been used around the world to affect educational policy. We believe that such studies should be continued in the future. We also believe that the TIMSS design was quite efficient and appropriate for gathering international data and the design should be the starting place for future studies.

The data are available for anyone to see and use. TIMSS has produced a very large, publicly available database that is available through the International Study Center at its website: www.timss.org. The data, the accompanying documents and all of the TIMSS reports are available. This is a very rich and carefully prepared database that provides an excellent opportunity for researchers to explore various research questions. This volume is an example of how many persons with different points of view can sift the data for important policy information. This policy of having the data and the reports freely available for data and analysis should be maintained to get the maximum benefits from future studies.

REFERENCES

Martin, M. O., & Kelly, D. L. (1996). *TIMSS technical report—Volume I: Design and development.* Chestnut Hill, MA: Boston College.

Martin, M. O., & Kelly, D. L. (1997). *TIMSS technical report—Volume II: Implementation and analysis, primary and middle school years.* Chestnut Hill, MA: Boston College.

Martin, M. O., & Kelly, D. L. (1998). *TIMSS technical report—Volume III: Implementation and analysis, final year of secondary school.* Chestnut Hill, MA: Boston College.

Martin, M. O., & Mullis, I. V. S. (1996). *Quality assurance in data collection.* Chestnut Hill, MA: Boston College.

Schmidt, W. H., McKnight, C. C., Valverde, G. A., Houang, R. T., & Wiley, D. E. (1997). *Many visions, many aims, Volume 1: A cross-national investigation of curricular intentions in school mathematics.* Dordrecht: Kluwer Academic Publishers.

APPENDIX

Contributors to the Volume

Carl Angell
Department of Physics
University of Oslo
P.O.Box 1048 Blindern
N-0316 Oslo
Norway

John A. Dossey
College of Education
Illinois State University
Normal, IL 61790-4520
USA

A. Geoffrey Howson
University of Southampton
Southampton, SO17 1BJ
UK

Chancey O. Jones
School and College Services Test
        Development
Educational Testing Service
Princeton, NJ 08541
USA

Dana L. Kelly
Education Statistics Services Institute
American Institutes for Research
1990 K Street, Suite 500
Washington, DC 20006
USA

Albert E. Beaton
Center for the Study of Testing
        Evaluation and Educational
Policy
Lynch School of Education
Boston College
Chestnut Hill, MA 02467
USA

Kadriye Ercikan
Department of Educational &
        Counselling Psychology, &
Special Education
Faculty of Education
University of British Columbia
Vancouver, Canada V6T 1Z4

Thomas Kellaghan
Educational Research Centre
St. Patrick's College
Dublin 9
Ireland

Edward Kifer
College of Education
University of Kentucky
Lexington, KY 40506-0001
USA

Marit Kjaernsli
Department of Teacher Education and
    School Development
University of Oslo
P.O.Box 1099 Blindern
N-0316 Oslo
Norway

Haggai Kupermintz
School of Education
University of Colorado
Campus Box 249
Boulder CO 80309-0249

Min Li
Department of Educational Psychology
College of Education, Box 353600
University of Washington
Seattle, WA 98195-3600
USA

Mary Lindquist
College of Education
Columbia State University
Columbus, GA 31901
USA

George Madaus
Center for the Study of Testing
    Evaluation and Educational Policy
Lynch School of Education
Boston College
Chestnut Hill, MA 02467
USA

Galina Kovalyova
Center for Evaluating the Quality of
    Education
Institute of General Secondary
    Education
Russian Academy of Education
Pogodinskaia st.8, Moscow 119905
Russia

Nancy Law
Faculty of Education
University of Hong Kong
Pokfulam Road
Hong Kong SAR
China

Svein Lie
Department of Teacher Education and
    School Development
University of Oslo
P.O.Box 1099 Blindern
N-0316 Oslo
Norway

Tanya McCreith
Department of Educational &
    Counselling Psychology, & Special
    Education
Faculty of Education
University of British Columbia
Vancouver, Canada V6T 1Z4

Tami S. Martin
Mathematics Department
Illinois State University
Normal, IL 61790-4520
USA

Ina Mullis
International Study Center
Lynch School of Education
Boston College
Chestnut Hill, MA 02467
USA

Eizo Nagasaki
Curriculum Research Center
National Institute for Educational Policy
    Research (NIER)
Tokyo, 153-8681
Japan

Jana Paleckova
Department of Educational
    Measurement
Institute for Information on Education
Senovazne nam. 26
100 00 Praha 1
Czech Republic

Tjeerd Plomp
University of Twente
Faculty of Educational Science and
    Technology
Department of Curriculum Technology
PO Box 217
7500 AE Enschede
The Netherlands

Maria Araceli Ruiz-Primo
School of Education
Stanford University
Stanford, CA 94305-3096
USA

Natalia Naidenova
Institute of General Secondary
    Education
Russian Academy of Education
Pogodinskaia st.8, Moscow 119905
Russia

Laura M. O'Dwyer
Center for the Study of Testing,
    Evaluation and Educational Policy
Lynch School of Education
Boston College
Chestnut Hill, MA 02467
USA

Willem J. Pelgrum
University of Twente
Faculty of Educational Science and
    Technology
Department of Curriculum Technology
PO Box 217
7500 AE Enschede
The Netherlands

David Robitaille
Department of Curriculum Studies
Faculty of Education
University of British Columbia
Vancouver, BC V6T 1Z4
Canada

Hanako Senuma
Curriculum Research Center
National Institute for Educational Policy
    Research (NIER)
Tokyo, 153-8681
Japan

Richard J. Shavelson
School of Education
Stanford University
Stanford, CA 94305-3096
USA

Jana Strakova
Department of Educational
    Measurement
Institute for Information on Education
Senovazne nam. 26
100 00 Praha 1
Czech Republic

Kenneth J. Travers
Department of Curriculum and
    Instruction
College of Education
University of Illinois
Champaign, IL 61820
USA

Jesse L. M. (Jay) Wilkins
Department of Teaching and Learning
Virginia Polytechnic Institute and State
    University
Blacksburg, VA 24061
USA

Michalinos Zembylas
Department of Teacher Education
325 Erickson Hall
Michigan State University
East Lansing, MI 48824
USA

Steven Stemler
PACE Center
Yale University
New Haven, CT 06511
USA

Alan Taylor
Applied   Research   and   Evaluation
    Services
Faculty of Education
University of British Columbia
Vancouver, BC V6T 1Z4
Canada

Hans Wagemaker
International Association for the
    Evaluation of Educational
    Achievement (IEA)
487 Herengracht
1017 BT Amsterdam
The Netherlands

Richard Wolf
Department of Human Development
Teachers College
Columbia University
New York, NY 10027
USA